THE BLOOMSBURY HANDBOOK OF SUSANNE K. LANGER

Also available from Bloomsbury:

The Bloomsbury Handbook of Fichte, edited by Marina F. Bykova
The Bloomsbury Companion to Arendt, edited by Peter Gratton and Yasemin Sari
The Philosophy of Susanne Langer, by Adrienne Dengerink Chaplin
The Selected Writings of Eva Picardi, by Eva Picardi and edited by Annalisa Coliva

THE BLOOMSBURY HANDBOOK OF SUSANNE K. LANGER

Edited by Lona Gaikis

BLOOMSBURY ACADEMIC
LONDON • NEW YORK • OXFORD • NEW DELHI • SYDNEY

BLOOMSBURY ACADEMIC
Bloomsbury Publishing Plc
50 Bedford Square, London, WC1B 3DP, UK
1385 Broadway, New York, NY 10018, USA
29 Earlsfort Terrace, Dublin 2, Ireland

BLOOMSBURY, BLOOMSBURY ACADEMIC and the Diana logo are
trademarks of Bloomsbury Publishing Plc

First published in Great Britain 2024

Published with the support of the Austrian Science Fund (FWF): [PUB 1012-G]

FWF Austrian Science Fund

A catalogue record for this book is available from the British Library.

A catalog record for this book is available from the Library of Congress.

ISBN: HB: 978-1-3502-9464-6
ePDF: 978-1-3502-9467-7
eBook: 978-1-3502-9465-3

Series: Bloomsbury Handbooks

Typeset by Integra Software Services Pvt. Ltd.
Printed and bound in Great Britain

To find out more about our authors and books visit www.bloomsbury.com
and sign up for our newsletters.

CONTENTS

LIST OF FIGURES

LIST OF CONTRIBUTORS

Randall E. Auxier is Professor of Philosophy and Communication Studies at Southern Illinois University Carbondale. He teaches courses in Western rhetorical theory, the philosophy of culture (including popular culture), the phenomenology of communication, semiotics, and other topics pertaining to meaning, language, and being. He is an activist, a radio DJ, and musician. He is the author of the Foreword to this book.

Carolyn Bergonzo is a poet and scholar based in Massachusetts. She holds an MFA in Literary Arts from Brown University and works at the Harvard Radcliffe Institute. With a focus on the creative and scholarly practices of women, Bergonzo has a specific interest in Susanne Langer's life and work. She has performed research on the Susanne Langer Papers at Houghton Library at Harvard University, co-curated an exhibition on Langer—with Tereza Hadravová and Linda Legassie—at Connecticut College in fall 2022, and is currently working on a biography of the philosopher.

Adrienne Dengerink Chaplin is a visiting research fellow at King's College London and a research associate at the Margaret Beaufort Institute in Cambridge. She works on the interface of philosophy, theology, and art and is the author of *The Philosophy of Susanne Langer: Embodied Meaning in Logic, Art and Feeling* (2020).

Donald Dryden is a visiting scholar in the department of Philosophy at Duke University. His research focuses on the history and philosophy of psychology and the biological sciences, as well as on the life and work of Susanne Langer, a subject on which he has done biographical research and has published a number of articles.

Giulia Felappi is a Lecturer in Philosophy at the University of Southampton. She works primarily on the intersection between philosophy of language and metaphysics. In particular, she focuses on the nature of contents, propositions, and facts, and the accounts of propositions and facts in the history of Early Analytic Philosophy, with particular reference to the work of Susanne K. Langer and Dorothy Wrinch.

Lona Gaikis is a Lecturer at the University of Art and Industrial Design in Linz and a researcher at the University of Applied Arts in Vienna. She offers lectures and seminars on current debates in philosophy, society, and the arts, and contributes as an author to exhibition catalogs, journals, and anthologies. Lona obtained her doctoral degree in Philosophy, Art and Cultural Studies from the Academy of Fine Arts in Vienna in 2017. Her research focused on the "new key" in Susanne K. Langer and the meaning of music(s) in artistic practice. She runs the platform *czirp czirp—sonic and experimental arts* in Vienna, and is editor of this book.

Christian Grüny is Professor for Aesthetics at the State University of Music and the Performing Arts in Stuttgart and senior researcher at the Max Planck Institute for Empirical Aesthetics in Frankfurt. After studying philosophy and linguistics in Bochum, Prague and Berlin, he received his PhD in 2003 in Bochum and was an assistant professor at the University of Witten/Herdecke from 2008–14. He was a visiting professor at the Music Academy of Hamburg, and an interim professor at the Art Academy of Düsseldorf and the Technical University Darmstadt. He has published on aesthetics, the philosophy of music, phenomenology, symbol theory, and the philosophy of culture.

Robert E. Innis is Professor Emeritus of Philosophy at the University of Massachusetts, Lowell. He has been Visiting Professor of Semiotics at Tartu University and Visiting 'Professor of Excellence' at the Pontifical Catholic University of São Paulo. In 1977 he was Humboldt Foundation Fellow at the University of Cologne. His interests are broad-based philosophical semiotics, aesthetics, ways of knowing, and philosophical dimensions of cultural psychology. Among his publications are *Semiotics: An Introductory Anthology* (1985), *Pragmatism and the Forms of Sense* (2002), *Susanne Langer in Focus: The Symbolic Mind* (2009), and *Between Philosophy and Cultural Psychology* (2020).

Rolf Lachmann is an independent scholar from Cologne. He is widely known for his research on Susanne K. Langer and process philosophy, as well as for introducing her philosophy to German-speaking scholars with his book *Susanne K. Langer: Die lebendige Form menschlichen Fühlens und Verstehens* (2000). Lachmann studied philosophy in Marburg, Washington D.C., and Düsseldorf, obtaining his PhD in 1990. From 1993 to 1999 he was a research assistant at Humboldt University, Berlin. His research focuses include ethics and applied ethics, philosophy of culture, and philosophical anthropology.

Thomas Leddy is Professor of Philosophy at San José State University. He has published one book, *The Extraordinary in the Ordinary: The Aesthetics of Everyday Life* (2012). His most recent publications are "Kant and Everyday Aesthetics" in *Kant's Critique of Aesthetic Judgment in the Twentieth Century: A Historical and Critical Comparison of Its Main Interpretations* (2020), and "Resolving the Tension of Everyday Aesthetics in a Deweyan Way," in *American Aesthetics Today: Theory and Practice* (2020). Tom has a blog, *Aesthetics Today* (http://aestheticstoday.blogspot.com/).

Eva Kit Wah Man is Chair Professor in Humanities at Hong Kong Baptist University. She publishes widely on comparative aesthetics, comparative philosophy, women's studies, feminist philosophy, cultural studies, and art and cultural criticism. Eva was a Fulbright scholar and conducted research at the University of California, Berkeley in 2004. She contributes public services to the Hong Kong Arts Development Council, Hong Kong Museums Advisory Committee, Hong Kong Public Libraries, Home Affairs Bureau of Hong Kong Government committees, and Hong Kong Jockey Club's Arts and Cultural Heritage projects.

Brian Massumi, University of Montreal, is the author of numerous works across philosophy, political theory, and art theory. His publications include *Couplets: Travels in Speculative Pragmatism* (2021), *99 Theses for the Revaluation of Value* (2018), and *Parables for the Virtual: Movement, Affect, Sensation* (augmented 2nd edn, 2021). With Erin Manning and 3Ecologies Project, he participates in the collective exploration of new ways of bringing philosophical and artistic practices into collaborative interaction.

Adam Nocek is an Associate Professor of Philosophy of Technology and Science and Technology Studies at the School of Arts, Media and Engineering, Arizona State University. He is also the Founding Director of ASU's Center for Philosophical Technologies. Adam recently published *Molecular Capture: The Animation of Biology* (2021).

Anne Pollok is a historian of philosophy (eighteenth- to twentieth-century aesthetics and philosophy of culture). She was an associate professor of Philosophy at the University of South Carolina, Columbia, SC (2013–20) and a teaching assistant at Stanford University (2008–13). In 2020 she joined the faculty at Gutenberg University Mainz, Germany, as a research assistant. Anne's numerous essays on Moses Mendelssohn, Ernst Cassirer, and women philosophers of the eighteenth to twentieth centuries, as well as her book, *Facets of Humankind: On Moses Mendelssohn's Anthropology* (2010) attest to her research interests.

Eldritch Priest is an Associate Professor at the School for the Contemporary Arts, Simon Fraser University. He writes on sonic culture, experimental aesthetics, and the philosophy of experience from a "pataphysical" perspective. Eldritch is also a composer and improviser, as well as a member of the experimental theory group "The Occulture" (http://www.theocculture.net/).

Martina Sauer is a senior editor of *Art Style | Art & Culture International Magazine* and a scientific advisor of the Society of Interdisciplinary Image Science and the German Society of Semiotics. She runs the Institute of Image and Cultural Philosophy in Bühl (Baden), Germany (https://www.bildphilosophie.de/en-gb). Martina has been a scientific associate in philosophy of art and culture, aesthetics, and design in Basel, Zürich, Bremen, Witten, and at Bauhaus-University, Weimar. She currently holds seminars on the fundamentals of perception at the Fresenius University of Applied Sciences, Akademie Mode & Design in Düsseldorf and Hamburg.

Helen Thaventhiran is a Lecturer in Modern Literature at the University of Cambridge, UK, and fellow of Robinson College. Her research areas are twentieth-century literature, literary criticism, aesthetics, intellectual history, and philosophy of language. Her monograph, *Radical Empiricists: Five Modernist Close Readers*, was published in 2015. She is currently working on a book titled *Margins of Philosophy*, offering an alternative history of twentieth-century philosophy of language, whose central characters include Victoria Welby, Vernon Lee, Alain Locke, Susan Stebbing, Susanne Langer, and Margaret Masterman. She is a member of the Editorial Board and Reviews Editor for the *Cambridge Quarterly*.

Christophe Van Eecke is a philosopher, art historian, filmmaker, and Visiting Professor at LUCA School of Arts (Genk, Belgium). He is the author of *Pandaemonium: Ken Russell's Artist Biographies as Baroque Performance* (2015), which was awarded the Research Prize of the Praemium Erasmianum Foundation, and *Only Connect: Five Exercises in Aesthetics* (2011), which took a Langerian approach to contemporary art and film. Christophe is currently preparing a book on Ken Russell and Neo-Victorianism.

Iris van der Tuin is Professor of Theory of Cultural Inquiry in the department of Philosophy and Religious Studies at Utrecht University, the Netherlands, where she is also Dean for Interdisciplinary Education. Iris is interested in humanities scholarship that traverses the "two cultures" and reaches beyond the boundaries of academia. Her recent book, *Critical Concepts for the Creative Humanities* (2022), is a glossary co-authored by Nanna Verhoeff. Iris first encountered Susanne K. Langer in a footnote to a student paper and has been interested in Langer and processes of philosophical canonization ever since.

Sander Verhaegh is Assistant Professor of Philosophy of Science at Tilburg University, the Netherlands. He is the principal investigator of "Exiled Empiricists: American Philosophy and the Great Intellectual Migration," funded by the European Research Council. The aim of this project is to reconstruct the American reception of logical empiricism in the years before the Second World War, when dozens of European philosophers sought refuge in the United States. Verhaegh is a member of the Susanne K. Langer Circle steering committee. His publications include *Working from Within: The Nature and Development of Quine's Naturalism* and the volume *Women in the History of Analytic Philosophy* (with Jeanne Peijnenburg).

FOREWORD

RANDALL E. AUXIER

When I first had the opportunity to teach a graduate course in the philosophy of art, which was in 2004, I believe, I chose Langer's *Feeling and Form* as the primary text, supplemented by Danto's *Transfiguration of the Commonplace*. Together these texts make for a rather complete survey of the best philosophies of art of the second half of the twentieth century, and, in 2004, the twentieth century had not really been replaced by the twenty-first, as it has now. I think that the situation for art is very different now than it was eighteen years ago, since the digital revolution has changed everything.

Danto's philosophy of art, which capped the analogue age of such philosophizing, is a nice and sensitive continuation of Langer's philosophy of art. Danto generalized her theory in several directions and confronted questions Langer left unaddressed. At that time and for the ensuing nine years, I was fortunate to work with Danto on the book that became *The Philosophy of Arthur Danto* (2013),[1] and I had many opportunities to discuss Langer with him. He always regarded Langer as his most important teacher, and he said she was clearly the finest mind among philosophers at Columbia University when he was a student, and of course she was treated badly, to the lasting shame of that university. They do not seem especially eager to claim Langer today either, but perhaps there has been some progress due to the long-term presence of Danto himself and Lydia Goehr, and a few others who read Langer sympathetically. From Danto I learned a good bit about those days in the 1950s when he was a student, and about how and why the philosophy of art and philosophy in general was ignoring her, while people in other disciplines were giving her due attention. So much the worse for philosophy. It seems an unfortunate twist of fate that placed Langer in a time and place in which she was led to see herself as a "philosopher" in the very narrow academic sense that word was used in her lifetime. That world had no place for women, and it distorted and twisted human thought into pigeonholes that made it nearly impossible for anyone, least of all women, to think across these artificial boundaries.

Yet Langer saw herself as a "philosopher," and strove for a career (or at least a living) under this narrow and artificial heading. I have always seen my own study of Langer, which began in 1988, as a contribution to rectifying something that went terribly wrong in the philosophy discipline. I was in search of an explanation for my own musical experiences, both as a musician and as a listener, and surely philosophers would shed some light on such questions. A young fool and his delusions are not easily parted. When I first read *Philosophy in a New Key*, I thought, "well, here it is, this is what I have been looking for." And then I slowly began to realize something. Yes, the public had embraced this book and theory, and had assimilated it, a bit, and moved on, as the public will do. But philosophers, including philosophers of art and aestheticians, basically ignored it. I could not understand why. Clearly, she was on to something important. The process of symbolization had not been well understood. The obvious fact is this: our most elevated symbolizations, scientific, in math, logic, or expressive, as in poetry and literature, were but the outer crust of a natural process by which human beings (and, indeed, even animals), negotiate with the passage of time in the physical world. Our symbols make the world stand still long enough for us to act on it more successfully. Was this not obviously the most important insight of the twentieth century? To me it still seems so. Perhaps it was too much to take in. We would have to rearrange the way we thought about, well, *everything* in order to pursue this insight to its full implications. We would have to change many things about our practices and attitudes as well as our mental habits. We did not wish to do that, nor be told that we needed to do it.

I could not at first discern the continuity of Danto's work with Langer's. He almost never mentioned her in writing. Only when I got to know him and he explained her profound influence did I begin to see how he was continuing her work. I asked him why he did not reference her or do any explication of her work, so as to bring attention to it. He said that he had learned as a graduate student that she was poison to a career among analytic philosophers.[2] I have since had this assessment confirmed by others who were around at that time, such as Joe Margolis and Marjorie Grene, the latter of whom was Langer's student at Radcliffe. Danto pretended to be an analytic philosopher, admitting he liked "the minimalism of analytic philosophy" for aesthetic reasons, not because it was the best way to do philosophy.[3] So he mastered the techniques of analytic philosophy as any artist would, and then presented his (and surreptitiously, Langer's) ideas in this form. He said to me that people were "very doctrinaire" in those days and that he did not dare try to buck the norms of the discipline. He had a family to feed.

Even now it is not without risk *in the dying philosophy discipline* to take Langer seriously. I have experienced the sideways glances myself. But I think it is worth enduring the sneers, then and now, to set aside the ridiculous popularity contests of the endangered sub-species *homo academicus philosophicus* in the twentieth century. Richard Rorty exposed the hollowness of these pretensions in 1979 and spent the rest of his life making good on his unpopular viewpoint.[4] I think he won. The rest of academia has gradually weighed in on the habits of this endangered creature and deemed it unfit in the struggle for survival. Sometimes extinction is a kindness done by history rather than nature. Goodbye *philosophicus*, varieties analytic and continental, which somehow took themselves to be enemies but were so closely related as to defy differentiation by the anthropologists of tomorrow. Thus, we stand on the brink of new movements in thought, unburdened of dross and chaff and the other excrescences that buried the thought of Langer beneath heaps and mounds of self-serious but irrelevant and mediocre fads and trends.

Now we look at a future in which, whatever philosophy was, it will not be that in the decades to come. And Langer is surviving the change of eras, attracting new readers and interpreters as she never has before. Finally, her rich collection of ideas and forms of analysis and imaginative thought are being discussed as they should be. And perhaps the very real influence she has exerted on our cultural development, from a thousand tiny hilltops, will be felt in the mountains and valleys of our further development. This volume is a clear indication that her contributions will get a hearing. I hope readers will consider and apply the ideas they take from it.

NOTES

1. Randall E. Auxier and Lewis E. Hahn, *The Philosophy of Arthur Danto*, Library of Living Philosophers 33 (Chicago: Open Court, 2013).
2. I believe this was his choice of words, but memory can play tricks on us. We were in a car and I was driving, and perhaps otherwise distracted. In any case, this was his meaning.
3. See my preface to *The Philosophy of Arthur Danto* for a complete account of this claim. Danto read this preface and affirmed its claims.
4. I allude to Rorty's *Philosophy and the Mirror of Nature* (Princeton, NJ: Princeton University Press, 1979), but he published many dozens, if not hundreds, of subsequent attacks and analyses of the basic claims in that book.

BIBLIOGRAPHY

Auxier, Randall E., and Lewis E. Hahn. *The Philosophy of Arthur Danto*, Library of Living Philosophers 33. Chicago: Open Court, 2013.

Rorty, Richard. *Philosophy and the Mirror of Nature*. Princeton, NJ: Princeton University Press, 1979.

PREFACE

Susanne K. Langer's philosophy popped up rather suddenly when, as a graduate student in about 2010, I discovered the *affective turn* in the media theory of Brian Massumi. I remember how it was his mentioning of Langer in the context of what he coined thinking-feeling that turned my attention to her work. His take on her chapter "Semblance" in *Feeling and Form* (1953)—Langer's discussion of how ornaments evoke a virtual sprawling of living form as an added factor to the perception of fixed shapes—intrigued me. Ever since, my new *real* has been endowed with the notion of those virtual worlds. Vision became something dynamic, objects turned into *events*. Music and media became the matter I would study for the next ten years, always remembering that there was this one female thinker who had placed a special emphasis on music's form in her theory of the arts. Much time has passed since then. My focus on artists who transcend media and channels of perception has grown, and reading Langer's philosophy lent me tools to unlock these kinds of artistic import. It preserved my enchantment for art's ability to open worlds of feeling through sound, gesture, and expressiveness. This lure for the arts, and for Langer, substantiated in more solid research that traced back her concepts and the roots of her enigmatic philosophy. It led to this book, and it will certainly continue in my writing on music and future forms of artistic media.

The idea for this collection of essays was formed during a panel on Susanne Langer, held at the American Society for Aesthetics' 2019 annual meeting in Phoenix, Arizona, to which Prof. Eva Kit Wah Man had invited me, along with Prof. Iris van der Tuin and Prof. Thomas Leddy. The goal was to redirect attention to Langer's art philosophy, which had remained rather untouched since last being debated at a Society meeting in 2010. The same year, Adrienne Dengerink Chaplin's book *The Philosophy of Susanne Langer* (2019) was published. Eva Man was also working on her publication *Cross-Cultural Reflections* (2020),[5] a collection of essays that accessed aesthetic discourses and body politics in Chinese art from a Langerian perspective. The research on and curiosity for Langer's work seemed to be gaining momentum. Shortly after, at the initiative of Adrienne, Iris, along with historian of analytic philosophy Sander Verhaegh, and longtime Langerian philosopher Randall E. Auxier, the Susanne K. Langer Circle formed—an international network of transdisciplinary scholars, both old and new researchers of Langer's work. One might smile and consider Whitehead's view that "the whole antecedent world conspires to produce a new occasion"[6] and how, on this occasion at least, the academic world has conspired to compile a collection of new perspectives on Langer's legacy. This holds true only in part. This book is—indeed—the culmination of the continued inspiration and growing interest in Langer's philosophical work, brought into being by the serendipity of individual initiatives. But beyond that, it is the fruit of a targeted and collective effort among the dawning of a new, post-pandemic reality. To keep one's mind clear in these anxious times is not the easiest of challenges, which is why I am so grateful for the contributions of almost all authors whose participation I requested. I believe that by connecting and collecting ourselves through these essays, new channels for discussion will open, possibly making Langer's late wish come true—for her thoughts to "eventuate in a parade of projects for young thinkers with long ways to go."[7]

This book brings together the work of both established and new scholars of Susanne Langer's philosophy. Those long familiar with her work developed new perspectives. Those rather new to her work felt challenged to continue the intriguingly contemporary aspects of her generative ideas. Overall, this book assembles essays from writers I have long admired for their lucid analyses and thought-provoking outlook on Langer's legacy. It also speculates new links and exciting applications of her philosophy, thereby reviving her audacity. I wish to express my heartfelt thanks to all authors who lent their individual perspectives to this exciting endeavor.

In the course of compiling this book, my special thanks must go to researcher and author Donald Dryden, who on many occasions shared material from his profound archival research, as well as personal anecdotes on Langer. He provided the portrait of Susanne Langer in the introduction pages to this collection, which was taken by her niece Susanne Dunbar Barrymore in 1954 in Ann Arbor. When inquiring as to the origin and copyright holder of this image, he put me in contact with Mrs. Dunbar Barrymore, who was delighted to revisit this memory of her "Tante Susy," standing by the log cabin she inhabited during a teaching term in Michigan. Donald's research goes back admiringly far, to when Langer was still alive, and he had reached out to her in a "fan letter" in 1968 to express his excitement upon the publication of the first volume of *Mind: An Essay on Human Feeling* (1967). On one occasion, a few years later, and after the appearance of the second volume, they even arranged a meeting on June 20, 1975 in Old Lyme, as he was visiting relatives in Boston and Langer had taken a break from what she called "a writing jag"—those periods in which she would put aside "everything else—letters, household, social life, and even music"[8]—to concentrate entirely on writing. At that time, Donald recounts, Langer would cruise about with a canoe strapped to her car, for she had found a technique to mount and dismount the heavy boat all by herself—ready for a canoe trip at any time—and on this occasion he joined her.[9] Langer was always close to nature and to culture and art—in her theory and her style of living. The growing interest in her thought might also account for a more eco-critical future.

A novelty of this publication is a series of previously unpublished photographs excavated from The Susanne K. Langer Papers at the Houghton Library in Harvard and from Donald's private research. I thank Leonard W. Langer and Stephen Langer, representatives of the Dr. Susanne K. Langer Estate, for kindly giving permission to publish these images. Thanks to the efforts of Donald and Iris, this material was made accessible. The digitalization of images was realized in part by Donald's private initiative of scanning much of his own research material. In the case of the second photograph of Langer's desk in Iris van der Tuin's Chapter 5, "The Horizontal, Vertical, and Transversal Mechanics of Susanne K. Langer's Card-Index System" (Figure 5.2), the archival material was digitalized by the Harvard Library with the kind support of Utrecht University.

I am particularly pleased to have been able to ascribe a name to the anonymous Māori Man with moko of Plate III in Susanne K. Langer's *Feeling and Form* (1953). Thomas Leddy's Chapter 18 "Susanne K. Langer, Everyday Aesthetics, and Virtual Worlds" (Figure 18.1), shows the photograph of Wiremu Pātara Te Tuhi, dated around 1880, cousin of King Tāwhiao, Waikato leader, and secretary to the Māori king, Ngāti Mahuta. He was editor and chief writer for the Māori King Movement's newspaper, *Te Paki-o-Matariki* in the late 1800s in New Zealand.[10]

All these resources and newly opened discussions about Susanne Langer's persona, her philosophy, personal thoughts, and methodology, uncover unique and hidden aspects of her great mind. Above all, they stimulate new questions to further the study of a thinker who was so open to the world and curious about the future.

NOTES

5. Eva Kit Wah Man, *Cross-Cultural Reflections on Chinese Aesthetics, Gender, Embodiment and Learning* (Berlin: Springer Nature, 2020).
6. Alfred N. Whitehead, *Modes of Thought* (Toronto: The Free Press, 1968), 164.
7. Susanne K. Langer, *Mind: An Essay on Human Feeling*, vol. 3 (Baltimore: Johns Hopkins University Press, 1982), 201.
8. Susanne K. Langer, letter to Donald Dryden, December 30, 1973.
9. Donald Dryden, email to author, October 15, 2022.
10. William H. Oliver, ed., "Te Tuhi, Wiremu Pātara," *Dictionary of New Zealand Biography*, vol. 1: 1769–1869 (Wellington, NZ: Allen & Unwin, 1990), 510–11.

BIBLIOGRAPHY

Langer, Susanne K. *Mind: An Essay on Human Feeling*, vol. 3. Baltimore: Johns Hopkins University Press, 1982.

Oliver, William H., ed. "Te Tuhi, Wiremu Pātara." In *Dictionary of New Zealand Biography*, vol. 1: 1769–1869. Wellington, NZ: Allen & Unwin, 1990.

Whitehead, Alfred N. *Modes of Thought*, 1938. Toronto: The Free Press, 1968.

NOTE ON THE COVER ARTIST

Barbara Kasten's art has intrigued me for several years. Just as the preparations for this collection of essays began, I was struck by a photograph from her series "Architectural Site 17, August 29, 1988," which shows a young female figure framed in a complex of dark windows and colorfully illuminated beams, walls, and rails—a speculative topology. She harks for a sound. This image epitomized the whispering futures that I recognized in Susanne K. Langer's philosophy.

Barbara Kasten (born in 1936, living and working in Chicago, Illinois) fractures space with mirrors, shadows, and light. Her photographs capture instant moments in what she calls "sets," thereby revealing the *slipperiness* in the perception of things. Kasten represents a generation of postmodern and postminimal North American artists, who engage in institutional critique and play with the ephemeral nature of the *real*. Her crystalline imagery oscillates between the familiar and the yet unfathomed, which parallels the potential of Langer's philosophy.

As a theorist for the arts, Langer was deeply interested in the epistemological import of artifacts. She maintained close friendships with many artists, among them avant-gardists, who found in her a kindred spirit as a philosopher and art theorist. Art does not represent anything, it *presents* an idea, is the thought that Langer lent to them. As one of few accounts for the influence Langer possibly had on artistic practice, avant-garde film maker Jonas Mekas, in his *Scrapbook of the Sixties* (2015), quotes Langer on the vogue of generative ideas and the problems they generate,[11] to become "the germ of a complete reorientation in metaphysics, or at least the 'Open Sesame' of some new positive science."[12] Indeed, her visionary thinking unites with and continues the pioneering spirits of the art world in its quest to expand the scope of knowing.

The female figure in the photograph is a replica of *Nydia, the Blind Flower Girl of Pompeii* (1853–4; carved in 1867), a neoclassical sculpture by American sculptor Randolph Rogers, pictured folded into the postmodern architecture of Richard Meier, in its home at the High Museum of Art, Atlanta, GA.

Courtesy of the artist Barbara Kasten and Bortolami Gallery, New York.

NOTES

11. Jonas Mekas, "Notes after Reseeing the Movies of Andy Warhol," 1970, in *Scrapbook of the Sixties*: *Writings 1954–2010* (Leipzig: Spector Books, 2015), 370.
12. Susanne K. Langer, *Philosophy in a New Key* (Cambridge, MA: Harvard University Press, 1993), 23.

BIBLIOGRAPHY

Langer, Susanne K. *Philosophy in a New Key*. Cambridge, MA: Harvard University Press, 1993.

Mekas, Jonas. *Scrapbook of the Sixties: Writings 1954–2010*. Leipzig: Spector Books, 2015.

ACKNOWLEDGMENTS

This book collects the individual research of twenty scholars from Europe, the US, Canada, and China. Each chapter sheds light on Susanne K. Langer's philosophy from a different perspective, and I must express my awe at the beauty of this engagement in its revival of a philosopher who had bridged so many divides.

Many people, to whom I am indebted, have been part of this book coming into life. I would like to thank my editor Colleen Coalter and her assistant Suzie Nash for guiding this project and helping me with their expertise and advice. At the core of this book's making, translator Mỹ Huê McGowran put much care into translating Rolf Lachmann's Chapter 6 "Susanne K. Langer's Foray into Art as a 'Phenomenology of Feeling,'" great job! Furthermore, I am very much obliged to copyeditor Nina Cook, who has been a great support in editing and formatting the chapters with me. I also thank the authors Adrienne Dengerink Chaplin, Sander Verhaegh, and Donald Dryden for their invaluable comments in the process of completing the manuscript, and Robert E. Innis for sharing his experience and kind advice on broader matters when pursuing the adventure of compiling such a collection of essays.

We kindly received text permissions for Sander Verhaegh's Chapter 1 "Susanne K. Langer and the Harvard School of Analysis" from Springer International Publishing, and for Eldritch Priest's Chapter 13 "Thinking Non/Humanly with Susanne K. Langer" from Duke University Press, which was published in parts as "What it's like to think what it's like to think" in his book *Earworm and Event* (Durham, NC: Duke University Press, 2022). Special thanks to Sander, who shares exciting insights into the intellectual atmosphere at Harvard in the 1920s/1930s from his chapter "Susanne Langer and the American Development of Analytic Philosophy," in *Women in the History of Analytic Philosophy*, ed. Jeanne Peijnenburg and Sander Verhaegh (Cham: Springer, 2022).

I would like to express my gratitude to Barbara Kasten and Bortolami Gallery, New York, for giving permission to reproduce her photograph "Architectural Site 17, August 29, 1988" as the cover image of this book. All photographic material of Susanne K. Langer is reproduced with permission of her Estate, represented by Leonard Langer and Stephen Langer. The 1954 portrait of Susanne K. Langer in Ann Arbor is courtesy of her niece Susanne Dunbar Barrymore. I would like to thank the James Lord Estate, represented by Harold Ober Associates, New York, for giving permission to reproduce the photograph of Susanne K. Langer at her desk in Old Lyme Connecticut in Iris van der Tuin's Chapter 5. It is taken from Lord's article "A Lady Seeking Answers," published May 26, 1968 in *The New York Times Book Review*. Thanks to Beinecke Library at Yale for giving us a lead in our search for the copyright holders of Lord's estate. The digitalization of photographs from Susanne K. Langer's archive was accomplished by Houghton Library at Harvard, but largely by Donald Dryden, who holds an extensive collection of digitalized images, papers, and notes. Thank you, Donald, for being so generous and open to collaboration.

The image of the dance theater work *Soledad* by Helen Lai (2015) in Eva Kit Wah Man's Chapter 16 "Virtual Powers in Susanne K. Langer's Theory of Dance" is published courtesy of CCDC City Contemporary Dance Company, Hong Kong. Its photographer

Ringo Chan is deceased, and I thank Eva for connecting me in this matter with the director of the CCDC, Kevin Wong, to receive the image permission. We are very grateful for permission to reproduce the portrait of Wiremu Pātara Te Tuhi, Waikato leader and Māori newspaper editor, from the Alexander Turnbull Library, Wellington, New Zealand. It is thanks to the library's search that we found the original film negative in the archives, which now enlivens the story of a man.

FIGURE 0.1 "Photograph of Dr. Susanne K. Langer in Ann Arbor, Michigan," by permission of the Estate of Susanne K. Langer. Photographer: Susanne Dunbar Barrymore (Spring 1954). Contributed by Donald Dryden.

Introduction

Logician and philosopher of art Susanne Katherina Langer (1895–1985), née Knauth, is a remarkable figure of twentieth-century thought. She devoted her philosophical engagement to mathematical and linguistic questions, and had a deeply ingrained curiosity for the meaning of forms of art, and the variety of cultural expressions. Working energetically for more than fifty years, the promising female logician was named alongside Abraham Kaplan and Charles Morris as one of "[t]he chief American theorists of the semantic school,"[1] yet traces of her philosophical legacy are rather scarce. This proverbially *unsung*[2] philosopher, whose book *Philosophy in a New Key* (1942), at that time a bestseller in popular and academic circles, remains surprisingly under-studied.

This book comprises essays by both European and North American scholars, by the researchers who spearheaded archival research in the 1990s and initiated the first substantive analyses of Langer's body of work, and by researchers who, more recently, have evolved their own ways of applying and furthering her ideas. It challenges long-established philosophical and conceptual divides on various levels. Langer herself—the offspring of nineteenth-century German emigrants—is regarded as a hybrid of continental and analytic traditions, and her study of philosophy was greatly influenced by her bilingualism in German and English. She was well versed not only in the ideas following German idealism, specifically neo-Kantianism, and aesthetics, but also in North American pragmatism and the logical empiricists. In fact, Langer was a central figure in what had initially been termed the "phenomenology of meaning,"[3] seeding and shaping the later American analytic tradition. The study of Langer's life and work reveals the many transatlantic conversations that influenced philosophy at its turning point in the short period between World Wars I and II. It also sheds new light on the personal and academic history of philosophers in exile during and after the wars. While Langer's philosophy, characteristically for this period in the US, was influenced by the newly formulated demand for an independent discipline of the logical analysis of meaning posed by early analytic philosophers and the protagonists of the Vienna circle—early Wittgenstein and Rudolf Carnap—her semiological approach was nuanced, and countered the predominating scientistic and positivistic orientation of her colleagues. In retrospect, Langer's approach seems even more determined, in the sense that she maintained an interdisciplinarity and considered art to have a unique kind of epistemological import, and therefore to represent, or in her words, "present", a class of its own. Essentially, she refused to leave the otherwise "unlogicized"[4] forms of mental life to the realm of the "ineffable," and instead proposed a complimentary *presentational form*, which she thought to underlie all forms of symbolic articulation. Langer backed up her individual challenge of developing a semiology that could encompass all human expression—propositional as well as pseudo-propositional—with Ernst Cassirer's theory of cultural symbolization. In his three volumes of *The Philosophy of Symbolic Forms*,[5] Cassirer offered a differentiated symbolic analysis for language, mythical thought, and the phenomenology of knowledge. His approach deviated from the mainstream in structuralism, which reduced all forms of human expression to linguistic schemes, instead forging the idea of cultural and mythical proto-linguistic matrices from which meaning emerges. However, Langer's philosophy was equally, if not more greatly, influenced by Alfred N. Whitehead's later process ontology. Indeed, in 1926 she graduated under his supervision, with her doctoral thesis "A Logical Analysis of Meaning,"[6] from Radcliffe University, Harvard's partner college for women's education. Langer's pervasive conceptual backdrop implies that Whitehead's cosmology enticed and provoked her philosophical development.

Reflecting on these heterogenic conceptual influences—on the one hand Cassirer's synthesis of Kant and Hegel, and on the other, Whitehead's lean towards the mathematization of the principles of inherence, culminating in an organicist philosophy—are the two most comprehensive investigations in anglophone academia to render the roots and their advancement in Langer's body of work: Robert E. Innis's *Susanne Langer in Focus: The Symbolic Mind* (2009) and Adrienne Dengerink Chaplin's *The Philosophy of Susanne Langer: Embodied Meaning in Logic, Art and Feeling* (2019). Both authors contributed to the current collection of essays. While Prof. Innis's first English monograph on Langer contextualizes her philosophy in a prism of the semiotic turn and her aspiration to challenge then-contemporary doctrines, Dutch philosopher Dr. Dengerink Chaplin offers a thorough revision of Langer's transatlantic roots in philosophy, showing how she had already drawn a blueprint to conceptualize an "embodied mind" long before it was coined under the cognitive turn.[7] Characteristic of both these authors is their affirmation of Langer's essential concepts, and their opening of perspectives that continue the visionary in her propositions.[8] Langer's ideas, particularly her critique of modern logocentrism, resonate with the poststructuralist philosophies of Gilles Deleuze, Jacques Derrida, and even phenomenologist François Merleau-Ponty; with postmodernist approaches to science such as those of Thomas Kuhn, Paul Ricœur, and John Searle; and with the neuro- and cognitive-scientific views of Antonio Damasio and Mark Johnson.[9]

Interestingly, the philosophy of Susanne K. Langer was appreciated and analyzed much earlier in German-speaking academia. Dr. Rolf Lachmann's monograph *Susanne K. Langer. Die lebendige Form menschlichen Fühlens und Verstehens* (2000)[10] called attention to her philosophy of mind almost a decade before Innis. As one of the earliest researchers of the Susanne Langer Papers at Houghton Library at Harvard in 1991–3,[11] Lachmann's careful excavation and analysis of core concepts was among the first to highlight in which ways process metaphysics had entered Langer's thought, and how this was essential for her later conceptualization of a theory of mind based on organic activity. A special 1997 issue of the journal *Process Studies*, devoted to Langer and edited by Lachmann, who is the German translator of Whitehead's *Symbolism: Its Meaning and Effect* (*Kulturelle Symbolisierung*, Frankfurt a. M.: Suhrkamp, 2000), introduced two other important scholars who also contribute to the present collection: Donald Dryden and Randall E. Auxier.[12] All three can be regarded as pioneers of Langer research, they have published widely, and continue to research and develop her work. However, even before 1997, a short volume titled *Gefühl, Abstraktion, symbolische Transformation*[13] was published in 1993 by art philosopher Barbara Kösters, preceding their tracing of Langer's covert conceptual foundation back to Whitehead. Kösters, too, discusses the epistemic implications of Langer's art theory regarding her orientation towards process. She also acknowledges Rolf Lachmann as her source for secondary literature and biographical dates.[14]

Interest in Susanne K. Langer's philosophy seemed to come in waves. Despite her presence in America, there was an unusual blind spot concerning her life and work in anglophone research. The reasons behind this are complex. For a start, women philosophers at that time were not equally promoted in the pursuit of academic careers. Co-education in universities was pretty much unheard of during Langer's time at Radcliffe University, and female voices rarely made it into the master narratives of philosophy. Genealogies were mainly dominated by men, and Langer simply did not appear in anyone's footnotes. Despite numerous awards and praises, she was deprived of an established career path in academia. Her male peers were placed front and center, leaving the reception of Langer's symbol theory wide open to misrepresentation.[15] Langer did, however, challenge these social constraints by forging alliances and continuing to do what she felt was her calling: the practice of philosophy. She studied with two significant thinkers of her time, Henry M. Sheffer and Alfred N. Whitehead. During her early career as a tutor of philosophy in Cambridge, Massachusetts, she embedded herself in the academic circles at Harvard—pioneers of the analytical branch of philosophy. Here, she liaised with visiting scholars from Europe. One of them was Herbert Feigl, philosopher from the Vienna Circle, who was granted a Rockefeller Fellowship at Harvard in 1930 and emigrated to the US that same year.[16] Langer went on to establish her own philosophical circle.[17] During this period, she corresponded with Edmund Husserl and even met him at his home in Schluchsee, Germany, while visiting Europe with her then husband in summer 1933.[18]

Langer's fluency in English and German gave her a great advantage over her American colleagues. She later became the English translator of Ernst Cassirer's *Language and Myth* (1946), which eased the way in US academia for the reception of the German scholar in exile. Despite Langer's dedication, it seems that her analytical venture into the arts was neither of interest nor of use in an intellectual atmosphere that Charles P. Snow characterized as "The Two Cultures," where scientific advancements hardly intersected with the interests and aspirations of the arts. Langer opposed the scientistic paradigm that banished aesthetic questions to the margins of philosophy, and aimed to overcome this artificial segregation. Moreover, she did not subordinate herself to mainstream philosophy, which at that time was all about specialization. Her interdisciplinary approach was therefore considered rather obscure in its holism. Langer was an avant-gardist, whose ideas anticipated later philosophic turns, but her peers seemed unready to grasp them, especially coming from a female philosopher.[19]

Yet, the exclusion of Langer was not solely due to sexism. An unconventional and determined thinker, Langer did not shy away from occasional polemics and controversies, including stark criticisms of her peers in logical positivism.[20] This already shimmers through in her first book *The Practice of Philosophy* (1930), whose first chapter is tellingly titled "Philosophy, the Scandal of Science." Here, Langer writes of the epochal shifts taking place with the emergence of analytic philosophy, as well as the rising issue of philosophy retaining its practice of being "committed to the method of universal doubt."[21] Equal to her fascination with the new possibilities is her unease with the new type of philosophic investigation that verbatim "rules out ethics and aesthetics, and disavows all metaphysics; in short, it refuses to make its subject-matter 'everything' and its view 'synoptic.'"[22] Langer recognized that philosophy's competence—to conceptualize the mythological and speculative meta-structures by which reality is formed—was at stake. This resulted in a deep ambivalence towards pragmatism and logical positivism. Langer underpinned her skepticism with recurring quips against the prevailing American philosophical traditions, noting the difficulty of tackling the problems of art by means of behaviorism and pragmatism alone.[23] This stance is upheld throughout Langer's work, provoking her to reach for the *art symbol*, for which she scaffolds a semantic theory in *Philosophy in a New Key* (1942). Her semiological relaunch continues in her genuinely unconventional and seemingly mystical approach to the virtual materials of the arts in *Feeling and Form* (1953), up until her trilogy of *Mind* (1967, 1972, 1982), which attempts to ground the philosophical megalith Mind in a vibrant biological matrix, and, paradoxically, returns to empirical analysis. Langer immersed herself in an abundance of philosophical adventures. She countered controversies by determinedly pushing forward her ideas and refining her concepts. Her consistency in her arguments, her shadowy existence, as well as her fearlessness of critique, lent Langer a kind of maverick reputation, which possibly invited her rediscovery in light of recent updates in poststructuralist philosophy, such as the call for *new materialism(s)*,[24] and made her more interesting to European scholars.

In addition to the monographic accounts of Langer's achievements by Lachmann, Innis, and Dengerink Chaplin, numerous scholars from philosophy, musicology, the arts, and theology have produced essays with very different assessments of her work—and discussed its controversies. While not all can be included in this book, their contributions must be praised for keeping the dialectic of Langer's thought alive. Philosophy is a polyphonic chorus, and this book's agenda is to inspire and encourage new waves of scholars to rediscover and refresh Langer's thinking.

PLACING SUSANNE K. LANGER

Starting off in symbolic logic, passing through an intricate semiology of the arts, towards a naturalistic theory of an embodied mind, Langer's intellectual path confronted various strands of philosophy. She equally listened to artists' voices and extended her aesthetic quest to possible intersections with empirical research. Langer's works culminate in detailed discussions on the meaning and effects of language and poetry, sculpture and painting, music, ritual, and dance, as she carves out their individual epistemological import. These idiosyncratic amalgamations of art analysis and speculations on how formalism could be reconciled with the biology of living organisms make her philosophy particularly intriguing. Part I of this book illuminates the influences most significant to Langer's philosophical practice, chief among them being Wittgenstein, Cassirer, and Whitehead, and traces her concepts back to these diverse perspectives. Readers will find the chapters organized according to Langer's philosophical development, which can be segmented into four phases paralleling her books *The Practice of Philosophy* (1930), *Philosophy in a New Key* (1941), *Feeling and Form* (1953), and culminating in the *Mind* trilogy (1967, 1972, 1982). The content of these chapters relates to historical, conceptual, material, and speculative aspects of Langer's philosophy.

The scene is opened by a historical positioning of the young logician, and the ways in which the intellectual atmosphere at Harvard—a central hub of analytic philosophy in the 1910s and 1920s—had influenced her. Sander Verhaegh's chapter, "Susanne K. Langer and the Harvard School of Analysis," places Langer among the thinkers of twentieth-century symbol theory and within the burgeoning of the new tradition. He sheds light on cross-continental influences that shaped the discipline during the early twentieth century. Langer's conceptual background is traced to the teachers Henry M. Sheffer, C. I. Lewis, and Alfred N. Whitehead, who examined her dissertation "A Logical Analysis of Meaning" in 1926. Verhaegh discusses the influence that Langer's research, her doctoral thesis, and her subsequent papers "Form and Content: A Study in Paradox" (1926) and

"A Logical Study of Verbs" (1927) had in the shaping and promotion of the method of logical analysis. His chapter features archival notes on how Langer, though not a Harvard graduate at that time, participated in pivotal philosophical discussions by founding her own private philosophical circle. Her active involvement made her a known philosopher even on the other side of the Atlantic, as her book *The Practice of Philosophy* (1930) circulated among logical empiricists in Vienna. A first edition of this rare print can still be found in the library collection at the University of Vienna's Faculty of Philosophy. Langer's position in relation to the Vienna Circle is addressed in more detail in Adrienne Dengerink Chaplin's Chapter 2, "Scientific Models and Artistic Images: Susanne K. Langer and the Early Wittgenstein." She dives into Langer's early fascination with Wittgenstein's *Tractatus*, examining the multi-layered issues concerning her interpretation of what she called his "prophetic gospel."[25] As one of the first American philosophers to engage seriously with Wittgenstein's early work, Langer's philosophy was unfortunately reduced to his supposed picture theory of representation as copy or mirror of the world.[26] This chapter clears up the distortions concerning his influence and challenges the claim that Langer's analysis of art intended to be an emotivist or representational theory of art and emotions. Langer, in fact, countered the common reading (by US logical positivists) of Wittgenstein as his being on a quest for a "logically perfect language."[27] She regarded him as part of a much wider symbolistic turn in philosophy, in proximity to Whitehead's approach to symbolization.[28] Dengerink Chaplin's attentive inspection of Wittgenstein's original use of the term "picture" (German, *Bild*) traces its spatial and relational scope, which distinguishes between picture as model, and image as abstraction.[29] Wittgenstein's impact on Langer's thought is also addressed in the subsequent Chapter 3, "Susanne K. Langer, Wittgenstein and the Profundity of Fairy Tales," in which literary theorist Helen Thaventhiran makes his speculative paradigm apparent. Instead of reaching primarily for Langer's academic texts, Thaventhiran introduces a rather unknown facet of Langer's biography—that of a children's book author. During her time at Radcliffe, shortly before stepping into the arena of analytic philosophy, Langer, in 1924, published the book *The Cruise of the Little Dipper, and Other Fairy Tales*, a collection of fables written by Langer and illustrated by her friend, artist and illustrator Helen Sewell (1896–1957). Contrasting her poetic—or mythological—narration with topics that had occupied Langer's early academic publications on the logic of symbols, Thaventhiran presents an inspiring complex of Langer's earliest responses to the *Tractatus*. Fairy tales, or *märchen* to Langer, were a "technique of our language,"[30] as Thaventhiran highlights, a method for transcending the ineffable in our abundant emotional lives. This chapter insightfully explores the meaning of myth and the special role of beliefs in the making of facts. It also draws attention to Wittgenstein's pedagogical writings, as well as Langer's third main influence: Cassirer's studies on the mythological import of symbols. Thaventhiran's portrayal of Langer as a writer of *märchen* sheds an intimate light on her role as a mother and educator, who always aimed to make the method of logical analysis an accessible skill.

Langer's achievements in logic come full circle in philosopher Giulia Felappi's Chapter 4, "Susanne K. Langer on Logic as the Study of Forms and Patterns of Any Sort." Highlighting Langer's unorthodox yet far-reaching understanding of logic, as a tool to encompass propositions and propositional forms, and then detect logical form in anything that follows a pattern, Felappi excavates the very base of Langer's departure. Influenced by the heterogenic traditions of not only Russell, Whitehead, Wittgenstein, and Moore, but also Sheffer and Lewis, Felappi reviews Josiah Royce as the source of Langer's doubt in only one general form of logic. There is no such thing as *the* logical form of any thing. Rather, logical analysis bears the possibility of exemplifying radically different forms in any matter.[31] And in Royce's sense, every thing, "from dialogues to dinners,"[32] can represent a pattern of its own, offering Langer a tool, which is "to the philosopher what the telescope is to the astronomer: an instrument of vision."[33] In tracing back Langer's foundations in logic, a discipline that at this time was itself a subject of study, Felappi explains her motivations for scouting its application in the field of arts, and primarily in music.[34]

A quite different subject area, and highly informative in understanding how Langer's heterogeneity of influences might have come about, is addressed by Iris van der Tuin's Chapter 5, "The Horizontal, Vertical, and Transversal Mechanics of Susanne K. Langer's Card-Index System." This chapter looks at the material and structural aspects of Langer's work process. Langer's lifelong systematic collecting and organizing of research notes accumulated in what one might call a *thinking-machine*: a system to remember, draw links, and write with, expounding the broader methodological aspect of ways of processing knowledge. This chapter is developed from van der Tuin's research visits to the Langer Archive at Houghton Library Harvard, which gave her insight into the horizontal, vertical, and transversal paradigms of Langer's card-index system. A closer look at the implementation of card-indexing and knowledge-storing architecture shows how filing systems were used not only to organize and archive, but also to facilitate algorithmic techniques that enabled cross-sectional references. Placed against the backdrop of the history and progress of card systems, e.g., Niklas Luhmann's *Zettelkasten*, van der Tuin points out that in Langer's case the filing system became a somewhat enlivened assistant that shaped her thinking and writing process—an external mechanism introducing new, diffractive patterns to her thought.

Equipped as she was with a knowledge management system that allowed nonlinear connections, and with her being well versed in both Continental and American philosophy, the impression arises that Langer had good reason in seeing the conceptual parallels in the diverse philosophies of Wittgenstein, Whitehead, and Cassirer as philosophers heralding a *symbolic turn*. Owing to the fact that Langer had, early on, achieved a synoptic view of philosophy, and could well contextualize the German-speaking philosophers Ludwig Wittgenstein and Ernst Cassirer, she must have been one of very few thinkers at the time to engage with the scope of Cassirer's three-part volume *Philosophie der Symbolischen Formen*, published from 1923 to 1929, on the symbolic structure and development of language, mythic thought, and the phenomenology of knowledge. Cassirer's prolific career as a polymath, one of the leading figures in German philosophy, however, ended abruptly. The Davos dispute in 1929 with Heidegger crushed his standing in academia,[35] and Cassirer's fate as Jewish scholar was sealed when Hitler came to power on January 30, 1933. He emigrated to Sweden on March 12 the same year. Expelled from German-speaking academia, his theory of symbols was strongly criticized,[36] and was only revised after a lengthy delay,[37] leaving his magnum opus to be translated and published in English, in 1955, ten years after his death.[38] In this sense, Langer's loyalty to symbolization meant bridging the so-called Continental Divide from the very beginning. Touching on the influence of continental philosophy on Langer's works, German researcher Rolf Lachmann presents the under-studied aspect of Langer possibly working on a variation of phenomenology, and branching a Hegelian and Husserlian legacy outside their respective European contexts. His Chapter 6, "Susanne K. Langer's Foray into Art as a 'Phenomenology of Feeling'" reconstructs the phenomenal character of Langer's core concept "feeling," from scattered references to phenomenology throughout her works.

Lachmann investigates her visit to Edmund Husserl in 1933 and their infrequent but existing written correspondence. Unlike his earlier Whiteheadian approach to the *Mind* trilogy, Lachmann's current cross-reading traces the phenomenological root in Langer's distinguished notion of *presentational form*, which quite possibly goes back to the phenomenological dual connotation of (re-)presentification, or (re-)presentation—the *real* as being sensually endowed (or *concrete*), and *abstract* at the same time. This link to German idealism comes up again in Anne Pollok's Chapter 7, "Susanne K. Langer's Theory of Self-Liberation through Culture," which looks closely at the Cassirerian influences on Langer's conception of art symbols in *Feeling and Form*. Pollok highlights the problematic context of symbols and their (re-)mythification in culture, especially regarding their abuse and politicization during the Nazi regime in Germany. According to Pollok, Langer should be regarded not only as one of very few scholars who studied Cassirer, but as possibly one of the most important, as she set forth core aspects of his neo-Kantian approach. Pollok gives particular emphasis to ethical and political issues as well as to the potential of rendering symbols through cultural expressions, which then lead to the further construction of civilizations. Such observations are uncannily contemporary in the current age of post-truth, and medially amplified social upheaval.

A different chord is struck in linking of Susanne K. Langer's philosophy to Ernst Cassirer in Christian Grüny's Chapter 8, "The Systematic Position of Art in Susanne K. Langer's and Ernst Cassirer's Thinking." German philosopher of music and translator of Langer's *Feeling and Form* (*Fühlen und Form: Eine Theorie der Kunst*, Hamburg: Meiner, 2018), Grüny offers a comparative analysis of Langer and Cassirer. Specifically, he looks at how they place art within their respective philosophic systems—Langer within the context of feeling, Cassirer in regard to his register of symbols. Broaching the issue of Cassirer having never completed his grand idealistic system of symbolic forms, which would have included the idea of an independent symbolic form of the arts—a relatively progressive concept for a nineteenth-century philosopher—Grüny emphasizes that asserting a metaphysics of symbols had never been Langer's thing.[39] He argues that Langer's intention in introducing a new, "symbolic" key was indeed to integrate into analytic philosophy Cassirer's phenomenology of (re-)presentation, which he so neatly packaged as *Darstellungsform*,[40] but without being confined by a metaphysical superstructure. Langer saw more potential in elaborating a naturalistic and non-reductive theory of mind by means of rendering art phenomenologically, as a twofold symbol, thereby fusing substance and function into one. Grüny points out the common thread of music running through her philosophy, and how the musical metaphor in her *new key* guides Langer's profound reconfiguration of the concept of artistic import—a model which would become formative for her later treatises on mind as a phenomenon emerging from the vital and vibrant matrix of life.

American philosopher Donald Dryden was among the vanguard of Langer researchers in the 1990s. He proposes a psychological approach to the concepts and philosophy of Susanne Langer. His Chapter 9, "The Meaning of 'Feeling' in Susanne K. Langer's Project of *Mind*" reads Langer's use of the term "feeling" in relation to William James's in *The Principles of Psychology* (1890), where James favored "thought" and "consciousness" as generic terms for all mental phenomena in his pioneering work on what we would call the "phenomenology of conscious experience,"[41] though his psychology was never about mental activity exclusively. Dryden sparks his discussion with a rather inconspicuous footnote on James early in Langer's *Mind: Essay on Human Feeling* (1967),[42] which leads the issue back to a detailed analysis of what Langer may have meant by feeling when, with the completion of *Feeling and Form* in 1953, it had become central to her understanding of the meaning of art, while also looking forward to the development of her philosophy of mind. This central concept, equally formative to her theory of art as well as her philosophy of mind, also led to various conflicting assessments of her philosophy. Throughout her work, Langer remains, indeed, rather vague when it comes to a definition, though she gives clear indications that point to her use of feeling in the sense of James'—as inclusive of all that is felt in conscious experience. This gives reason to assume that she only secondarily uses it as a generic term to describe moods, emotions, sensations, or desires. Dryden highlights that this ambiguity proves equally problematic and enticing when attempting to trace Langer back to one main philosophical strand. Dryden's focus here, on the psychological aspects in Langer, in retrospect, allows her late philosophy of mind to almost resonate as a variation of experimental psychology.

The discourse on Langer's psychology is broadened by Robert E. Innis's Chapter 10, "Psychological Dimensions, Cultural Consequences, and their Breakings in Susanne K. Langer's Symbolic Mind," which drives forth the reading of Langer's philosophy of mind as more of an empirical endeavor. Seen in connection to Dryden's previous analysis of feeling, this chapter points to Langer's original take on psychology by conceiving of the life of feeling as "a stream of tensions and resolutions."[43] Innis's Chapter 10 reads Langer as Langer, without restoring her concepts to any of her major influences, except when referencing her purpose of demonstrating a theory of the human as *animal symbolicum*, as introduced by Cassirer. This chapter elucidates Langer's philosophy in its uniqueness, and easily comes to terms with her bio-psychological and semiotic framework, touching upon the concepts of "individuation" and "involvement" from Langer's later process-oriented ontology. Innis's discourse transitions seamlessly between the topics of nature and culture, illuminating the extent to which "breakings," as intra-psychic and intra-social conflicts, take effect. Core to this discussion is Langer's analysis of the roots of the symbolic mind in her last volume of *Mind III* (1982), and the interrelation of imaginary and material fabrics that culture and civilization are embedded in. It carries as its imperative the normative task of giving form to the perceived chaos that is the avalanche of breakings in our collective fabrics.

Chapter 11, "Music as the DNA of Feeling, and some Speculations on Whitehead's Influence on Susanne K. Langer's Philosophy"—by Lona Gaikis—is my own contribution to this collection of essays, and picks up the rather invisible threads between Susanne Langer and her teacher Whitehead. I speculate whether Langer's key term "feeling" stands in any connection to Whitehead's frequent use of "feeling" or "feelings" in *Process and Reality* (1929). As a graduate student of Whitehead, Langer continued visiting his course on the philosophy of nature from 1927 to 1928. My research is semi-archival, as it refers to Langer's published notes in *Process Studies*, vol. 26 (1997, edited by Rolf Lachmann), but conjectures on the interrelations of their respective terminology. These notes show how Whitehead, in the course Langer attended, was developing his concept of "vector feeling"—the stream of, or undulating waves of, feeling occurring between entities—as preliminary work for his later book *Process and Reality* (1929). This discussion covers Langer's generative idea of a tonal "new key" in philosophy, which grounds *meaning-making* in musical form, respectively a matrix of "dynamic sound-patterns." I give a deeper analysis of Langer's functional understanding of symbols and its overall effect on her conceptual framework. This chapter intends to present Langer's parallels with—but also her determined deviation from—Whitehead's metaphysics. It speculates upon the direct relations between Langer and Whitehead, and intends to open the scene for the generative ideas in Part II of this book.

GENERATIVE IDEAS

Susanne Langer's practice of philosophy was unbound by convention. Her work is wide-encompassing, ranging from phenomenology to philosophy of language and a metaphysics of symbols, from logical positivism and shared issues in pragmatism to flirting with radical empiricism and psychology, and reaching for a process metaphysics in her quest to develop adequate concepts that could render the abundant, yet hard to realize, proto-linguistic realms of mental life. A growing understanding of her extraordinary endeavor and an appreciation for the transversality of her ideas and insights—in the wake of newly discovered frontiers in the sciences and the arts—have led to a renaissance of Langer's theory. She foresaw developments that would go on to become elemental towards the end of the millennium. Her theoretical hybridity in developing a theory of art that emphasizes the symbolization processes as an inter-organic flow is enticing newer generations of both philosophers and artists engaged in new media, as well as those interested in the creation of new research fields. Some of Langer's generative ideas on embodied cognition and biologically informed processes of symbolization might seed novel approaches in postmodern sciences, cultural studies, and philosophy's branching into unexpected planes. Part II of this book gives space to thinkers continuing the avant-garde in Langer. Still, their contributions are not uncritical of Langer's ideas, and read her in the context of the scientific, cultural, and technological advancements that have taken place since her time. They take her thinking into new directions.

The Whiteheadian lineage in Susanne Langer's philosophy, last discussed in Part I, is taken up again in philosopher Adam Nocek's Chapter 12, "Susanne K. Langer and Philosophical Biology." Expanding Langer's development of core concepts, from "model" to "image," and from "primary and secondary illusions," to her later "act" model, Nocek addresses the possibility of a Langerian "philosophical biology" that he finds seeded in her trilogy of *Mind*. Deviating from the usual trails that discuss Langer from a theory of art, Nocek flips the perspective by rooting his analysis in biology. He argues that theoretical biology faces difficulties with the conceptualization of organic development. Science's bias toward standard quantification, generalization, and formalistic abstraction often stands in the way of grasping the full complexity of biological phenomena. Langer's late devising of a philosophy of mind rooted in the physiological and biochemical processes of organic activity could alleviate this conceptual gridlock, as she does not resort to a reductive materialism or to metaphysical dualism.[44] As Langer states, "'Life' is obviously not easy to define."[45] The implications of Langer's Whiteheadian roots reflect many current efforts to renew concepts in both science and the arts, as it caters to eco-critical research, and theories exploring the bonds and boundaries of anthropomorphism.

Continuing these efforts, Eldritch Priest's Chapter 13, "Thinking Non/Humanly with Susanne K. Langer" tackles Langer's contribution to the addressing and understanding of animal mentality. Departing from her later work in *Mind*, and reading Langer as a philosopher of process, Priest's contribution is concerned with the representation of animal thought. Even though Langer's biologically informed theory of the human mind conceives of animal cognition as foundational to its evolutionary emergence, this chapter points out that Langer makes a clear distinction between human and non-human thinking. Central to Priest's argument is Langer's understanding of symbolization processes in humans as acts of organic compulsions, satisfied through an essentially "impractical enthusiasm" for expression, whereas animal mentality is confined to practical demands. This qualitative difference, however, is suspended in the paradox of play, where simians, too, exhibit symbolic activity. Priest's musings take this chapter into the outer limits of what we define as animalistic, and speculates, as well, on the permeable contours of the human as *animal symbolicum*. In line with the previous two chapters that address Langer's foray into biology and the study of animal mentality, Brian Massumi's Chapter 14, "From Aesthetic Frights to the Politics of Unspeakable Thought with Susanne K. Langer" also enters the discussion from Langer's study of animal behavior, iterating the similarities between human and non-human mental substrata, as she reasons in *Philosophy in a New Key* about the basic conditions that could facilitate language in anthropoids. The media theorist thus continues his preceding Langerian discourse on *thinking-feeling* from 2008 and 2011,[46] which involved her notion of virtuality and movement in static forms—formative to his coining of an "affective turn" in the mid-2000s.[47] Massumi now applies his approach to an analysis of the utilitarian and expressive (and suggestive) functions of language. Pointing at how literal meaning in language is turned on its head in the development of human rationality, Massumi takes the "granular level of feeling"—that which determines perception at both conceptual and conceptional levels (vital import)—as the main point of friction and the location of far-reaching errors that twist meaning and lead to derangements in discourse. Expanding on Langer's brief discussion of errors,[48] systematically planted "mis-takes," according to Massumi, distort the texture of social and political discourse today. He diagnoses a "deficit of symbolic thinking-feeling" in the case of raging conspiracy cultures, and calls for a counter-politics of "minor gestures" on the level of meaning's vital import in artistic and philosophical activism.[49] Massumi directs the impact of Langer's symbolism back to *more than* human needs.

In Chapter 15, "Towards Vitality Semiotics and a New Understanding of the *Conditio Humana* in Susanne K. Langer," Martina Sauer puts forth her own theory of cultural psychology. Developed from cultural, anthropological, semiotic, and empirical research, Sauer's proposition of a Vitality Semiotics builds on the Langerian conjecture that artifacts are congruent with a vibrant network of vital experience. Sauer demonstrates Langer's biosemiological engagements from *Mind*, vol. 1 and *Mind*, vol. 2 as well as her empirically oriented configuration, the *act* model, and she drives this further, towards broader issues concerning the accountability of art in the building of social structures. From Langer's act model Sauer establishes a call for action: "a new understanding of the *conditio humana*." Inasmuch as Langer's philosophy provides keystones to Vitality Semiotics, Sauer critically engages the transcendental ideals ingrained in Langer's concept of art as a purposeless affair. She sees Langer's empirical import as a fruitful link for bringing in aspects from developmental psychology in developing a cultural theory.

The final chapters of this book belong to authors who continue the art-theoretical approach of Langer's philosophy. Susanne Langer has indispensable connections not only with the sciences and psychology but also with the epistemological import of the arts. Chapter 16, "Virtual Powers in Susanne K. Langer's Theory of Dance and its Application in Post-Colonial Hong Kong" by Canadian-Chinese theorist Eva Kit Wah Man focuses on Langer's thoughts on dance in *Feeling and Form* (1953). She discusses the fecundity of applying Langer's multi-leveled conception of art perception to the occurrent arts, such as dance and choreography. This chapter analyzes Langer's idea of primary and secondary illusions, thereby pointing at the seemingly mystic forces unleashed by bodies in motion. Langer defines the primary illusion of dance as a "virtual realm of Power,"[50] which strongly shows her Cassirerian influence, coming from Langer's translation of *Language and Myth* (1946)—his first book published in English. This deep engagement with the differences of mythological and linguistic import shapes the greater part of how

Langer conceptualizes the arts. It is this influence that lets her theory appear rather mystical, or obscure, for it emphasizes art's animistic appeal. Eva Man transfers the insights gained from Langer's treatment of the power of dance to contemporary dance choreography in Hong Kong, pointing at the currency and need for mythico-subversive stage performance in climates of political constraint. The political in artistic expression is one of several topics in the following chapter, which expands Langer's theory regarding the radical changes that post-war art has undergone in the global North since Langer's *Feeling and Form* (1953). Philosopher and art historian Christophe Van Eecke proposes a Langerian theory of performance art in his Chapter 17, "'Virtual Acts' as a Langerian Approach to Performance Art." According to Van Eecke, the flourishing of performance and body-related arts that currently blur with various forms of media calls for a reconfiguration of terminologies to render what thus far have been materialistically labeled "Body Art" or "fleshworks." Van Eecke sees the need for a robust concept of performance art that goes beyond these generalities. His chapter reaches for Langer's authenticity regarding art's *real* material—its virtuality. Idiosyncratic of the Langerian semantic approach is her avoidance of reductive schemes. She neither stylizes art as divine inspiration, as a metaphysical prerequisite of reality, nor does she exclusively access the meaning of art from the purely formal analysis of composition, material, or context. This offers rather unorthodox, yet surprisingly effective models for capturing art's virtual expressivity without resorting to the limitations of confining terms. Thus, Van Eecke's chapter unfolds a theory of "virtual acts" in performance. He elucidates the nature and functionality of Langer's concepts with examples from Chris Burden, Marina Abramović, and theater *enfant terrible* Christoph Schlingensief.

Philosopher Thomas Leddy wraps up the discussions that continue Langer's theory of the arts with Chapter 18, "Susanne K. Langer, Everyday Aesthetics, and Virtual Worlds." Leddy's approach in seeking the extraordinary in the ordinary in everyday aesthetics engages the implications of Langer's "virtual worlds" in *Feeling and Form* as an overlaying and illusory level of perception intrinsic to artifacts—impressions of kinetic volume, virtual time and gesture, power, and memory. Alluding only indirectly to the media theory potential in Langer's coining of a virtual realm (Langer is mentioned in the birth of what we call *virtual realities* today[51]), Leddy's line of reasoning considers the heterotopy of the analogue world. Departing from a Deweyan standpoint and in the wake of a ubiquitous mediatization of perception—on the screens in our hands—Leddy sees a need to expand the theory of everyday aesthetics. For Langer, engaging aesthetically in the virtuality of everyday objects meant a consciously induced estrangement from actuality—a "disengagement from a belief"[52]—not the constructing of a *new real.* Deweyans advocate for bringing art and life back together, and while Leddy senses an unfortunate discontinuation of virtuality from its "mundane environment" in Langer's approach, he acknowledges her as an ally in conceptualizing an everyday aesthetics, as her notion of "virtual worlds" offers potential advancements to the Deweyan stance. Alluding to the old philosophical problem of the world and its copies, Leddy sees a new challenge to his discipline in the bifurcation of the screen world and actual world.

NOTES

1. Max Rieser, "The Semantic Theory of Art in America," *The Journal of Aesthetics and Art Criticism* 15, no. 1 (September 1956): 12.
2. Antonio Damasio, *The Feeling of What Happens: Body and Emotion in the Making of Consciousness* (San Diego, CA: Harcourt Brace, 1999), 287.
3. Susanne K. Langer's dissertation, "A Logical Analysis of Meaning," (PhD diss., Radcliffe College, Cambridge, MA, 1926), references Reinhold F. A. Hoernlé's "Plea for a Phenomenology of Meaning" published in 1921 to systematize the perceived chaos in defining a general systemization of the meaning of meaning. At that time, Langer was continuing the early efforts of Frege, Pierce, Lady Welby, Husserl, and Meinong to adequately define this discipline's materials.
4. Langer herself borrows this term from Harvard professor Henry M. Sheffer, who had a substantial impact on Langer's philosophy and introduced her to formal logic. Susanne K. Langer, "The Art Symbol and the Symbol in Art," in *Problems of Art: Ten Philosophical Lectures* (New York: Charles Scribner's Sons, 1957), 125.
5. Cassirer's treatises on *Language* (1923), *Mythical Thought* (1925), and *The Phenomenology of Knowledge* (1929) were a yet-incomplete catalog of symbolic forms, as Cassirer continued to study newly emerging modern art, and would realize a fourth volume to engage the greater idea of a *Metaphysics of Symbolic Forms*. This material was compiled and translated posthumously in Ernst Cassirer, *The Philosophy of Symbolic Forms: Metaphysics of Symbolic Forms*, vol. 4 (New Haven, CT: Yale University Press, 1996).
6. Langer, "A Logical Analysis of Meaning." Langer's thesis was examined and accepted by professors C. I. Lewis, Alfred N. Whitehead, and Henry M. Sheffer.
7. See Adrienne Dengerink Chaplin, ch. 10, "Mind as Embodied Meaning," in *The Philosophy of Susanne Langer: Embodied Meaning in Logic, Art and Feeling* (London/Oxford: Bloomsbury Academic, 2019).
8. In this spirit, Innis highlights Langer's contribution to philosophy as a "'double-bladed' approach," joining a clearly organized categorical and conceptual system with a phenomenological assessment of empirical material. Her crossing of methodological boundaries makes "Langer's way of doing philosophy," as Innis remarks, "so fruitful and exciting" (Robert E. Innis, *Susanne Langer in Focus: The Symbolic Mind* [Bloomington: Indiana University Press, 2009], 255). Langer wanted to show that the manifestation of meaning was not only bound to the discursive forms of languages, but could find its equivalent in the viscerality of the body—and possibly even beyond. His assessment corresponds with Dengerink Chaplin, who emphasizes Langer's compatibility with the neurosciences and the fact that she actually "*anticipated* many major turns in philosophy," despite the lack of demonstrable references (Adrienne Dengerink Chaplin, *The Philosophy of Susanne Langer: Embodied Meaning in Logic, Art and Feeling* [London/Oxford: Bloomsbury Academic, 2019], 261). See also Donald Dryden, "Susanne K. Langer and American Philosophic Naturalism in the Twentieth Century," *Transactions of the Charles S. Peirce Society* 33, no. 1 (1997): 161–82, and "The Philosopher as Prophet and Visionary: Susanne Langer's Essay on Human Feeling in the Light of Subsequent Developments in the Sciences," *The Journal of Speculative Philosophy*, New Series, 21, no. 1 (2007): 27–43.
9. Cf. Dengerink Chaplin, "Bridging Divides," in *The Philosophy of Susanne Langer*, 258–61.
10. Rolf Lachmann, *Susanne K. Langer. Die lebendige Form menschlichen Fühlens und Verstehens* (Munich: Fink Verlag, 2000).
11. Lachmann, *Susanne K. Langer*, acknowledgment (Cologne, 30 November 1999).

12. This edition with special focus on Susanne K. Langer and Whitehead contains Donald Dryden's article, "Whitehead's Influence on Susanne Langer's Conception of Living Form," 62–85; Randall E. Auxier, "Susanne Langer on Symbols and Analogy: A Case of Misplaced Concreteness?" 86–106, and Rolf Lachmann's "From Metaphysics to Art and Back: The Relevance of Susanne K. Langer's Philosophy for Process Metaphysics," 107–25, as well as a transcript of "Susanne K. Langer's Notes on Whitehead's Course on Philosophy of Nature," 126–50. In *Process Studies* 26, no. 1–2 (1997).
13. Barbara Kösters, *Gefühl, Abstraktion, Symbolische Transformation*, ed. Wolfgang Hogrebe (Frankfurt a.M: Peter Lang, 1993).
14. Cf. Kösters, *Gefühl, Abstraktion, Symbolische Transformation*, 16.
15. In chapter 3, "Philosophy and Women" of her book *The Philosophy of Susanne Langer*, 51–53, Adrienne Dengerink Chaplin describes the main reasons for Langer's invisibility in comparison to her male peers, and contrasts the 1926 graduate Langer with Willard V. Quine's educational and early academic career path only four years later in 1930. Both were comparable in their academic steps, the field of research, and their achievements. The main difference was that Langer's pioneering writings on the basics of logic were greatly ignored by her contemporaries. Rather than being recognized as a radical departure from the standard path, her unconventional conception of the symbol was judged as "natural confusion."
16. Sander Verhaegh's chapter in this volume, "Susanne K. Langer and the Harvard School of Analysis," 21–34, discusses these intellectual ties and Langer's lasting impression on the European branch of logic.
17. Verhaegh, Chapter 1 this volume, "Susanne K. Langer and the Harvard School of Analysis," 27–8.
18. Rolf Lachmann discusses Langer's relationship and appraisal of phenomenology in his Chapter 6 in this volume, "Susanne K. Langer's Foray into Art as a 'Phenomenology of Feeling,'" 93–103.
19. Dengerink Chaplin summarizes five main reasons for Langer's neglect: 1. Her field of research in the arts; 2. Her boundary-crossing approach to understanding the "meaning of meaning"; 3. The fact that she was ahead of her time; 4. Her European sources in philosophy were mostly unfamiliar to her colleagues in US academia; and 5. Langer was marginalized by a "general and deep-rooted cultural prejudice against women in philosophy," which prevented her for a long time from procuring a non-tenured position and the academic influence this entailed. *The Philosophy of Susanne Langer*, 2–5.
20. Langer commentator Stefan Morawski, for example, remarks upon her "reiterated polemics with empiricism, naturalism, and with neo-positivist philosophers of science." Stefan Morawski, "Art as Semblance," *The Journal of Philosophy* 81, no. 11 (November 1984): 655. Langer, *The Practice of Philosophy* (New York: Henry Holt, 1930), 13; *Philosophy in a New Key*, 14–16; or *Feeling and Form: A Theory of Art Developed from Philosophy in a New Key* (New York: Charles Scribner's Sons, 1953), 384.
21. Susanne K. Langer, *The Practice of Philosophy* (New York: Henry Holt, 1930), 9.
22. Langer, *The Practice of Philosophy*, 18.
23. On this matter, Lachmann, for instance, quotes from Langer's *Feeling and Form*, 34: "Probably under the pressure of the psychologistic currents that have tended, for the last fifty years at least, to force all philosophical problems of art into the confines of behaviorism and pragmatism, where they find neither development nor solution, but are assigned to vague realms of 'value' and 'interest,' in which nothing of great value or interest has yet been done ... The whole tenor of modern philosophy, especially in America, is uncongenial to serious speculation on the meaning and difficulty and seriousness of art works." Lachmann, *Susanne K. Langer*, 22.

24. Iris van der Tuin and Adam Nocek, "New Concepts for Materialism: Introduction," *Philosophy Today* 63, no. 4 (2020): 815–22.
25. Langer, *The Practice of Philosophy*, 108.
26. See e.g., Ernest Nagel, "Review of Philosophy in a New Key by Susanne K. Langer," *The Journal of Philosophy* 40, no. 12 (June 1943): 323–9; Melvin Rader, "Review: Feeling and Form," *The Journal of Aesthetics and Art Criticism* 12, no. 3 (March 1954): 396–8; Morris Weitz, "Symbolism and Art," *The Review of Metaphysics*7, no. 3 (March 1954): 466–81. Charles Varela and Lawrence Ferrara highlight the latent depreciative tone with which Langer's then-novel conception of symbols was received. See "The Nagel Critique and Langer's Critical Response," *Journal for the Anthropological Study of Human Movement* 2, no. 2 (Autumn 1982): 99–111.
27. Introduction by Bertrand Russell. In Ludwig Wittgenstein, *Tractatus Logico-Philosophicus*, ed. Charles K. Ogden, tran. Frank P. Ramsey (London: Kegan Paul, Trench, Trubner, 1922), 8.
28. Cf. Langer, *The Practice of Philosophy*, 108–9.
29. See Dengerink Chaplin's Chapter 2 in this volume "Scientific Models and Artistic Images: Susanne K. Langer and the Early Wittgenstein," 35–47.
30. Langer, *The Cruise of the Little Dipper*, 12, 36–8.
31. Cf. Giulia Felappi, Chapter 4 in this volume, "Susanne K. Langer on Logic as the Study of Forms and Patterns of Any Sort," 63–77.
32. Langer, "A Logical Study of Verbs," 124. Cited in Giulia Felappi's Chapter 4 in this volume, "Susanne K. Langer on Logic," 67.
33. Susanne K. Langer, *An Introduction to Symbolic Logic* (New York: Allen and Unwin, 1937), 41. Cited in Felappi, Chapter 4 in this volume, "Susanne K. Langer on Logic as the Study of Forms and Patterns of Any Sort," 63.
34. Langer's earliest engagement with a structural analysis of the arts was a paper that sought to ground musical form in Boolean algebra. See Susanne K. Langer, "A Set of Postulates for the Logical Structure of Music," *The Monist* 39, no. 4 (October 1929): 561–70.
35. See Peter E. Gordon, *Continental Divide: Heidegger, Cassirer, Davos* (Cambridge, MA: Harvard University Press, 2010).
36. Ernst Cassirer shares a discussion with Swedish philosopher Marc-Konrad Wogau, in "Zur Logik des Symbolbegriffs," who implied a relapse to sensualism with Cassirer's twofold concept of symbols. In *Wesen und Wirkung des Symbolbegriffs* (Darmstadt: Wissenschaftliche Buchgesellschaft, 1938), 202–50.
37. One of the first researchers to excavate Ernst Cassirer's philosophy in anglophone academia was John Michael Krois in *Cassirer: Symbolic Forms and History* (New Haven, CT: Yale University Press, 1987).
38. Cassirer died in New York in May 1945, mere days before the end of the Second World War. His works were translated into English by Ralph Manheim. See Ernst Cassirer, *The Philosophy of Symbolic Forms*, vols. 1–3 (New Haven, CT: Yale University Press, 1955/7).
39. Cf. Christian Grüny's Chapter 8 in this volume, "The Systematic Position of Art in Susanne K. Langer's and Ernst Cassirer's Thinking," 128.
40. Ibid. 129 fn. 10.
41. Donald Dryden, Chapter 9 in this volume, "The Meaning of 'Feeling' in Susanne K. Langer's Project of Mind," 136.
42. Susanne K. Langer, *Mind: An Essay on Human Feeling*, vol. 1 (Baltimore: Johns Hopkins University Press, 1967), 21 fn. 36.
43. Langer, *Feeling and Form* (New York: Charles Scribner's Sons, 1953), 372.

44. See Adam Nocek's Chapter 12 in this volume, "Susanne K. Langer and Philosophical Biology," 183–199.
45. Langer, *Mind*, vol. 1 (Baltimore: Johns Hopkins University Press, 1967), 258.
46. See Brian Massumi's discussion on "semblance" in "The Thinking Feeling of What Happens: Putting the Radical Back in Empiricism," in *Semblance and Event: Activist Philosophy and the Occurrent Arts* (Cambridge, MA: The MIT Press, 2011), 39–86. This chapter had previously been published online as a conversation between Brian Massumi and Arjen Mulder, "The Thinking-Feeling of What Happens: A Semblance of a Conversation," *INFLeXions*, no. 1—How is Research-Creation? (May 2008): 1–40.
47. Masumi references Langer's notion of semblance in describing the phenomena of embodied perception, or the affectively endowed double sight involved when we perceive objects. Ornament, as Langer describes in *Feeling and Form*, 64–5, evokes, e.g., a sense of virtual movement. See Massumi, "The Thinking-Feeling of What Happens" (2008), 4–5.
48. See e.g., "Prototypes of Error" in Massumi's Chapter 14 in this volume, "From Aesthetic Frights to the Politics of Unspeakable Thought with Susanne K. Langer," 216, referencing Langer, *Philosophy in a New Key*, 29–30.
49. See Brian Massumi in this volume, "From Aesthetic Frights to the Politics of Unspeakable Thought with Susanne K. Langer," 218–19.
50. Langer, *Feeling and Form*, 175.
51. Cf. Jaron Lanier, *Dawn of the New Everything: Encounters with Reality and Virtual Reality* (New York: Henry Holt, 2017), 42.
52. Cf. Langer, *Feeling and Form*, 49.

BIBLIOGRAPHY

Cassirer, Ernst. "Zur Logik des Symbolbegriffs." In *Wesen und Wirkung des Symbolbegriffs*, 202–50. Darmstadt: Wissenschaftliche Buchgesellschaft, 1938.

Cassirer, Ernst. *The Philosophy of Symbolic Forms*, vols. 1–4: (*Language* [1955], *Mythical Thought* [1955], *The Phenomenology of Knowledge* [1957], *Metaphysics of Symbolic Forms* [1996]). Translated by Ralph Mannheim (vols. 1–3) and John Michael Krois (vol. 4). New Haven, CT/London: Yale University Press.

Damasio, Antonio. *The Feeling of What Happens: Body and Emotion in the Making of Consciousness*. San Diego, CA: Harcourt Brace, 1999.

Danto, Arthur C. "Mind as Feeling; Form as Presence; Langer as Philosopher." *The Journal of Philosophy* 81, no. 11 (November 1984): 641–7.

Dengerink Chaplin, Adrienne. *The Philosophy of Susanne Langer: Embodied Meaning in Logic, Art and Feeling*. London/Oxford: Bloomsbury Academic, 2019.

Dryden, Donald. "The Philosopher as Prophet and Visionary: Susanne Langer's Essay on Human Feeling in the Light of Subsequent Developments in the Sciences." *The Journal of Speculative Philosophy*, New Series, 21, no. 1 (2007): 27–43. Available online: https://www.jstor.org/stable/25670642 (accessed July 18, 2023).

Dryden, Donald. "Susanne K. Langer and American Philosophic Naturalism in the Twentieth Century." *Transactions of the Charles S. Peirce Society* 33, no. 1 (1997): 161–82. Available online: https://www.jstor.org/stable/40320624 (accessed April 13, 2022).

Dryden, Donald. "Susanne K. Langer." In *Dictionary of Literary Biography*, vol. 270: *American Philosophers before 1950*, edited by Philip B. Dematteis and Leemon B. McHenry, 189–99. Farmington Hills, Michigan: Gale, 2003.

Gordon, Peter E. *Continental Divide: Heidegger, Cassirer, Davos*. Cambridge, MA: Harvard University Press, 2010.
Innis, Robert E. *Susanne Langer in Focus: The Symbolic Mind*. Bloomington: Indiana University Press, 2009.
Kösters, Barbara. *Gefühl, Abstraktion, Symbolische Transformation*. Edited by Wolfgang Hogrebe for Studia Philosophica et Historica. Frankfurt a.M: Peter Lang, 1993.
Krois, John Michael. *Cassirer: Symbolic Forms and History*. 1st edn. New Haven, CT: Yale University Press, 1987.
Lachmann, Rolf, ed. *Kulturelle Symbolisierung*. Frankfurt a. M.: Suhrkamp, 2000.
Lachmann, Rolf, ed. *Process Studies: Special Focus on Susanne K. Langer and Whitehead*, 26, no. 1–2 (1997). Available online: https://scholarlypublishingcollective.org/uip/ps/issue/26/1-2 (accessed July 17, 2023).
Lachmann, Rolf. *Susanne K. Langer. Die lebendige Form menschlichen Fühlens und Verstehens*. Munich: Fink Verlag, 2000. Available online: http://digi20.digitale-sammlungen.de/de/fs1/object/display/bsb00043407_00001.html (accessed July 18, 2023).
Langer, Susanne K. "A Logical Analysis of Meaning." PhD diss., Radcliffe College, Cambridge, MA, 1926.
Langer, Susanne K. "A Logical Study of Verbs." *Journal of Philosophy* 24, no. 5 (March 3, 1927): 120–9. Available online: http://www.jstor.org/stable/2015082 (accessed July 18, 2023).
Langer, Susanne K. "A Set of Postulates for the Logical Structure of Music." *The Monist* 39, no. 4 (October 1929). Available online: http://www.jstor.org/stable/27901211 (accessed July 18, 2023).
Langer, Susanne K. *Feeling and Form: A Theory of Art Developed from Philosophy in a New Key*. New York: Charles Scribner's Sons, 1953.
Langer, Susanne K. *Philosophy in a New Key: A Study in the Symbolism of Reason, Rite and Art*. 17th printing. Cambridge, MA: Harvard University Press, 1993.
Langer, Susanne K. *The Cruise of the Little Dipper, and Other Fairy Tales*. New York: Graphic Society, 1924.
Langer, Susanne K. *The Practice of Philosophy*. New York: Henry Holt, 1930.
Langer, Susanne K., Paul Henle, and Horace M. Kallen, eds. *Structure, Method, and Meaning: Essays in Honor of Henry M. Sheffer*. New York: Liberal Arts Press, 1951.
Lanier, Jaron. *Dawn of the New Everything: Encounters with Reality and Virtual Reality*. New York: Henry Holt, 2017.
Massumi, Brian. "The Thinking Feeling of What Happens: Putting the Radical Back in Empiricism." In *Semblance and Event: Activist Philosophy and the Occurrent Arts*, 39–86. Cambridge, MA: The MIT Press, 2011.
Massumi, Brian, and Arjen Mulder. "The Thinking-Feeling of What Happens: A Semblance of a Conversation." *INFLeXions* no. 1—How is Research-Creation? (May 2008): 1–40. Available online: http://inflexions.org/n1_The-Thinking-Feeling-of-What-Happens-by-Brian-Massumi.pdf (accessed July 18, 2023).
Morawski, Stefan. "Art as Semblance." *The Journal of Philosophy* 81, no. 11 (November 1984): 654–63. Available online: https://www.jstor.org/stable/2026172 (accessed July 18, 2023).
Nagel, Ernest. "Review of Philosophy in a New Key by Susanne K. Langer." *The Journal of Philosophy* 40, no. 12 (June 10, 1943): 323–9. Available online: https://doi.org/10.2307/2018224.

Rader, Melvin. "Review: Feeling and Form." *The Journal of Aesthetics and Art Criticism* 12, no. 3 (March 1954): 396–8. Available online: http://www.jstor.org/stable/426984 (accessed July 18, 2023).

Rieser, Max. "The Semantic Theory of Art in America." *The Journal of Aesthetics and Art Criticism* 15, no. 1 (September 1956): 12–26. Available online: https://doi.org/10.2307/427490.

Russell, Bertrand. "Introduction." In Ludwig Wittgenstein, *Tractatus Logico-Philosophicus*, edited by Charles K. Ogden, translated by Frank P. Ramsey, 7–19. London: Kegan Paul, Trench, Trubner, 1922.

van der Tuin, Iris, and A. J. Nocek. "New Concepts for Materialism: Introduction." *Philosophy Today* 63, no. 4 (2020): 815–22. Accessible online: https://doi.org/10.5840/philtoday2019634294.

Varela, Charles, and Lawrence Ferrara. "The Nagel Critique and Langer's Critical Response." *Journal for the Anthropological Study of Human Movement* 2, no. 2 (Autumn 1982): 99–111. Available online: https://jashm.press.uillinois.edu/2.2/index.html (accessed July 18, 2023).

Weitz, Morris. "Symbolism and Art." *The Review of Metaphysics* 7, no. 3 (March 1954): 466–81. Available online: http://www.jstor.org/stable/20123390 (accessed July 18, 2023).

Susanne K. Langer's body of work emerged from a web of key movements in twentieth-century philosophy. The essays presented in this book excavate Langer's work from various angles to reveal the multifaceted nature of her thinking. Part I situates Langer within the intellectual atmosphere and philosophical influences of her time. It presents conceptual, structural, and material aspects of her work, and highlights several instances of her unique and visionary approach to philosophy. Part II expands upon Langer's central ideas, showcasing her insights and transdisciplinary reach. These central ideas are discussed transversally across chapters, stimulating further interest in Langer's philosophy from any point of engagement.

PART ONE

Placing Susanne K. Langer

CHAPTER ONE

Susanne K. Langer and the Harvard School of Analysis

SANDER VERHAEGH

INTRODUCTION

Susanne Langer was a student at Radcliffe College between 1916 and 1926—a highly transitional period in the history of American philosophy. Intellectual generalists such as William James, John Dewey, and Josiah Royce had dominated philosophical debates at the turn of the century but the academic landscape gradually started to shift in the years after the First World War. Many scholars of the new generation adopted a more piecemeal approach to philosophy—solving clearly delineated, technical puzzles using the so-called "method of logical analysis."[1] Especially at Harvard, the intellectual climate rapidly changed. The department hired several philosophers who had contributed to the development of symbolic logic—H. M. Sheffer, C. I. Lewis, and A. N. Whitehead—and Harvard quickly began to be viewed as a central hub for analytic philosophy in the United States.

This chapter contextualizes Langer's earliest work by reading it through the lens of this shifting academic environment. Though Harvard did not allow women to take its courses until 1943, Langer is one of the most significant fruits of this period. Her dissertation "A Logical Analysis of Meaning" and her first publications are all illustrations of the approach that came to dictate the American philosophical conversation. By exploring the increased focus on the logical-analytic method and Langer's attempts to expand the new approach to what she later called "non-discursive" symbolisms,[2] I situate her publications in the intellectual context of the 1920s.

THE METHOD OF LOGICAL ANALYSIS

In 1914, two years before Langer entered college, Bertrand Russell spent a semester at Harvard University. The philosopher-logician was already an academic celebrity at the time. He and Whitehead had just completed their three-volume *Principia Mathematica*, he had been one of the first to employ the new logic to tackle philosophical problems, and he was one of the leading opponents of James's theory of truth. Naturally, the

I thank Lona Gaikis for her helpful comments and suggestions. Parts of sections 4 and 7 of this chapter are taken from my paper "Susanne Langer and the American Development of Analytic Philosophy," in *Women in the History of Analytic Philosophy*, ed. Jeanne Peijnenburg and Sander Verhaegh (Cham: Springer, 2022).

department was eager to acquire the Englishman who started to receive "more attention than any logician since Aristotle."[3] According to Harvard philosopher George Santayana, there was "no one whom the younger school of philosophers" was "more eager to learn of" than Russell. And T. S. Eliot, who attended the latter's logic seminar, even wrote a poem—*Mr. Apollinax*—about the way the Harvard community behaved in the presence of the forty-one-year-old philosopher.[4]

Russell visited "the other Cambridge" in a crucial period in the history of Harvard philosophy. The local department had, in Russell's view, been "the best in the world"[5] but it had lost three of its intellectual leaders in a few years' time. James had died, Santayana had moved to Europe, and Royce had suffered a mild stroke. Though he had received Harvard's first invitation when all three were still active, the department was in a state of deep crisis when he first entered Emerson Hall in March 1914. Not surprisingly, Russell made full use of the opportunity to fill the void created by the unit's intellectual decapitation. He persuaded the department's chairman "that logic is *the* important thing" in philosophy, taught courses on epistemology and the *Principia*, and argued that logic is "the essence of philosophy"[6] to a crowd of over five hundred people.[7] Symbolic logic had always played a role in Harvard's philosophy curriculum due to Royce and the indirect influence of C. S. Peirce but its impact had been rather limited in a department that valued methodological pluralism.[8] Russell's Lowell Lectures *Our Knowledge of the External World* were explicitly designed to make amends and illustrated "by means of examples, the nature, capacity, and limitations of the logical-analytic method in philosophy."[9]

The method of logical analysis means many things in Russell's philosophy. On the most general level, it asks philosophers to adopt a scientific attitude and to view their discipline as a collaborative, objective enterprise, aiming "at results independent of [their] tastes and temperament."[10] Russell maintained that philosophy should aim at "piecemeal, detailed, and verifiable results" instead of "large untested generalities recommended only by a certain appeal to imagination."[11] On a more detailed level, Russell's method involved what Michael Beaney calls a "transformative" approach to analysis.[12] Whereas traditional philosophers had tried to analyze complex ideas and propositions by *dissecting* them into component parts, Russell advocated *rephrasing* them into their proper, logical form. The best-known illustration of this method is Russell's theory of descriptions, often heralded as a "paradigm of philosophy."[13] In "On Denoting," Russell (1905) aimed to dissolve ontological questions about non-referring descriptions such as "the present King of France" by analyzing them away. Rather than dissecting a sentence such as "The present King of France is bald" into a subject (the present King of France) and a predicate (is bald), he proposed to rephrase the sentence as "There is one and only one King of France, and whatever is King of France is bald," arguing that, on such an analysis, there is no longer any puzzle about the sentence's truth value.[14]

A key component of Russell's approach is the notion of "logical form." The method just sketched presupposes that every sentence has an underlying logical structure that may be masked by its grammatical appearance. In his Lowell Lectures, Russell characterized logical form as whatever remains unchanged when the constituents of a sentence are altered. In a series of propositions such as (1) "Socrates drank the hemlock," (2) "Coleridge drank the hemlock," (3) "Coleridge drank opium," and (4) "Coleridge ate opium," the constituents of (1) are altered one by one while the logical form of the propositions remains the same. The logical form of a proposition, in other words, "is not another constituent, but is the way the constituents are put together."[15] According to Russell, philosophy can be defined as the discipline which is "concerned with the analysis and enumeration of logical forms." Whereas the special sciences aim to answer questions that are decided by empirical evidence, philosophy is "the science of the possible" and concentrates attention upon the investigation of the "logical forms" that allow us to meaningfully talk about the world.[16]

TRANSITION

Russell's attempts to sell his logical-analytic method were successful. After his 1914 visit, he became the faculty's prime candidate to become James's successor. In his final days as the department's chair, Ralph Barton Perry wrote that they had to try "by hook or crook [to] attach [Russell] to ourselves"; and when Royce died in 1916, the department's new chair James Haughton Woods acted swiftly, offering Russell a position.[17] And though Russell never became a Harvard professor because of a conviction during the First World War, the department immediately started to search for people with a similar profile, hoping that they could help Harvard attract "many of the cleverest of the youth with predilections for logic."[18] The department hired Sheffer and Ralph Monroe Eaton as logic instructors, both of whom would stay at Harvard until the end of their careers. And a few years later, Woods managed to obtain the funds to offer positions to Lewis (1921) and Whitehead (1924), appointing two of the best-known logicians in the anglophone world. Lewis had just published his seminal *A Survey of Symbolic Logic* (1918), presenting his system of strict implication; Whitehead was the co-author of *Principia Mathematica*. Sheffer, finally, had studied with Russell in Cambridge before the latter's 1914 visit and was viewed as "Russell's most enthusiastic representative at Harvard."[19]

As a result of these changes, Harvard quickly became a central hub for technical philosophy in the United States. Whereas James had once confessed that he was "a-logical, if not illogical, and glad to be so," there was "an unmistakable drift in the direction of logic" among graduate students by the late 1920s.[20] Roy Wood Sellars wrote about the "efflorescence of mathematical logic so characteristic of Harvard"[21] and Palmer and Perry[22] boasted about the department's "unquestioned leadership" in the field in an article about the evolution of Harvard philosophy.[23] Bruce Kuklick's study of Harvard philosophy doctorates confirms these conclusions about the rapid transformation of the Harvard intellectual climate. The proportion of dissertations on technical subjects (logic, methodology, epistemology, and philosophy of science) increased from 0 percent in the 1890s to a stunning 54.8 percent in the 1920s.[24] The new generation of Cambridge's best and brightest—e.g., Susanne Langer, William Parry, Henry Leonard, W. V. Quine, and Nelson Goodman—produced dissertations that fell squarely in line with Russell's plea for the use of the logical-analytic method in philosophy. Whether or not it was a direct consequence of Russell's suggestion to invest more in logic, the department had quickly become a frontrunner in the analytic approach that would come to dominate American philosophy after the Second World War.

LANGER'S DICTUM

Whitehead, Sheffer, and Lewis were Langer's most prominent teachers. It is hardly surprising, therefore, that Langer became a strong advocate of the method of logical analysis. She explicitly adopted the approach in her dissertation "A Logical Analysis of Meaning" (1926) and early publications such as "Confusion of Symbols and Confusion of Logical Types" (1926) and "A Logical Study of Verbs" (1927).[25] In fact, Langer was already well known for her adoption of the method when she was still a student. Unpublished lecture notes of Sheffer's 1924 seminar on philosophic methods make mention of "Mrs. Langer's dictum that the analytic is the only method in philosophy."[26] Naturally, Sheffer himself also favored the approach. His seminar discussed "dialectic," "intuitional," "pragmatic," and "phenomenological" methods but it was clear that Sheffer was committed to the method of logical analysis. He was convinced that "logic *is* philosophy."[27] Indeed, in reflecting on the period, Langer would later describe Sheffer as the "intellectual guide of a small group of perceptive, serious students … who looked forward to a new philosophical era, that was to grow from logic and semantics."[28]

Langer, like Russell, presupposed a transformative conception of analysis. In her dissertation, Langer set herself the task to find the "logical form of all meaning-situations."[29] And in her textbook *An Introduction of Symbolic Logic*, one of the first logic handbooks published in the United States, she offered students a host of examples to teach them the importance of the distinction between a statement's grammatical appearance and its underlying logical form.

> [I]n "Jones killed his wife" the word which a grammarian would call the direct object does more than a direct object should, namely to denote the element to which Jones stood in the relation of killing; it also conveys that this element stood in the relation "wife" to Jones. In other words, "Jones killed his wife" means more than "A kd B," though that is its grammatical form; it signifies "A kd B and B wf A." Here we see how … it is … the easiest thing in the world to miss [a statement's] logical form completely.[30]

The notion of logical form, in other words, played an important role in Langer's work. In explaining the notion, Langer explicitly relied on Russell's account, extensively quoting from the 1914 lectures in which the British philosopher had characterized logical form as the way "constituents are put together" using a series of propositions starting with "Socrates drank the hemlock."[31] According to Langer, she could not have done "better than to quote Bertrand Russell's admirably lucid exposition of logical forms."[32] Langer repeated this strategy in her above-mentioned textbook, published seven years later. After defining logic as "a science of forms," she again used Russell's account to explain to students what she meant when she talked about the "logical form of our language."[33]

PLURALISM

Although Russell is the most-cited philosopher in *The Practice of Philosophy* and some of Langer's first journal publications, it would be a mistake to conclude that she was Russell's disciple, except in the broad sense of adopting a logical-analytic approach and a transformative conception of analysis. There are at least two important differences between Russell's and Langer's accounts, both of them inspired by her *direct* teachers.[34] Rather than thanking Russell, Langer often expressed her indebtedness to Sheffer in her earliest publications. In her dissertation, Langer noted that her analysis of meaning is a "philosophical application of the purely formal work done by … Dr. Sheffer"[35] and in her logic textbook, Langer thanked Sheffer for the insight that logic is the "science of forms."[36]

A first key difference is that Langer advocated a pluralistic stance toward logical structures. Most early twentieth-century philosophers and logicians, including Russell, had defended a universalist conception of logic, culminating in Wittgenstein's thesis that propositions are pictures of facts and that facts and their corresponding propositions have the same logical form.[37] For Langer, however, there is no such thing as *the* logical form of reality. In "Form and Content: A Study in Paradox" (1926), Langer argued that the world can be symbolized by different logical systems:

> The false premise ... is the supposition that there is such a thing as *the* form of anything. A logical form is always relative to a system; a logical term or complex of terms without reference to any particular system is as meaningless as a word or phrase without reference to any particular language.[38]

Whereas Russell presupposed that logic is absolute, Langer accepted a pluralistic philosophy of logic. She repeated her thesis in her first monograph *The Practice of Philosophy*, adding that we can pragmatically choose between logics by selecting the system that best suits our purposes:

> There is no such [thing] as *the form* of a real thing, or of an event ... there are many patterns possible within the same reality ... This means, in the end, that *all understanding is selective*, and that the great work of science is to find out those ways of conceiving an object which shall be most appropriate to certain purposes ... Originality and genius in science consist mainly in the ability to recognize the configurations which are important for a given purpose.[39]

In advocating such a pluralistic conception, Langer was clearly influenced by her Harvard background. For Lewis and Sheffer, too, had defended variants of pluralism and were known for their pragmatic conception of the *a priori*[40] and theory of notational relativity.[41] In a co-authored paper on the development of American philosophy, Langer credits both her teachers for the innovation and traces the idea back to Royce, who first conceived of logic as the study of abstract *forms* (plural):

> Royce's logic belongs to the fertile new inquiry. His conception of logic *as the study of abstract forms* is exemplified in the analysis of formal systems developed by H. M. Sheffer, demonstrating the relativity of abstract structures themselves to the notation by which they are rendered ... The technical development of Royce's logic ... led to the free construction of 'logics' by C. I. Lewis—systems of inference all somehow related to the classical pattern, but 'queer' in their assumptions and ... formal appearance.[42]

Whitehead, finally, helped Langer develop a *diachronic* perspective on logical systems. Following the latter's *Science and the Modern World* (1925) and moving from individual propositions to systems of thought, Langer maintained that the history of philosophy should not just be viewed as a succession of different theories but as a series of logical languages or conceptual frameworks.[43]

NON-DISCURSIVE SYMBOLISMS

Langer does not only defend a *pluralistic* conception of logical form; she also *generalizes* it. Though she often uses Russell's 1914 account to explain the distinction between a sentence's logical form and its grammatical appearance, Langer moves beyond the English philosopher-logician in using the term in a much broader sense, including for example "musical form," "physical, grammatical, social forms," and "norms of conduct"; in short, anything that "follows a pattern of any sort, exhibits order, internal connection."[44] Russell and many of his contemporaries employed the method of logical analysis exclusively to study what Langer would later call "discursive" symbolisms,[45] relegating all other types of expressions to the realm of emotion or the "unspeakable."[46] Langer, however, maintained that the approach can also be used to study non-discursive or "presentational" symbolisms (e.g., art, myths, and dreams):

> This logical "beyond," which Wittgenstein calls the "unspeakable," both Russell and Carnap regard as the sphere of subjective experience, emotion, feeling, and wish ... The study of such products they relegate to psychology, not semantics. And here is my point of radical divergence from them ... We are dealing with symbolisms here ... The field of semantics is wider than that of language.[47]

Langer would not develop her seminal distinction between discursive and presentational symbolisms until the 1940s but much of her early work can be also read as an attempt to break with the more restricted conception of logical form. Her dissertation aimed to show that "Mr. Russell's system of 'propositional' logic" is not sufficiently general to account for all "possible meaning-situations";[48] one of her first journal publications aimed to develop a set of postulates to reveal the "logic of music";[49] and *The Practice of Philosophy* argued that a theory of meaning which fails to incorporate the significance of art, "commits exactly the sins of narrowness which logical philosophy is supposed to avert."[50]

Langer's attempts to move beyond the presuppositions of early analytic philosophy were, again, inspired by her Harvard teachers. In an essay written for a *Festschrift* for Sheffer, Langer argued that Russell failed to "see the entire potential range of philosophical studies built on the study of relational logic," emphasizing that "Whitehead came nearer to it," that "Peirce and Royce saw it" but that "the actual development of systematic abstraction" had been the accomplishment of her most valued teacher.[51] It was Sheffer who had shown her Russell's "error of treating logic as essentially a study of *propositional* forms" and taught her that logic should concern itself "with *all* sorts of forms." [emphasis added][52]

NEW DIRECTIONS

The previous sections sketch some of the ways in which Langer was a child of her time. In the earliest stages of her career, Langer liberally combined influences from Peirce, Royce, Russell, Lewis, Whitehead, and especially Sheffer. A more complete account of Langer's intellectual context would also have included German influences on her thought: philosophers such as Ernst Cassirer, Edmund Husserl, and Ludwig Wittgenstein all played a significant role in the formation of Langer's ideas.[53] Langer's admiration for German intellectual movements was somewhat unusual in the early 1920s since Germany had been widely viewed as the aggressor during the First World War. By "labeling a conception, a policy, or a mode of conduct 'German,'" Frank Thilly wrote a few years after the end of the conflict, philosophers were able "to put the quietus on it: whatever was German was wrong."[54] Still, her reading of the Germanophone literature—which came naturally to her since her parents were German immigrants[55]—significantly influenced Langer's development.

In arguing that Langer was partly a product of her scholarly environment in the earliest stages of her career, I do not wish to suggest that she was just a passive recipient of the views of her teachers. On the contrary, Langer developed these views into new directions and played an active role in shaping the course of American analytic philosophy throughout the 1930s. She was probably the first American philosopher to use the term "analytic philosophy" in print[56] and her books were widely reviewed in US philosophy journals.[57] When the New York philosopher Leo Abraham, a few years after the publication of *The Practice of Philosophy*, made a list of the philosophers who had given "considerable impetus to the development of a distinct science of symbolism," he included Langer on his list, along with Russell, Wittgenstein, and Carnap.[58] Similarly, when American philosophers such as Charles Morris and Ernest Nagel were asked about the development of logic and scientific philosophy in the United States on their trips to Europe, they both mentioned Langer as one of the main representatives.[59] Langer's early publications made her one of the few American experts in a field that was quickly becoming more popular in the 1930s, as is evinced by the role she played in institutionalizing the new approach through her activities for organizations such as the Association of Symbolic Logic and the Unity of Science movement.

Langer's book also had quite an impact *outside* the United States. When Quine, who was also a student of Sheffer, Lewis, and Whitehead, attended meetings of the Vienna Circle in 1933, for example, he was surprised to discover that they were reading *The Practice of Philosophy*.[60] And it was definitely not the first time that members of the Circle had been studying Langer's book. Moritz Schlick, the group's leader, had already written about it two years before, when he had first received a copy from its publisher, Henry Holt and Company. In his reply to the publisher, Schlick had praised the book in exceptionally strong terms:

> I have thoroughly enjoyed reading it. There have been very few philosophical books indeed during the last years that have given me a similar pleasure. The book is certainly excellently written. The author's exquisite style, lucid, fluent and brilliant, has been a source of real joy for me, and must be, I am sure, for every reader. But what is more important: the philosophy expounded in the book is the true kind of philosophy: its method, the method of logical analysis, will be the only method of future philosophizing.[61]

Carnap, too, appears to have been impressed by Langer's work. Although he did not attend the 1933 meetings of the Vienna Circle, he listed Langer as one of the people he would like to work with if he were to obtain a Rockefeller Fellowship to move to the United States.[62] And when Carnap finally did arrive in the United States approximately two years later, Langer was one of the first people he met. Carnap's diary reveals that he, Langer, and a few Harvard academics had tea at Quine's place on December 26, 1935, about a week after he arrived in the United States. In his diary, Carnap notes that Langer used her first meeting with the by-then famous German philosopher to ask him about Frege.[63]

It was Herbert Feigl, however, who was most influenced by Langer in the early 1930s. When the Viennese philosopher visited Harvard for a year on Rockefeller Fellowship, Langer was one of his most frequent contacts. In letters to Schlick, Feigl regularly mentions meetings of a discussion group on logic and philosophy organized by the Radcliffe philosopher. Feigl called it the "*Langer Zirkel*" and told Schlick that it reminded him of the *Wiener Kreis*:

> I was delighted to meet Susanne Langer, who is a professor here at Radcliffe College … She is an excellent woman and her versatility is admirable … We (i.e. a group of young people who are interested in logic and philosophy …) meet at her place every Monday evening for discussions on the Viennese model.[64]

> In the Langer Circle we almost always discuss logic ... it is almost like in Vienna! Mrs. Langer is always very interesting; amusingly, she reminds me a bit of Wittgenstein in her demeanor, in her intuitive determination, and in the biblical conciseness of her statements.[65]

Feigl's letters reveal that Langer, despite her junior position, played an important role in Cambridge's philosophical circles. Not only did she host meetings for young people "interested in logic and philosophy," she also invited senior professors to her circle. When her group started to discuss work in the philosophy of physics, for example, the meetings were attended by the later Nobel laureate Percy W. Bridgman, who had recently published his influential *The Logic of Modern Physics* (1927). And when Russell spent another period at Harvard in 1929, he also visited Langer's "cozy attic studio" to discuss philosophy.[66]

Considering Feigl's high opinion of Langer, it is hardly a surprise that his positivist manifesto "Logical Positivism: A New Movement in European Philosophy" prominently mentions her book as one of the three "American publications" that exhibit tendencies related to the approach that had been developed in Vienna,[67] the other two being Lewis's *Mind and the World-Order* and Bridgman's *The Logic of Modern Physics*. At the time, however, this would have been a remarkable list: Bridgman and Lewis were both established professors and were viewed as some of the most influential scholars of the country; Langer was a thirty-five-year-old analytic philosopher who had just started developing the views of her teachers into exciting, new directions.

NOTES

1. Susanne K. Langer, "A Logical Analysis of Meaning," (PhD diss., Radcliffe College, Cambridge, MA, 1926), 2.
2. Susanne K. Langer, *Philosophy in a New Key: A Study in the Symbolism of Reason, Rite and Art* (Cambridge, MA: Harvard University Press, 1942), 75.
3. Royce, cited in Victor F. Lenzen, "Bertrand Russell at Harvard, 1914," *Russell: The Journal of Bertrand Russell Studies* 3 (1971): 4.
4. Santayana to Russell, February 8, 1912, cited in Bertrand Russell, *Autobiography* (London: George Allen & Unwin, 1967–9/1975), 259–60; Thomas S. Eliot, *Prufrock and Other Observations* (London: The Egotist Press, 1917).
5. Russell, *Autobiography*, 205.
6. Bertrand Russell, *Our Knowledge of the External World as a Field for Scientific Method in Philosophy* (Chicago: Open Court, 1914), 26.
7. Russell to Morrell, March 19 and May 29, 1914, cited in Nicholas Griffin, *The Selected Letters of Bertrand Russell* (Boston: Houghton Mifflin, 1992), 497, 508.
8. Between 1878 and 1915, only two out of the 103 Harvard philosophy dissertations were on "Logic and Methodology" (see George H. Palmer and Ralph B. Perry, "Philosophy, 1870–1929," in *The Development of Harvard University since the Inauguration of President Eliot, 1869–1929* (Cambridge, MA: Harvard University Press, 1930], 30). Most likely, Palmer and Perry are referring to the dissertations of Henry M. Sheffer, "A Program of Philosophy Based on Modern Logic," PhD diss., Harvard University, 1908, and Clarence I. Lewis, "The Place of Intuition in Knowledge," PhD diss., Harvard University, 1910.
9. Russell, *Our Knowledge of the External World as a Field for Scientific Method in Philosophy*, xv.

10. Ibid.
11. Ibid., 2.
12. Michael Beaney, *The Analytic Turn: Analysis in Early Analytic Philosophy and Phenomenology* (London: Routledge, 2007), 2.
13. Frank P. Ramsey, *The Foundations of Mathematics* (London: Routledge, 1931), 263.
14. See Beaney, *The Analytic Turn: Analysis in Early Analytic Philosophy and Phenomenology*, 2–3.
15. Russell, *Our Knowledge of the External World as a Field for Scientific Method in Philosophy*, 34.
16. Bertrand Russell, "Scientific Method in Philosophy," 1914, in *Mysticism and Logic and Other Essays* (London: George Allen & Unwin, 1917), 75–93.
17. Perry to Bernard Berenson, March 20, 1914, cited in Bruce Kuklick, *The Rise of American Philosophy: Cambridge, Massachusetts, 1860–1930* (New Haven, CT: Yale University Press, 1977), 409; Woods to Russell, September 23, 1916, The Bertrand Russell Archive, McMaster University (hereafter BRA), RA1, Box 5.57.
18. Woods to Russell, January 5, 1916, BRA, RA1, Box 5.57.
19. Juliet Floyd, "Sheffer, Lewis, and the 'Logocentric Predicament,'" in *C. I. Lewis: The A Priori and the Given* (New York: Routledge, 2021), 33. Incidentally, this is also the period in which Harvard philosophers helped generate new attention for Peirce's contributions to the development of symbolic logic. Harvard acquired the latter's papers in 1914 and reserved funds to have a group of scholars organize and catalog them, leading to the six-volumed *Collected Papers of Charles Sanders Peirce*. Peirce's work was to have a significant influence on Langer's development. See Robert E. Innis, *Susanne Langer in Focus: The Symbolic Mind* (Bloomington: Indiana University Press, 2009).
20. James to Peirce, December 24, 1909, cited in Cheryl Misak, *Cambridge Pragmatism: From Peirce and James to Ramsey and Wittgenstein* (Oxford: Oxford University Press, 2018), 104; Palmer and Perry, "Philosophy, 1870–1929," 31.
21. Roy W. Sellars, "Current Realism in Great Britain and United States." *The Monist* 37, no. 4 (1927): 513.
22. Palmer and Perry, "Philosophy, 1870–1929," 27.
23. The fact that Harvard became widely viewed as center for technical philosophy does not imply that the department could keep up with the developments in Europe. Indeed, Quine would later complain that although "American philosophers *associated* Harvard with logic because of Whitehead, Sheffer, Lewis, and the shades of Peirce and Royce ... the action was in Europe," where the work of Ackermann, Bernays, Gödel, Herbrand, Löwenheim, Skolem, and von Neumann was revolutionizing the field (Willard V. Quine, "Autobiography of W. V. Quine," in *The Philosophy of W. V. Quine*. Library of Living Philosophers (La Salle, IL: Open Court, 1986), 9, emphasis added).
24. See Kuklick, *The Rise of American Philosophy: Cambridge, Massachusetts*, ch. 24; approx. 3. PhDs in psychology are not included in these numbers.
25. Langer's dissertation was supervised by Whitehead, who had then recently arrived at Harvard. The dissertation is viewed as one of the best Harvard theses in logic from that period. Juliet Floyd, who studied all Harvard dissertations on logic up through 1932, describes Langer's dissertation as "the finest thesis I saw." Juliet Floyd, "Recent Themes in the History of Early Analytic Philosophy," *Journal of the History of Philosophy* 47, no. 2 (2009): 199.
26. Unidentified author, "Secretary's Note Book: Seminary in Philosophic Method 1924," Harvard University, Houghton Library, Susanne K. Langer Papers, Box 29. See also Kris

McDaniel, "Ontology and Philosophical Methodology in the Early Susanne Langer," in *Innovations in the History of Analytical Philosophy* (London: Palgrave Macmillan, 2017), fn. 3.

27. Floyd, "Sheffer, Lewis, and the 'Logocentric Predicament,'" 34.
28. Susanne K. Langer, "Henry M. Sheffer 1883–1964," *Philosophy and Phenomenological Research* 25, no. 2 (1964): 306.
29. Langer, *A Logical Analysis of Meaning*, 2.
30. Susanne K. Langer, *An Introduction to Symbolic Logic* (New York: Allen and Unwin, 1937), 53.
31. See section "The method of logical analysis" in this chapter.
32. Susanne K. Langer, *The Practice of Philosophy* (New York: Henry Holt, 1930), 91.
33. Langer, *An Introduction to Symbolic Logic*, x, 31.
34. I say "at least two important differences" because there are many smaller, more technical, points of disagreement. See, e.g., Susanne K. Langer, "Confusion of Symbols and Confusion of Logical Types," *Mind* 35, no. 13 (1926): 222–9; or Susanne K. Langer, "The Logical Perspectives of the World," *Journal of the Philosophy* 30, no. 7 (1933): 178–87.
35. Langer, *A Logical Analysis of Meaning*, iii.
36. Langer, *An Introduction to Symbolic Logic*, x. Adrienne Dengerink Chaplin also notes that Sheffer was the "first and foremost significant influence on Langer's thinking," in *The Philosophy of Susanne Langer: Embodied Meaning in Logic, Art and Feeling* (London: Bloomsbury, 2020), 61.
37. Ludwig Wittgenstein, "Tractatus Logico-Philosophicus" 2.2., in *Annalen der Naturphilosophie* 14 (1921): 195–262.
38. Susanne K. Langer, "Form and Content: A Study in Paradox," *Journal of Philosophy* 23, no. 16 (1926): 437.
39. Langer, *The Practice of Philosophy*, 135–6, 142 (Langer's emphasis). See also Giulia Felappi, "Susanne Langer and the Woeful World of Facts," *Journal for the History of Analytical Philosophy* 5, no. 2 (2017): 38–50, §2, and McDaniel, "Ontology and Philosophical Methodology in the Early Susanne Langer," 272–83 for more detailed analyses of Langer's pluralism.
40. See Clarence I. Lewis, "A Pragmatic Conception of the A Priori," *The Journal of Philosophy* 20, no. 7 (1923): 169–77.
41. See Henry M. Sheffer, "Notational Relativity," *Proceedings of the Sixth International Congress of Philosophy* (1927): 348–51. For a reconstruction of Lewis's and Sheffer's pluralisms, see Floyd, "Sheffer, Lewis, and the 'Logocentric Predicament.'"
42. Susanne K. Langer and Eugene T. Gadol, "The Deepening Mind: A Half-Century of American Philosophy," *American Quarterly* 2, no. 2 (1950): 126 (original emphasis).
43. See Langer, *The Practice of Philosophy*, 170, 193. The related idea that every "epoch" has a "generative idea" or "key," which Langer would later employ in *Philosophy in a New Key*, was also influenced by *Science and the Modern World*. See Langer (*Philosophy in a New Key*, ch. 1) and Dengerink Chaplin (*The Philosophy of Susanne Langer*, 109–12).
44. Langer, *An Introduction to Symbolic Logic*, 23–4.
45. See Langer, *Philosophy in a New Key*, ch. 4.
46. Wittgenstein, "Tractatus Logico-Philosophicus," 4.115.
47. Langer, *Philosophy in a New Key*, 70 (original emphasis).
48. Langer, *A Logical Analysis of Meaning*, iv.
49. Susanne K. Langer, "A Set of Postulates for the Logical Structure of Music," *The Monist* 39, no. 4 (1929): 562.

50. Langer, *The Practice of Philosophy*, 152.
51. Susanne K. Langer, "Abstraction in Science and Abstraction in Art," in *Structure, Method, and Meaning: Essays in Honor of Henry M. Sheffer* (New York: The Liberal Arts Press, 1951), 179.
52. Susanne K. Langer, "A Logical Study of Verbs," *Journal of Philosophy* 24, no. 5 (1927): 129.
53. See, e.g., Dengerink Chaplin, *The Philosophy of Susanne Langer*, chs. 5 and 7; Chapter 2, this volume; and Pollok, Chapter 7 in this volume.
54. Frank Thilly, "Book Review: The Present Conflict of Ideals," *The Philosophical Review* 29, no. 2 (1920): 185. In fact, it may even be argued that Russell's plea for logical analysis was partly successful because German speculative philosophy and psychology, which had strongly influenced the previous generation of Harvard philosophers, became suspect during the war. Many Harvard philosophers had passionately contributed to public debates about the war and some of them had explicitly made a connection between German thought and the German war effort. George Santayana, in *Egotism in German Philosophy* (New York: Charles Scribner's Sons, 1916), and Ralph B. Perry, in *The Present Conflict of Ideals: A Study of the Philosophical Background of the World War* (New York: Longmans, Green & Co., 1918), for example, drew a clear link between German militarism and the development of nineteenth-century speculative philosophy.
55. Donald Dryden, "Susanne K. Langer 1895–1985," in *American Philosophers before 1950* (Farmington Hills, MI: Gale, 2003), 190.
56. See Langer, *The Practice of Philosophy*, 17; Greg Frost-Arnold, "The Rise of 'Analytic Philosophy': When and How did People Begin Calling themselves 'Analytic Philosophers'?" in *Innovations in the History of Analytical Philosophy* (London: Palgrave Macmillan, 2017).
57. For example, in Charles A. Baylis, "Book Review: The Practice of Philosophy," *Journal of Philosophy* 28, no. 12 (1931): 326–9; Max H. Fisch, "Book Review: The Practice of Philosophy," *Philosophical Review* 43, no. 3 (1934): 321–6; Herbert Feigl, "Book Review: An Introduction to Symbolic Logic," *American Journal of Psychology* 51 (1938): 781; Susan Stebbing, "Book Review: An Introduction to Symbolic Logic," *Philosophy* 13, no. 52 (1938): 481–3.
58. Leo Abraham, "What is the Theory of Meaning About?" *The Monist* 46, no. 2 (1936): 229n5.
59. See Charles W. Morris, "The Relation of the Formal and Empirical Sciences within Scientific Empiricism," *Erkenntnis* 5 (1935): 6–16, and Ernest Nagel's "list of recent publications in America which stand at somewhat the same place as the logical-positivists"; Nagel to Neurath, October 14, 1934, "Moritz Schlick Papers," 275, Haarlem, Wiener Kreis Archiv.
60. Quine to Sheffer, February 16, 1933. W. V. Quine Papers, item 981, Harvard University, Houghton Library.
61. Moritz Schlick, "Moritz Schlick Papers," March 22, 1931, 103/Holt-1. Haarlem, Wiener Kreis Archiv.
62. Carnap to Kaufmann, September 27, 1933, Rudolf Carnap Papers (hereafter RCP), 024-22-07, Archives of Scientific Philosophy, University of Pittsburgh. Cf. Christoph Limbeck-Lilienau, "Rudolf Carnap und die Philosophie in Amerika. Logischer Empirismus, Pragmatismus, Realismus," in *Vertreibung: Transformation und Rückkehr der Wissenschaftstheorie* (Vienna: LIT Verlag, 2010), 130. Carnap's correspondence with Feigl suggests that Carnap was planning on reviewing *The Practice of Philosophy*. See Feigl

to Carnap, April 29, 1934, Herbert Feigl Papers, 02-69-03, Minneapolis, University of Minnesota Archives.
63. Carnap, "Rudolf Carnap Papers," 025-82-01. Langer was very interested in Frege's work at the time. See Felappi, "Susanne Langer and the Woeful World of Facts," 45. Carnap had taken several courses with Frege at Jena in the 1910s.
64. Feigl, December 6, 1939, Moritz Schlick Papers, 99/Fei-17 (translation by author).
65. Feigl, April 5, 1931, Moritz Schlick Papers, 99/Fei-19 (translation by author).
66. Feigl to Schlick, December 6, 1930, Moritz Schlick Papers, 99/Fei-17.
67. Albert E. Blumberg and Herbert Feigl, "Logical Positivism: A New Movement in European Philosophy," *The Journal of Philosophy* 28, no. 11 (1931): 281–96.

BIBLIOGRAPHY

Abraham, Leo. "What is the Theory of Meaning About?" *The Monist* 46, no. 2 (1936): 228–58. Available online: https://www.jstor.org/stable/27901483 (accessed February 21, 2022).

Baylis, Charles A. "Book Review: The Practice of Philosophy." *Journal of Philosophy* 28, no. 12 (1931): 326–9. Available online: https://doi.org/10.2307/2016149.

Beaney, Michael. *The Analytic Turn: Analysis in Early Analytic Philosophy and Phenomenology*. London: Routledge, 2007.

Blumberg, Albert E., and Herbert Feigl. "Logical Positivism: A New Movement in European Philosophy." *The Journal of Philosophy* 28, no. 11 (1931): 281–96.

Bridgman, Percy W. *The Logic of Modern Physics*. New York: Macmillan, 1927.

Dengerink Chaplin, Adrienne. *The Philosophy of Susanne Langer: Embodied Meaning in Logic, Art and Feeling*. London: Bloomsbury, 2020.

Dryden, Donald. "Susanne K. Langer 1895–1985." In *American Philosophers before 1950*, edited by Leemon B. McHenry and Philip B. Dematteis, 189–99. Farmington Hills, MI: Gale, 2003.

Eliot, Thomas S. *Prufrock and Other Observations*. London: The Egotist Press, 1917.

Feigl, Herbert. "Book Review: An Introduction to Symbolic Logic." *American Journal of Psychology* 51 (1938): 781.

Felappi, Giulia. "Susanne Langer and the Woeful World of Facts." *Journal for the History of Analytical Philosophy* 5, no. 2 (2017): 38–50.

Fisch, Max H. "Book Review: The Practice of Philosophy." *Philosophical Review* 43, no. 3 (1934): 321–6. Available online: https://www.jstor.org/stable/2179712 (accessed February 21, 2022).

Floyd, Juliet. "Recent Themes in the History of Early Analytic Philosophy." *Journal of the History of Philosophy* 47, no. 2 (2009): 157–200. Available online: https://muse.jhu.edu/article/262927 (accessed February 21, 2022).

Floyd, Juliet. "Sheffer, Lewis, and the 'Logocentric Predicament.'" In *C. I. Lewis: The A Priori and the Given*, edited by Quentin Kammer, Jean-Philippe Narboux, and Henri Wagner, 27–103. New York: Routledge, 2021.

Frost-Arnold, Greg. "The Rise of 'Analytic Philosophy': When and How Did People Begin Calling Themselves 'Analytic Philosophers'?" In *Innovations in the History of Analytical Philosophy*, edited by Sandra Lapointe and Christopher Pincock, 27–67. London: Palgrave Macmillan, 2017. Available online: https://link.springer.com/chapter/10.1057/978-1-137-40808-2_2 (accessed July 17, 2023).

Griffin, Nicholas. *The Selected Letters of Bertrand Russell*. Boston: Houghton Mifflin, 1992.

Innis, Robert E. *Susanne Langer in Focus: The Symbolic Mind*. Bloomington: Indiana University Press, 2009.

Kuklick, Bruce. *The Rise of American Philosophy: Cambridge, Massachusetts, 1860–1930*. New Haven, CT: Yale University Press, 1977.

Langer, Susanne K. "Abstraction in Science and Abstraction in Art." In *Structure, Method, and Meaning: Essays in Honor of Henry M. Sheffer*, edited by Paul Henle, Horace M. Kallen, and Susanne K. Langer, 171–82. New York: The Liberal Arts Press, 1951.

Langer, Susanne K. "Confusion of Symbols and Confusion of Logical Types." *Mind* 35, no. 13 (1926): 222–9. Available online: http://www.jstor.org/stable/2249354 (accessed February 21, 2022).

Langer, Susanne K. "Form and Content: A Study in Paradox." *Journal of Philosophy* 23, no. 16 (1926): 435–8. Available online: https://www.jstor.org/stable/2014823 (accessed February 21, 2022).

Langer, Susanne K. *An Introduction to Symbolic Logic*. New York: Allen and Unwin, 1937.

Langer, Susanne K. "A Logical Analysis of Meaning." PhD diss., Radcliffe College, Cambridge, MA, 1926. Available online: https://langercircle.sites.uu.nl/wp-content/uploads/sites/662/2021/02/Langer-Susanne-K.-1926-A-Logical-Analysis-of-Meaning-doctoral-dissertation-scanned-PDF-from-photocopy.pdf (accessed February 21, 2022).

Langer, Susanne K. "The Logical Perspectives of the World." *Journal of the Philosophy* 30, no. 7 (1933): 178–87. Available online: https://www.jstor.org/stable/2015751 (accessed February 21, 2022).

Langer, Susanne K. "A Logical Study of Verbs." *Journal of Philosophy* 24, no. 5 (1927): 120–9. Available online: http://www.jstor.org/stable/2015082 (accessed February 21, 2022).

Langer, Susanne K. *Philosophy in a New Key: A Study in the Symbolism of Reason, Rite and Art*. Cambridge, MA: Harvard University Press, 1942.

Langer, Susanne K. *The Practice of Philosophy*. New York: Henry Holt, 1930.

Langer, Susanne K. "A Set of Postulates for the Logical Structure of Music." *The Monist* 39, no. 4 (1929): 561–70. Available online: http://www.jstor.org/stable/27901211 (accessed February 21, 2022).

Langer, Susanne K. "Henry M. Sheffer 1883–1964." *Philosophy and Phenomenological Research* 25, no. 2 (1964): 305–7.

Langer, Susanne K., and Eugene T. Gadol. "The Deepening Mind: A Half-Century of American Philosophy." *American Quarterly* 2, no. 2 (1950): 118–32. Available online: https://doi.org/10.2307/3031449.

Lapointe, Sandra, and Christopher Pincock, eds. *Innovations in the History of Analytical Philosophy*. London: Palgrave, 2017.

Lenzen, Victor F. "Bertrand Russell at Harvard, 1914." *Russell: The Journal of Bertrand Russell Studies* 3 (1971): 4–6.

Lewis, Clarence I. *Mind and the World-Order*. New York: Charles Scribner's Sons, 1929.

Lewis, Clarence I. "The Place of Intuition in Knowledge." PhD diss., Harvard University, 1910.

Lewis, Clarence I. "A Pragmatic Conception of the A Priori." *The Journal of Philosophy* 20, no. 7 (1923): 169–77. Available online: https://www.jstor.org/stable/2939833 (accessed February 21, 2022).

Limbeck-Lilienau, Christoph. "Rudolf Carnap und die Philosophie in Amerika. Logischer Empirismus, Pragmatismus, Realismus." In *Vertreibung: Transformation und Rückkehr der Wissenschaftstheorie*, edited by Friedrich Stadler, 85–164. Vienna: LIT Verlag, 2010.

McDaniel, K. "Ontology and Philosophical Methodology in the Early Susanne Langer." In *Innovations in the History of Analytical Philosophy*, edited by Sandra Lapointe and

Christopher Pincock, 265–98. London: Palgrave Macmillan, 2017. Available online: https://link.springer.com/book/10.1057/978-1-137-40808-2 (accessed February 21, 2022).

Misak, Cheryl. *Cambridge Pragmatism: From Peirce and James to Ramsey and Wittgenstein*. Oxford: Oxford University Press, 2018.

Morris, Charles. W. "The Relation of the Formal and Empirical Sciences within Scientific Empiricism." *Erkenntnis* 5 (1935): 6–16. Available online: https://www.jstor.org/stable/20011733 (accessed February 21, 2022).

Palmer, George H., and Ralph B. Perry. "Philosophy, 1870–1929." In *The Development of Harvard University since the Inauguration of President Eliot, 1869–1929*, edited by Samuel Eliot Morison, 3–32. Cambridge, MA: Harvard University Press, 1930.

Perry, Ralph B. *The Present Conflict of Ideals: A Study of the Philosophical Background of the World War*. New York: Longmans, Green & Co, 1918.

Quine, Willard V. "Autobiography of W. V. Quine." In *The Philosophy of W. V. Quine*. Library of Living Philosophers, vol. 18, edited by Lewis E. Hahn and Paul A. Schilpp, 1–46. La Salle, IL: Open Court, 1986.

Ramsey, Frank P. *The Foundations of Mathematics*. London: Routledge, 1931.

Russell, Bertrand. *Autobiography*. London: George Allen & Unwin, 1967–9/1975.

Russell, Bertrand. On Denoting. *Mind* 14 (1905): 479–93. Available online: https://www.jstor.org/stable/2248381 (accessed February 21, 2022).

Russell, Bertrand. *Our Knowledge of the External World as a Field for Scientific Method in Philosophy*. Chicago: Open Court, 1914.

Russell, Bertrand. "Scientific Method in Philosophy." 1914. In *Mysticism and Logic and Other Essays*, 75–93. London: George Allen & Unwin, 1917.

Santayana, George. *Egotism in German Philosophy*. New York: Charles Scribner's Sons, 1916.

Sellars, R. W. "Current Realism in Great Britain and United States." *The Monist* 37, no. 4 (1927): 503–20. Available online: https://www.jstor.org/stable/27901129 (accessed February 21, 2022).

Sheffer, Henry M. "Notational Relativity." *Proceedings of the Sixth International Congress of Philosophy* (1927): 348–51.

Sheffer, Henry M. "A Program of Philosophy Based on Modern Logic." PhD diss., Harvard University, 1908.

Stebbing, Susan. "Book Review: An Introduction to Symbolic Logic." *Philosophy* 13, no. 52 (1938): 481–3. Available online: https://www.jstor.org/stable/3746393 (accessed February 21, 2022).

Thilly, Frank. "Book Review: *The Present Conflict of Ideals*." *The Philosophical Review* 29, no. 2 (1920): 184–91. Available online: https://www.jstor.org/stable/2178851 (accessed February 21, 2022).

Whitehead, Alfred N. *Science and the Modern World*. New York: Macmillan, 1925.

Wittgenstein, Ludwig. "Tractatus Logico-Philosophicus." In *Annalen der Naturphilosophie* 14 (1921): 195–262.

CHAPTER TWO

Scientific Models and Artistic Images: Susanne K. Langer and the Early Wittgenstein

ADRIENNE DENGERINK CHAPLIN

Langer's enthusiasm for the early Wittgenstein has often been an obstacle in her reception. After all, did Wittgenstein himself, in later life, not denounce the so-called "picture theory" of his *Tractatus*? This chapter shows that Langer did not read the *Tractatus* as a copy or mirror theory of representation but as part of a broader linguistic or, as she would call it, *symbolic* turn in philosophy. In this "turn" it is recognized that the world can be presented and re-presented through different angles of refraction, each highlighting different aspects or dimensions of reality according to perceived value, purpose, or need. The chapter also shows how Langer went *beyond* Wittgenstein by extending the range of types of forms in which those representations could occur, from discursive language and logic to non-discursive works of art. Drawing on her notions of expression, analogy, and intuition, the chapter shows how Langer saw both scientific models and artistic images as different but equally valid symbolic forms for humans to make sense of the world.

LANGER AS LOGICIAN

Although best known as a philosopher of art and, more recently, philosopher of mind, Langer started her career as a logician. As such, she not only contributed to the early development of symbolic logic but to the rise of analytic philosophy in America. She was one of the earliest American philosophers to critically engage with Russell and Whitehead's *Principia Mathematica* (1910–13) and wrote her PhD (1926) under Whitehead's supervision. In 1936, alongside C. I. Lewis, Alonzo Church, and W. V. Quine, she was the co-founder of the Association for Symbolic Logic and served as a consulting editor of its new publication, *The Journal of Symbolic Logic*. Versatile in German (her native language) as well as French and Italian, she served as the journal's reviewer of international publications and was, as such, often better informed about developments in Europe and elsewhere than many of her colleagues. She wrote one of the first textbooks on symbolic logic, *An Introduction to Symbolic Logic* (1937), which served for a long time as a standard textbook at many universities.[1]

In 1939, alongside P. W. Bridgman, W. V. Quine, Rudolf Carnap, and other pioneering philosophers, she was on the organizing committee of the Fifth International Congress for the Unity of Science that took place at Harvard and became a pivotal gathering point for American and European philosophers who had been forced to flee Europe after the rise of Hitler in 1933.[2] The congress became a launch pad of the new Anglo-American "analytic" philosophy. Langer had been one of the first to coin the world "analytic" to refer to the new philosophy before it became current in the 1950s. As she wrote in *The Practice of Philosophy* in 1930: "There is ... one type of philosophy based upon a rule of procedure and defining itself thereby—that is the so-called 'logical' or 'analytic' type. It is sometimes called by the misleading name, 'scientific philosophy.'"[3] Six years later Ernest Nagel was to use the term in his article "Impressions and Appraisals of Analytic Philosophy in Europe" to refer to the new philosophy he had encountered on his travels in Europe.[4]

Langer was recognized as an important thinker by members of the Vienna Circle where her book *The Practice of Philosophy* (1930) had been read at their main gatherings. The Circle's founder, Moritz Schlick, had praised it highly for its lucidity, fluency, brilliance, and style.[5] In an influential article authored by Schlick's student Herbert Feigl, that promoted the new movement of logical positivism, the authors list Langer as one of three American philosophers with kindred views: "Logical positivism's foremost philosophical exponents are R. Carnap (Vienna), H. Reichenbach (Berlin), M. Schlick (Vienna), and L. Wittgenstein (Cambridge, England). It is interesting to note that recent American publications by P. W. Bridgman, Suzanne K. Langer, and C. I. Lewis exhibit related tendencies."[6] Crucially, Schlick, Feigl, and Langer shared a similar reading of Wittgenstein's *Tractatus*.[7]

Langer's enthusiasm for the early Wittgenstein has often been an obstacle in her reception. After all, did Wittgenstein himself not denounce the so-called "picture theory" of his *Tractatus*? Morris Weitz, for instance, echoes many critics when he writes, "Mrs. Langer's whole theory of non-discursive symbolism seems to be rooted in her picture or mirror theory of language ... This whole conception has been refuted, and by not other more certainly than by the later Wittgenstein himself."[8]

Although there are indeed suggestions that the later Wittgenstein rejected his early work—in the preface to *Philosophical Investigations* he mentions "grave mistakes" and, in conversations recorded by Friedrich Waismann, its "dogmatism,"—on Langer's reading this may not have referred to the putative failings of a supposed mirror theory of representation.[9] Instead, Langer was one of the first American philosophers who recognized Wittgenstein's early work as part of a broader linguistic or, as she would call it, *symbolic* turn in philosophy. In this "turn" it is recognized that the world can be presented and re-presented through a plurality of symbolic forms that highlight different and irreducible dimensions of reality. At the same time, Langer went *beyond* Wittgenstein by extending the range of types of forms in which those representations could occur.

SEMINAL SOURCES

An important factor in Langer's understanding of the early Wittgenstein was her knowledge of the thought of three important philosophers: her mentor, the logician Henry Sheffer, the neo-Kantian philosopher Ernst Cassirer, and the process philosopher Alfred N. Whitehead.[10]

Sheffer is arguably best known for his economizing mathematical operator, referred to as the "Sheffer stroke."[11] But he had always been more interested in logic as a study of forms and their ways of meaning than in a refinement of its technical notational apparatus. For him, logic was not primarily a study of deductive reasoning and inference, but, following his mentor Josiah Royce, a study of *patterns and forms*. Sheffer recognized that the same reality could be perceived in terms of a plurality of different forms and configurations, each requiring a different system of representation or notation, something he referred to as "notational relativity."

Cassirer, too, highlighted the plurality of different prisms or "symbolic forms" through which the world could be seen, as reflected in his trilogy *The Philosophy of Symbolic Forms*.[12] He recast Kant's fixed and universal forms of intuition and cognition as historically evolving, culturally based symbolic forms of perception and experience each of which highlighting different aspects of reality. Cassirer directed Langer's attention to the presence of symbolic forms outside the realm of logic, to include the realms of myth, art, and ritual.

Whitehead, finally, warned against the imposition of one type of abstraction or symbolization on different kinds of aspects of the world or, alternatively, reducing the world to any one abstraction, such as science. In his books *Symbolism: Its Meaning and Effect* (1927) and *Process and Reality* (1929), he showed how our first encounters with the world are inevitably diffuse and indeterminate. In his analysis of the symbolic reference, he drew attention to the fact that the symbol–symbolized relation was reversible and interchangeable depending on context and user.[13]

Read through these lenses Langer recognized the *Tractatus* as a significant and inspiring work—an "extraordinary prophetic gospel" no less—even as she proceeded to extend its principles beyond those envisaged by Wittgenstein himself.[14]

THE *TRACTATUS*

As one of the most dense and enigmatic texts in the history of philosophy, the *Tractatus* (1922) has opened itself to a wide range of interpretations. Wittgenstein himself often felt misunderstood. As he wrote to Russell shortly after he finished the manuscript:

> The main point [of the *Tractatus*] is the theory of what can be expressed (gesagt) by prop[osition]s—i.e., by language—(and, which comes to the same, what can be *thought*) and what cannot be expressed by prop[osition]s, but only shown (gezeigt); which I believe is the cardinal problem of philosophy ... It is VERY hard not to be understood by a single soul![15]

And in the preface, he muses, "The book will perhaps only be understood by those who have themselves already thought the thoughts which are expressed in it—or similar thoughts."[16] This, I suggest, was indeed the case with Langer—as it had been with Schlick. Both had already "thought the thoughts" that were presented in the work.[17]

PICTURES AS FACTS

Langer's reading of the *Tractatus* zooms in on a select number of theses that she considers keys for unlocking the work as a whole. These include:

> 2.1 We make to ourselves pictures [*Bilder*] of facts [*Tatsachen*].
> 2.12 The picture [*Bild*] is a model [*Modell*] of reality.
> 2.13 To the objects correspond in the picture the elements of the picture.
> 2.131 The elements of the picture stand, in the picture, for the objects.
> 2.14 The picture consists in the fact that its elements are combined with one another in a definite way.

2.141 The picture is a fact [*Tatsache*].
2.15 That the elements of the picture are combined with one another in a definite way, represents that the things are so combined with one another. This connection of the elements of the picture is called its structure, and the possibility of this structure is called the form of representation (*Abbildung*) of the picture.
2.16 In order to be a picture a fact [*Tatsache*] must have something in common with what it pictures.
2.17 What the picture must have in common with reality in order to be able to represent it after its manner—rightly or falsely—is its form of representation.
2.171 The picture can represent every reality whose form it has.

Developing Wittgenstein's assertions that "we make ourselves pictures of facts" [2.1] and that "the picture is a fact" [2.141], Langer writes,

> [A] fact is an intellectually formulated event, whether the formulation be performed by a process of sheer vision, verbal interpretation, or practical response. A fact is an event as we see it, or would see it if it occurred for us.[18]

Put differently, a fact lends a *form* to the amorphous flux of sensations and renders it available for the conscious mind. As she put it in *The Practice of Philosophy*,

> "Facts" are the basic formulations of any system of apperception. They are not arbitrary logical constructions, neither are they "absolute" and stark in their own form—indeed, by pure sense-experience or intuition, if there could be such a thing, facts would not even be apparent.[19]

Langer rejects the usual distinction between what is symbolic and what is literal. Whatever might be designated as literal is simply another formulation of experience articulated in a symbolic form. One formulation of experience is not more "symbolic" than the other. For her, there are not two ontologically different realms—the world of symbols and the world of "brute facts"—but one world as perceived under different aspects depending on the focus and need of the perceiver. So, when Wittgenstein writes that "the world is the totality of facts, not of things" [1.1], Langer reads this in Cassirerian terms, that is, as the world being both mediated and constructed by particular symbolic forms or, to use Wittgenstein's term, "pictures." Most importantly, "there is no such thing as *the form* of a real thing, or of an event."[20] Instead, there can be "*several adequate descriptions* of reality."[21]

PICTURES AS MODELS

In order to understand Wittgenstein's meaning and Langer's reading of the word "picture" or *Bild*—Langer will have most likely read the *Tractatus* in its original German—it is important to know some of its German connotations. While often translated as "picture," "Bild" is equally used to refer to three-dimensional representations and statues—the German word for "sculptor" is "Bildhauer"—and is also used for "model." We know from Wittgenstein's 1914 notebooks that the idea of a "picture theory" had come to him when reading about a Parish court case in which small cars and dolls were being used in order to reconstruct a car accident.[22] In the same way that the relation between the objects in the model corresponded with the relations of the relevant cars, houses, and people in the situation of the car accident, so a proposition might serve as a model of a state of affairs by virtue of a comparable correspondence between its parts and those of the world.

Wittgenstein's interest in pictures understood as models goes back further to his years of studying engineering in Manchester. One of the textbooks that had deeply influenced him was *Principles of Mechanics* by the German physicist Heinrich Hertz and he twice refers to him in the *Tractatus*.[23] Hertz used the word *Bild* to refer to models as a way of understanding physical processes. When describing *Bilder* as particular forms of "representations" Hertz consistently uses the term *Darstellungen*—that is, Kantian cognitive constructs—rather than *Vorstellungen*—that is, passive "ideas," Humean "impressions," or Machian "sensations":

> We form for ourselves images (*Bilder*) or symbols of external objects; and the form which we give them is such that the necessary consequents of the images in thought are always the images of the necessary consequents in nature of the things pictured.[24]

The same phrase used by Hertz to explain the symbolism of physics will be used later by Wittgenstein to explain the symbolism of discursive language as such. Moreover, for Hertz, the world does not merely allow for one kind of modeling, it opens itself to a plurality of possible *Bilder*. As Hertz wrote:

> Various images (*Bilder*) of the same objects are possible, and these images may differ in various respects ... Of two images of equal distinctness the more appropriate is the one which contains, in addition to the essential characteristics, the smaller number of superfluous or empty relations—the simpler of the two. Empty relations cannot altogether be avoided; they enter into the images because they are simply images—images produced by our mind and necessarily affected by the characteristics of its mode of portrayal.[25]

Both Schlick and Cassirer had great admiration for Hertz's philosophy of science and cognition. In *The Phenomenology of Knowledge*, Cassirer refers to Hertz as "The first modern scientist to have effected a decisive turn from the copy theory of physical knowledge to a purely symbolic theory."[26] In the case of Wittgenstein, Hertz's notion of *Bild* prompted him to think of language and equivalent forms of representation as "models" that can function like charts, maps, graphs, and so on. As regards a piece of music, such models can be as different as a gramophone record, a musical thought, a score, or the waves of sound. For him these were all different ways in which the "logical form" or structure of the music could be represented or translated.

Reading these statements in the light of Hertz will preclude interpretations of Wittgenstein's picture theory in terms of a Russellian copy theory of knowledge based on empirical sense data. As his biographer Ray Monk observes, "[s]o ingrained in Wittgenstein's philosophy was this Hertzian and, ultimately Kantian, view that he found the empiricist view difficult even to conceive."[27]

PICTURES AS EXPRESSIONS

For Wittgenstein and Langer all propositions are "expressions." Wittgenstein does not tire of stressing this in theses 3.31 and beyond: "Every part of a proposition which characterizes its sense I call an expression (a symbol). (The proposition itself is an expression.) Expressions are everything—essential for the sense of the proposition—that propositions can have in common with one another. An expression characterizes a form and a content" [3.31].

Clearly, "expression" in this sense is not self-expression as the venting of emotions but the symbolic formulation of the way we sense and understand the world. Langer puts it thus in *The Practice of Philosophy*: a symbol is "not a reproduction of its object, but an *expression*—an exhibition of certain relevant moments, whose relevance is determined by the purpose in hand."[28] Even Schlick used the word "expression" to explain the relation between symbol and symbolized. As he wrote in his article "Form and Content" (1932), its title borrowed from an article by Langer in 1926:[29]

> We say that one fact (the arrangement of little black marks) *expresses* the other (the eruption of the volcano), so the particular relation between them is called Expression. In order to understand language, we must investigate the nature of Expression. How can certain facts "speak of" other facts? That is our problem.[30]

Schlick, it is worth noting, was in regular contact with Cassirer throughout the 1920s—they were both leading interpreters of Einstein's theory of relativity—and would most certainly have known Cassirer's view of symbols as forms of conception and expression.

STAYING SILENT

While Wittgenstein did *in principle* acknowledge the possibility of a multiplicity of forms of expression, *in practice* he did not conceive of any other than those of discursive language or explanatory models rooted in science. On that basis he did not think philosophy itself could ever say something of importance. If we were to follow "the right method of philosophy" as outlined by himself, this would be its inevitable implication:

> To say nothing except what can be said, i.e., the propositions of natural science, i.e., something that has nothing to do with philosophy: and then always, when someone else wished to say something metaphysical, to demonstrate to him that he had given no meaning to certain signs in his propositions. [6.53]

Based on this scientistic paradigm, the only valid philosophical propositions were those based on models derived from the physical sciences. Fully aware of the limitations of such a philosophy he lamented, "We feel that even if all possible scientific questions be answered, the problems of life have still not been touched at all" [6.52]. Bertrand Russell pointed to the same problem in relation to language in general:

> Our confidence in language is due to the fact that it ... shares the structure of the physical world, and therefore can express that structure. But if there be a world which is not physical, or not in space-time, it may have a structure which we can never hope to express or to know ... Perhaps that is why we know so much physics and so little of anything else.[31]

Perhaps it was not surprising that, based on his deep disillusionment with philosophy of this kind, Wittgenstein left the academic world and became a teacher in a remote mountain village. After all, as the last thesis of his work made clear: "Whereof one cannot speak, thereof one must be silent" [7].

LANGER'S EXTENSION OF WITTGENSTEIN

For Langer, however, this was not the last word. She did not share Wittgenstein's defeatist conclusion that "the limits of [scientific] language mean the limits of my world" [5.6]. Nor did she think, as did Russell and Carnap, that anything that could not be expressed in empirically sense-based language was merely subjective or fanciful. Langer went beyond Russell and Carnap, as well as Schlick and Wittgenstein, in recognizing a wider range of forms of representation than they had been able to envisage. In her words:

> The logical "beyond" which Wittgenstein calls the "unspeakable," both Russell and Carnap regard as the sphere of subjective experience, emotion, feeling, and wish, from which only *symptoms* come to us in the form of metaphysical and artistic fancies. The study of such products they relegate to psychology, not semantics ... [However], there is an unexplored possibility of genuine semantic beyond the limits of discursive language.[32]

The term "discursive" is used by Langer to refer to a type of symbolism in which its discrete elements contain independent meaning rooted in tacit convention or explicit agreement. Such elements can be arranged and rearranged in different configurations, such as is the case with, for example, words in a sentence, letters in mathematical formula, or marks or dots on a map or a graph. In each of these cases the elements can be understood to stand independently for something that can be known and identified separately by means of a dictionary, reference key, instruction, or translation. By contrast, the elements in non-discursive symbolisms derive their meaning predominantly if not purely from the context of the whole. Their discrete elements, such as a line in a picture or a note in a song, do not have any independent meaning outside the context in which they are used.

MODELS AND IMAGES

Langer's distinction between discursive and non-discursive symbolisms maps roughly on another distinction which she introduces in the first volume of *Mind*. This is the distinction between a model and an image. Whereas a model "illustrates a principle of construction quite apart from any semblance," an image "abstracts the semblance of its object, and makes one aware of what is there for direct perception."[33] Whereas a model shows how something *works*, an image shows how something *appears*. This makes images particularly suited to express how the world is sensed and felt, in short, the world of lived experience:

> An image ... abstracts [an object's] phenomenal character, its immediate effect on our sensibility or the way it presents itself as something of importance, magnitude, strength or fragility, permanence or transience, etc. It organizes and enhances the impression directly received.[34]

This applies specifically to works of art. In contrast to discursive models derived from the physical sciences, a non-discursive artistic image has the capacity to show human feelings as they occur in lived experience. As Langer puts it in volume 1 of *Mind*:

> The art symbol sets forth in symbolic projection how vital and emotional and intellectual tensions appear, i.e., how they feel. It is this image that gets lost in our psychological laboratories, where models from non-biological sciences and especially from intriguing machinery have taken the field, and permit us to analyze and understand many processes, yet lead us to lose sight of what phenomena we are trying to analyze and understand.[35]

Although Wittgenstein held that a picture could "represent every reality whose form it had" [2.171], he never considered the possibility of this kind of Langerian non-discursive image. All his "pictures" or "logical forms"—from linguistic propositions and graphs to musical scores and gramophone grooves—were conceived of in terms of discursive Hertzian models, in which "one name stands for one thing, and another for another thing, and they are connected together" [4.0311]. He explicitly refers to Hertz when he writes, "In the proposition there must be exactly as many things distinguishable as there are in the state of affairs, which it represents. They must both possess the same logical (mathematical) multiplicity (cf. Hertz's Mechanics, on Dynamic Models)" [4.041].

Langer's notion of logical or structural form allows for a broader range of symbolic forms than envisaged by Wittgenstein. Drawing on Sheffer's conception of logic as a study of patterns and forms *in general*, "logical forms" can be identified in any pattern or shape, whether in rock formations, clothing patterns, jelly molds, melodies, rituals, types of behavior, processes, or, indeed, inner feelings.

All symbolization, whether discursive or non-discursive, rests on the recognition of a common form between symbol and symbolized, sometimes referred to by Langer as "analogy." Analogy is not the same as resemblance. On the contrary, "it is only by analogy that one thing can represent another which does *not resemble* it."[36] As she had put it in *An Introduction to Symbolic Logic*:

> A logical picture differs from an ordinary picture in that it need not look the least bit like its object. Its relation to the object is not that of a copy, but of analogy. We do not try to make an architect's drawing look as much as possible like the house … All that the plan must do is to copy exactly the proportions of length and width, the arrangement of rooms, halls and stairs, doors and windows.[37]

That said, so-called "ordinary pictures" also make use of analogies, whether that be contour, size, shape, texture, intensity, color, and so on. A picture of a person or cat can share with reality just "a certain *proportion of parts*."[38] As she explains in *An Introduction to Symbolic Logic*:

> One must not make the mistake of associating "structure" always with something *put together* out of parts that were previously separate. A snowflake is a detailed construct of very recognizable individual parts, but these have not been "put together"; they crystallized out of one homogenous drop of water. They were never separate, and there has been no process of combination.[39]

Langer's conception of symbol in terms of a shared logical form echoes Peirce's definition of the "icon" as "any sign that may represent its object mainly by its similarity."[40] But since everything in the world always resembles something else *in some respect* Langer emphasizes that an "icon" always selects and highlights certain features while suppressing others. That is why she tends to shy away from using the term "isomorphism" with its suggestion of a unique sameness or correspondence of form, and, with Wittgenstein, preferred the less determinate term "expression" to refer to the symbolic reference. This is also hinted at by Wittgenstein when he says that a picture "reaches up to" or is "like a scale applied to" reality in which the configurations are "feelers of its elements with which the picture touches reality" [2.1511; 2.1512; 2.1515]. Such wordings suggest a deeper continuity between his early and later thought than is often assumed.

INTUITION

Because there is no vantage point outside the symbol–symbolized relation by which one can judge the relation itself, any analogy can ultimately only be grasped by intuition.

> The great value of analogy is that by it, and it alone, we are led to seeing a single "logical form" in things which may be entirely discrepant as to content. The power of recognizing similar forms in widely various exemplifications, i.e., the power of discovering analogies, is logical intuition.[41]

Intuition, for Langer, whether artistic or otherwise, is not an irrational mystical or metaphysical vision but a "fundamental intellectual activity, which produces logical or semantic understanding."[42] It is the basic capacity to see forms and configurations:

> There are certain relational factors in experience which are either intuitively recognized or not at all, for example, distinctness, similarity, congruence, relevance. These are formal characteristics which are protological in that they "must be seen to be appreciated."[43]

The same principle explains the essential role of metaphors in the formation of new language to name new phenomena and experiences. They highlight and express what is considered significant or relevant about the object to be named. Drawing on linguist Von Humboldt, Cassirer illustrated this idea of significance with the two different classical words for moon, either highlighting its role as providing illumination—as reflected in the Latin term "luna"—or its role in measuring time—the Greek term "mēn."[44] More recent examples might be the names for entities and operations related to the internet such as "cloud," "web," "browser," "host," "bookmark," "platforms," "highways," and so on. It would be difficult to pinpoint the exact kind of similarity between these phenomena and those from which they borrowed their names. Yet, they make obvious intuitive sense for the user community and are no longer recognized as metaphors. Interestingly, in his article on logical positivism, Feigl hints at the same difficulty when discussing the nature of similarities between a proposition and fact. As he writes:

> This similarity itself cannot be expressed in the same language. To express it we should require a different language in which a proposition would express this similarity in virtue of a similarity between it and the similarity. Russell, in his introduction to Wittgenstein's *Tractatus Logico-Philosophicus*, indicates that the difficulty might be met by an unending hierarchy of languages. Wittgenstein takes the position that the similarity is inexpressible. It shows itself.[45]

In her discussion of intuition Langer refers to John Locke's notion of "natural light." She quotes him as saying that it is through natural light that "the mind perceives, that white is not black, that a circle is not a triangle, that three are more than two, and equal to one and two."[46] The mind perceives such truths "by bare intuition, without the intervention of any other idea."[47] Both logic or science and art depend on "protological" intuition for insight into the "nature of relations whereby we recognize distinctions and identities, contradictions and entailments, and use."[48] According to Langer "[t]he scientist must have insight to convey intellectual knowledge, and the artists must have insight to inspire insight. This power which is called insight or intuition, is based upon our perception of patterns [and forms]."[49] Artistic intuition is not more mysterious or irrational than logical intuition. Both are "incommunicable, yet rational."[50] In Langer's words:

> A scientist of genius is a person who can apprehend a new concept through some natural medium, for whom there are unprobed patterns in nature, which catch his mind's eye so that he can see the general form of a system which becomes lucid for others and even for him only as he gives it literal expression. This is the logical process which in popular parlance is called "having a hunch."[51]

The difference between scientific and artistic intuition lies in the kind of forms that each aims to intuit or "abstract." Artists draw on nature for their symbolic forms to make a "hunch vaguely contagious."[52] They provide an elusive and transient experience a relatively stable and integrated formal unity which it lacks in the mind's stream of consciousness. For art to serve as the objectification of feeling, its forms have to be *intuited as* analogical to the dynamic tensions of sentient life. Works of art create images or "Gestalten" that are *recognized as* resembling the structure and texture of certain feelings. It is not an "objective" correspondence between symbol and symbolized but a *felt sense* of resemblance between "actual organic tensions and virtual perceptually created tensions."[53] Artistic perception, for her, is "the perception of expressiveness in works of art."[54] Only in that sense can music be said to have, in the words of Mark Johnson, "a structure and pattern of temporal flow, pitch contours, and intensity loudness/softness that is *analogous* to felt patterns of the flow of human experience" (emphasis added).[55] Music, Langer says, is "not the cause or the cure of feelings, but their logical expression."[56]

In conclusion, Langer's notions of expression, analogy, and intuition should prevent anyone interpreting her or, for that matter, Wittgenstein, as a copy or mirror theorist of representation. As Langer repeats over and over, symbolic forms, whether logic and science or ritual and art, are different modes or kinds of abstraction that apply to the whole of reality seen through different angles of refraction. They are prisms that highlight different aspects or dimensions of reality according to perceived value, purpose, or need. It was Langer's own power of logical intuition that enabled her to move effortlessly from the realm of symbolic logic in the 1920s and 1930s to the realm of art and aesthetics in the 1940s and 1950s and the biology of mind in the 1970s and 1980s. Always searching for shared forms and common patterns Langer can be said to have remained a Shefferean logician till the end of her life.

NOTES

1. Susanne K. Langer, *An Introduction to Symbolic Logic* (New York: Allen and Unwin, 1937); 2nd revised edn New York: Dover, 1953; 3rd edn New York: Dover, 1953. It was republished as recently as 2011.
2. In his autobiography Quine referred to the conference as "the Vienna Circle, with accretions." Willard V. Quine, *The Time of My Life: An Autobiography* (Cambridge, MA: MIT Press, 1985), 140.
3. Susanne K. Langer, *The Practice of Philosophy* (New York: Henry Holt, 1930), 17–18, 67.
4. Ernest Nagel, "Impressions and Appraisals of Analytic Philosophy in Europe," *The Journal of Philosophy* 33, no. 1 (1936a): 524; 33, no. 2 (1936b): 29–53.
5. Johannes Friedl and Heiner Rutte, eds., *Moritz Schlick Erkenntnistheoretische Schriften 1926–1936*, Kritische Gesamtausgabe Abteilung II, Band 1.2 (Vienna: Springer Verlag, 2013), 383, fn. 7.
6. Albert E. Blumberg and Herbert Feigl, "Logical Positivism: A New Movement in European Philosophy," *Journal of Philosophy* 28, no. 11 (1931): 281.
7. For a more detailed account of Langer's early period as a logician, see Sander Verhaegh's Chapter 1 in this volume, "Susanne K. Langer and the Harvard School of Analysis."
8. Morris Weitz, "Symbolism and Art," *Review of Metaphysics* 7, no. 3 (March 1, 1954): 470.
9. Ludwig Wittgenstein, *Philosophical Investigations*, ed. Rush Rhees and Gertrude E. M. Anscombe, trans. G. E. M. Anscombe (Oxford: Blackwell, 1967), preface; Brian F. McGuinness, ed., *Ludwig Wittgenstein and the Vienna Circle: Conversations Recorded by Friedrich Waismann* (Oxford: Blackwell, 1979), 182.

10. For more extensive accounts of the influence of these thinkers on Langer, see chapters 4–6 in Adrienne Dengerink Chaplin, *The Philosophy of Susanne Langer: Embodied Meaning in Logic, Art and Feeling* (London/Oxford: Bloomsbury Academic, 2019), 61–131.
11. The Sheffer Stroke replaced the two operations of a propositional negation and disjunction -p &-q or "neither p nor q" with one single operation of non-conjunction p | q.
12. Ernst Cassirer, *Philosophie der symbolischen Formen* (Berlin: B. Cassirer,1923–9); *The Philosophy of Symbolic Forms*, vols. 1–3 (New Haven, CT: Yale University Press, 1955–7).
13. Alfred N. Whitehead, *Symbolism: Its Meaning and Effect*, 1927 (New York: Fordham University Press, 1985), 10–13; *Process and Reality* (New York: Free Press, 1929), 180–3.
14. Langer, *The Practice of Philosophy*, 108.
15. Letter from Wittgenstein to Russell, August 19, 1919. In Georg H. von Wright, *Letters to Russell, Keynes and Moore* (Oxford: Blackwell, 1977), 71.
16. Ludwig Wittgenstein, *Tractatus Logico-Philosophicus* (London: Routledge, 1996), 27. Direct quotations will be referred to in the text with the thesis number.
17. For a more extensive account of Langer's reading of Wittgenstein, see chapter 7, "Wittgenstein: Expressive Form," in Dengerink Chaplin, *The Philosophy of Susanne Langer*, 133–56. Some sections draw directly on this chapter.
18. Susanne K. Langer, *Philosophy in a New Key: A Study in the Symbolism of Reason, Rite and Art* (Cambridge, MA: Harvard University Press, 1942), 269.
19. Langer, *The Practice of Philosophy*, 150–1.
20. Ibid., 135–6.
21. Ibid., 37, 138.
22. Ray Monk, *Ludwig Wittgenstein: The Duty of Genius* (London: Vintage Books, 1991), 118.
23. References to Hertz can be found in theses 4.04 and 6.361 in Wittgenstein's *Tractatus*.
24. Heinrich Hertz, *The Principles of Mechanics Presented in New Form* (New York: Dover, 1956), 1.
25. Ibid., 2.
26. Ernst Cassirer, *The Philosophy of Symbolic Forms: Phenomenology of Knowledge*, vol. 3 (New Haven, CT: Yale University Press, 1957), 20.
27. Monk, *The Duty of Genius*, 26.
28. Langer, *The Practice of Philosophy*, 141.
29. Susanne K. Langer, "Form and Content: A Study in Paradox," *Journal of Philosophy* 23 (August 5, 1926): 435–8.
30. Moritz Schlick, "Form and Content: An Introduction to Philosophical Thinking," in *Moritz Schlick: Philosophical Papers 1925–1936*, ed. Henk L. Mulder and B. F. B. Van de Velde-Schlick (Dordrecht: Reidel, 1979), 286.
31. Bertrand Russell, *Philosophy* (New York: W. W. Norton, 1927), 265. Quoted in Langer, *Philosophy in a New Key*, 88.
32. Langer, *Philosophy in a New Key*, 86–7.
33. Susanne K. Langer, *Mind: An Essay on Human Feeling*, vol. 1 (Baltimore: Johns Hopkins University Press, 1967), 67–8.
34. Langer, *Mind*, vol. 1, 59.
35. Ibid., xix.
36. Langer, *An Introduction to Symbolic Logic*, 30.
37. Ibid., 29.
38. Langer, *Philosophy in a New Key*, 69.
39. Langer, *An Introduction to Symbolic Logic*, 25.

40. Charles S. Peirce, "The Icon, Index, and Symbol," in *The Collected Papers of Charles S. Peirce*, vol. 2, ed. Charles Hartshorne and Paul Weiss (Cambridge, MA: Harvard University Press, 1931–58), 2.276, 2:157.
41. Langer, *An Introduction to Symbolic Logic*, 33.
42. Susanne K. Langer, *Problems of Art: Ten Philosophical Lectures* (New York: Charles Scribner's Sons, 1957), 66.
43. Langer, *Problems of Art*, 166.
44. Ernst Cassirer, *An Essay on Man: An Introduction to a Philosophy of Human Culture* (New Haven, CT: Yale University Press, 1972), 134.
45. Blumberg and Feigl, "Logical Positivism: A New Movement in European Philosophy," *Journal of Philosophy* 28, no. 11 (1931), 287.
46. Quoted in Langer, *Problems of Art*, 65.
47. Ibid.
48. Langer, *Problems of Art*, 65.
49. Langer, *The Practice of Philosophy*, 166.
50. Langer, *Problems of Art*, 69.
51. Langer, *The Practice of Philosophy*, 166.
52. Ibid.
53. Susanne K. Langer, "Abstraction in Art," *Journal of Aesthetics and Art Criticism* 22, no. 4 (Summer 1964): 382.
54. Langer, *Problems of Art*, 59.
55. Mark Johnson, *The Meaning of the Body, Aesthetics of Human Understanding* (Chicago: University of Chicago Press, 2007), 238.
56. Langer, *Philosophy in a New Key*, 218.

BIBLIOGRAPHY

Blumberg, Albert E., and Herbert Feigl. "Logical Positivism: A New Movement in European Philosophy." *Journal of Philosophy* 28, no. 11 (1931): 281–96.

Cassirer, Ernst. *An Essay on Man: An Introduction to a Philosophy of Human Culture*. New Haven, CT: Yale University Press, [1944] 1972.

Cassirer, Ernst. *The Philosophy of Symbolic Forms,* vols. 1–3. New Haven, CT: Yale University Press, 1955–7.

Cassirer, Ernst. *The Philosophy of Symbolic Forms*: *Phenomenology of Knowledge,* vol. 3. New Haven, CT: Yale University Press, 1957.

Dengerink Chaplin, Adrienne. *The Philosophy of Susanne Langer*. London/Oxford: Bloomsbury Academic, 2019.

Friedl, Johannes, and Heiner Rutte, eds. *Moritz Schlick Erkenntnistheoretische Schriften 1926–1936*, Kritische Gesamtausgabe Abteilung II, Band 1.2. Vienna: Springer Verlag, 2013.

Hertz, Heinrich. *The Principles of Mechanics Presented in New Form*. 1899. Translated by D. E. Jones and J. T. Walley. New York: Dover, 1956.

Johnson, Mark. *The Meaning of the Body, Aesthetics of Human Understanding*. Chicago: University of Chicago Press, 2007.

Langer, Susanne K. "Abstraction in Art." *Journal of Aesthetics and Art Criticism* 22, no. 4 (Summer 1964): 379–92.

Langer, Susanne K. "Form and Content: A Study in Paradox." *Journal of Philosophy* 23 (August 5, 1926): 435–8.

Langer, Susanne K. *An Introduction to Symbolic Logic*. New York: Allen and Unwin, 1937.
Langer, Susanne K. *Mind: An Essay on Human Feeling*, vol. 1. Baltimore: Johns Hopkins University Press, 1967.
Langer, Susanne K. *Philosophy in a New Key: A Study in the Symbolism of Reason, Rite and Art*. Cambridge, MA: Harvard University Press, 1942.
Langer, Susanne K. *The Practice of Philosophy*. New York: Henry Holt, 1930.
Langer, Susanne K. *Problems of Art: Ten Philosophical Lectures*. New York: Charles Scribner's Sons, 1957.
McGuinness, Brian F., ed. *Ludwig Wittgenstein and the Vienna Circle: Conversations Recorded by Friedrich Waismann*. Oxford: Blackwell, 1979.
Monk, Ray. *Ludwig Wittgenstein: The Duty of Genius*. London: Vintage Books, 1991.
Nagel, Ernest. "Impressions and Appraisals of Analytic Philosophy in Europe. I." *The Journal of Philosophy* 33, no. 1 (1936a): 5–24; 33, no. 2 (1936b): 29–53.
Peirce, Charles S. "The Icon, Index, and Symbol." In *The Collected Papers of Charles S. Peirce*, vol. 2, edited by Charles Hartshorne and Paul Weiss, 274–308. Cambridge, MA: Harvard University Press, 1931–58.
Quine, Willard V. *The Time of My Life: An Autobiography*. Cambridge, MA: MIT Press, 1985.
Schlick, Moritz. "Form and Content: An Introduction to Philosophical Thinking." In *Moritz Schlick: Philosophical Papers 1925–1936*, edited by Henk L. Mulder and B. F. B. Van de Velde-Schlick, 285–369. Dordrecht: Reidel, [1932] 1979.
von Wright, Georg H. *Letters to Russell, Keynes and Moore*. Oxford: Blackwell, 1977.
Weitz, Morris. "Symbolism and Art." *Review of Metaphysics* 7, no. 3 (March 1, 1954): 466–81.
Whitehead, Alfred N. *Process and Reality: An Essay in Cosmology*. Edited by David Ray Griffin and Donald W. Sherburne. New York: Free Press, 1929.
Whitehead, Alfred N. *Symbolism: Its Meaning and Effect*. New York: Fordham University Press, [1927] 1985.
Wittgenstein, Ludwig. *Philosophical Investigations*. Edited by Rush Rhees and Gertrude E. M. Anscombe. Translated by G. E. M. Anscombe. Oxford: Blackwell, 1967.
Wittgenstein, Ludwig. *Tractatus Logico-Philosophicus*. Translated by Charles K. Ogden. London: Routledge, [1922] 1996.

CHAPTER THREE

Susanne K. Langer, Wittgenstein, and the Profundity of Fairy Tales

HELEN THAVENTHIRAN

When Wittgenstein was ill in bed in 1935, he was awed by a fairy tale. "I remember him picking up the volume of Grimms' tales," his Russian teacher recalled, "and reading out with awe in his voice: 'Ach, wie gut ist daß niemand weiß Daß ich Rumpelstilzchen heiß.' ['Oh how good it is that nobody knows that I am called Rumpelstiltskin.'] 'Profound, profound,' he said."[1] What quality of profundity Wittgenstein found here might appear more unguessable than the name of that story's imp. While the stakes of the guess might be lower than for the miller's daughter, there is a powerful feeling that fairy or wonder tales carry just the sort of depth to be gauged in reading Wittgenstein's philosophy. When reading the tales that mattered to him, Wittgenstein's "voice and facial expression" were, in the words of another of his interlocutors, "unforgettable."[2] Fairy tales also figured in his wish to be remembered, when voice and face were lost. He names, in his will, his fairy tale volumes as a material legacy to his closest companion during his final years. "I make the following gifts of specific articles or chattels namely:—To Dr. Benedict Richards my French Travelling Clock my Fur Coat my complete Edition of Grimm's Fairy Tales."[3] But the physiognomy of his thinking here is now obscure: what was the wonder of the wonder tale for Wittgenstein? These tales, with their charms, spells, riddles, oaths, and promises certainly carry a possibility of philosophical wonder, an implicit linguistic philosophy of "what words do in the world."[4] Yet when they make occasional appearances across Wittgenstein's remarks, they tend to seem like only lightly symbolic "chattels," or distracting paraphernalia from the margins of a properly philosophical account of verbal behaviors. What, though, if it matters that "Wittgenstein's understanding of ethics can be tied to his sense of the power and profundity of that tale" of Rumpelstiltskin?[5] If fairy tales are among the "techniques of our language"[6] significant for understanding Wittgenstein and those he influenced? This chapter approaches these questions by attending to a writer of fairy tales and philosopher with debts to Wittgenstein, Susanne Langer.

It takes a capacious philosophy of symbolic forms to tie together ethics and imps. Just the kind for which Alfred North Whitehead made space in declaring that "Philosophy may not neglect the multifariousness of the world—the fairies dance, and Christ is nailed to the cross."[7] As Whitehead's student, Langer was similarly expansive in her vision of what philosophy should include and her reaction to reading Wittgenstein plays an important part in that expansiveness. Langer holds the strange position of a disciple of the *Tractatus*—"that extraordinary prophetic gospel," in her words—who nonetheless found its vision incomplete, and yet who did not, as she forged a less "despair[ing]" philosophy, pursue the kinds of post-Tractarian shifts in Wittgenstein's thought which sit well alongside her own work. Langer's first scholarly book, *The Practice of Philosophy* (1930), quotes more from the *Tractatus* than perhaps anyone, except Louis Zukofsky.[8] The long stretches of quotation (these occupy four full pages) include Wittgenstein's nod to the Brothers Grimm in proposition 4.014, where he illustrates his argument about internal relations with a parenthesis that looks to the tale of "The Gold Children." Illustration in the *Tractatus* is generally absent, impossible even. Here, this rare incursion of ordinary culture into the logically atomist universe, the fairy tale, hides within the brackets and behind the generality of reference to "the story," becoming so abstract that it has been easy to pass over in silence.

> The gramophone record, the musical thought, the score, the waves of sound, all stand to each other in that pictorial internal relation, which holds between language and the world.
>
> To all of them the logical structure is common.
>
> (Like the two youths, their two horses and their lilies in the story. They are all in a certain sense one.)[9]

Langer, after her intense passage of quotation from this middle section of the *Tractatus*, also goes silent about Wittgenstein. Or at least her early intervention in the transatlantic reception of Wittgenstein's thought does not continue in any explicit form, although her subsequent arguments for non-discursive symbolism can be seen as sustained articulations of resistance to the Tractarian terminus. The imprint of 4.014 on her thought perhaps also shows in the method of her 1937 book, *An Introduction to Symbolic Logic*, which borrows this technique of illustrating logical relations by fairy tale: Langer takes the royal children, enchanted then redeemed, of "The White Swans" to animate her exposition of the relations between members and classes.[10] Langer's early reading of the *Tractatus* was non-standard, refusing a view of it as merely a positivist correspondence theory; it was, for her, in company with her reading of Cassirer and Whitehead, part of a symbolic turn.[11] Wittgenstein's riddling book could, she writes, "show, more completely than any other in the literature of logic, the importance of configuration for any sort of meaning relation, from the simple denotation of names or suggestiveness of natural signs, to the most intricate symbolic expression, in literal notation or poetic metaphor."[12] But Wittgenstein was wrong, Langer suggests, to present his gospel of configuration and then "despair."[13] Where the *Tractatus* ultimately halts at the limits of discursive language and urges silence, Langer's work takes symbolic forms like the fairy tale out of parentheses and explores a more expansive sense of how philosophy can understand its subjects, including non-discursive or "presentational" symbolic forms.[14]

Langer's asymmetrical debts to early but not later Wittgenstein have been noted briefly and variously. Richard Wollheim offered an early dismissal of *Feeling and Form* as a "retrogressive step" in aesthetic philosophy, partly on the grounds of Langer's "unquestioning acceptance and then uncompromising generalization of a very odd and obscure view of language held by Wittgenstein in the *Tractatus*."[15] Later scholars note the magnitude of the debt, yet without the same censure. Garry Hagberg, for example, argues that the "influence of the *Tractatus* is central to all of Langer's work and is implicit on almost every page" of her oeuvre.[16] Adrienne Dengerink Chaplin's monograph, *The*

Philosophy of Susanne Langer: Embodied Meaning in Logic, Art and Feeling (2019) restores a fuller intellectual-historical context to Langer's Tractarian aesthetics, arguing that "Langer was one of the first—if not the first—American philosophers to recognize the significance of the *Tractatus* and to read Wittgenstein's work as part of a broader linguistic or, more precisely, symbolic turn in philosophy."[17] My chapter is not a contribution to this settling of intellectual debts. It does, however, take a prompt from the dominant recent characterizations of Langer as opening up, or going beyond a narrow version of philosophical analysis, in order to see what might be sayable about symbolic forms like fairy tales as techniques of language, with implications both for how we read some aspects of Wittgenstein and how Langer figures in histories of philosophy.[18] I also hope, as I think through their interest in fairy tales, to raise the question of how much Langer's opening up of the realm of meaning relevant to philosophical analysis owes to her response to Wittgenstein and how much, to understand certain puzzles in Wittgenstein's middle work, it may help to think about Langer, as one of his most stretching readers.

In what ways can fairy tales offer a window onto Langer's "expanded notion of logic, as a study of meaning in form"? In addressing this question, my concern here is not with Langer's extended discussion of fairy tale and its relation to myth in her best-known work, *Philosophy in a New Key* (*PNK*).[19] Nor do I focus on how her understanding of fairy tales was received by other scholars, although it did spark some response, even beyond the folklorists. When, for example, Ananda Coomaraswamy writes to Langer about her philosophy, he hones in on fairy tales: "I have been looking at your New Key. There is much for me to agree with and much I cannot agree with, notably in connection with the appraisement of fairy tales ... You ignore the point of view of those whose interest in fairy tales is essentially metaphysical."[20] Rather, this chapter aims to locate the fairy or wonder tale as somewhere around which some intriguing and philosophically significant thinking takes place, both for Langer and Wittgenstein, about names, metaphysics and magic, feeling and form. To consider this first involves some intellectual-historical work of establishing their convergence on the fairy or wonder tale or, as both these German-speakers named it, *märchen*.

Langer, who was mostly schooled at home, read and wrote *märchen* from childhood. Home was a realm of wonder tales, away from the Gradgrinds of the schoolroom.[21] During her degree at Radcliffe College, Langer sustained this fairy tale thread through the skein of her formal education. Her papers at the Houghton Library include work about Lady Wilde's Irish folk tales as well as imaginative compositions, alongside work in formal logic. From 1924–6, Langer then completed a doctorate with Whitehead, for whom educational theory was in lively dialogue with the maneuvers of process philosophy across this decade.[22] Whitehead published 'The Rhythm of Education' in 1929, arguing that, for the pupil, subjective receptiveness and emotional patterning should hold sway over the learning of mere information. As his pupil, Langer sustained an interest in receptiveness and how it can be fostered when we "illustrate to little children."[23] She explains, in *The Practice of Philosophy*, a book to which Whitehead provided a preface, the dangers of instructing children by analogy.

> [We teach] how the earth moves round the sun by letting an orange circle about a large balloon, and tell them that the cosmic process is the same, but impossible to observe under ordinary circumstances. The orange is the symbol, the earth is what is meant. Now, worlds and oranges, suns and balloons differ in so many ways that they do not present an obvious similarity. A fairly high degree of rationality is required to appreciate the analogy.[24]

Here, Langer's concerns parallel those of Wittgenstein, whose primary school teaching in Austria led him, in 1925, to produce a word book to offset the excessive abstraction required of children by current classroom dictionaries. As Wittgenstein explains in the preface to his *Wörterbuch für Volksschulen*, "the purely alphabetic order, which pushes a heterogeneous word between closely related ones, then in my opinion demands too much of the child's ability to abstract."[25] The *Wörterbuch* is not merely a nominalist accumulation of useful words; it is also a philosophical construct that intervenes in questions of language and cognition. How we "illustrate to little children" is, then, involved with the central aspects of Wittgenstein's and Langer's philosophies—abstraction, analogy, simile, symbol—as much as with their schoolroom or home routines.

Langer's first pupils in the art of imaginative illustration were her two sons, born in 1922 and 1926, and for whom she wrote *The Cruise of The Little Dipper and Other Fairy Tales* (1923). The titular tale in this collection of *märchen* is a tender, blithe fairy tale about a child "so poor he had no name." The protagonist is a "very poor boy" whose father has been lost at sea for some years. The boy lives with his cruel great-aunt, who denies him sugar or butter, soap bubbles or pretty shells. Each day, as he gathers wood for her fire, while the other children play on the beach, the boy whistles, "and since he had no name, they called him 'Birdling.' His great-aunt called him 'You!'."[26] Birdling remains calm at the viciousness of his aunt and is cocooned by a world of others who do not much interfere in his sufferings but are not quite indifferent. First, this is the other children, who lend Birdling some things to build toy boats in exchange for some whistled songs. Then, when Birdling shrinks to the size of a deflated balloon and sets sail on his toy boat, the "Little Dipper," it is a crew of quavering shrimps, who sail with him, encountering a sea monster, rescuing a fairy prince, and then, by an inversion of authority and power, rescuing his father. He navigates the fairy tale's characteristic division between ordinary malice (the great-aunt's spankings and deprivations) and supernatural malevolence (Shag, King of the Deep Sea).[27] Birdling is not proportionately frightened—he goes calmly where others would not, into Shag's lair—but he has a clear-sighted sense of fear of the minor punishments of his great-aunt. This obeys the depthless pattern of fairy tale composition, in which there is an oblique quality of the relation of event to feeling. Here, the extraordinary is received as though ordinary and there are not laws, only contingencies, surprises, changes, transmogrifications. "For the fairytale is irresponsible; it is frankly imaginary," as Langer writes in *PNK*.[28] Birdling's blitheness also embodies the principle that "[t]here is no psychology in a fairy tale," only behavior.[29] This may be part of the form's draw for Wittgenstein if we consider that "the achievement of *Philosophical Investigations* was to de-psychologize psychology,"[30] and for Langer, with her "antipsychologistic theory of 'significant form'."[31] Langer's writing of the *Little Dipper* is also part of her training in receptiveness to the patterns of thought about the virtual subject that Whitehead developed during his years at Harvard. This wonder tale of Birdling, who lives across elements and in slanted relation to human naming, with its companionable sailing Nautilus shells, shy shrimps, and protective mullein leaves,[32] inhabits the world of Whitehead's "reformed subjectivist principle," and its "breaking with the anthropocentric limitations of modern philosophy."[33]

It is worth pausing with the aspect of Langer's tale that most marks out its philosophical resonance, rather than, say, the ready availability it has for psychoanalytic readings: the insistence that Birdling has "no name." For having no name is a philosophical predicament, too, and Langer's framing of her fairy tale in these terms draws a line of connection through to her philosophical thought. Langer's preface to the 1951 edition of *PNK* remarks that "[r]eally new concepts, having no names in current language, always make their earliest appearance in metaphorical statements." For Langer, thinking about this metaphorical dynamic to the language of thought takes place not only through Wittgenstein and Whitehead but also in the company of Ernst Cassirer, whose understanding of human symbol-making remains one of her dominant influences. Cassirer sharpened Langer's sense of what she describes as "the hypostatic mechanism of the mind by which the world is filled with magical things—fetishes and talismans, sacred trees, rocks, caves, and the vague, protean ghosts that inhabit them."[34] Amid Langer's description of this non-discursive symbolization, which gives rise to myths, rituals, and religious thinking, as well as to the properties of the fairy tale universe, we again find names. "Names are the very essence of mythic symbols; nothing on earth is a more concentrated point of sheer meaning than the little, transient, invisible breath that constitutes a spoken word."[35]

Wittgenstein also worked through his sense of words as "mythic symbols" around 1930, as he made notes during readings of James Frazer's turn-of-the-century anthropological monolith, *The Golden Bough*. This reading led to a series of aphoristic comments, some typed up into a considered sequence, some remaining in a large manuscript book. Of the former kind, we find, "And magic always rests on the idea of symbolism and of language," and of the latter, "A whole mythology is deposited in our language."[36] These are not comfortably proximate aphorisms but rather restless rewordings; Wittgenstein was troubled by how to compose remarks about Frazer's "stupid" conflations of magic and scientific immaturity, and also by the *Tractatus*, whose method had become a source of disquietude. "For when I began in my earlier book to talk about the 'world' (and not about this tree or table)," Wittgenstein worries, "was I trying to do anything except conjure up something of a higher order by my words?" This remark comes from an introductory sequence for a possible final typescript, which Wittgenstein then crossed out, marking the margins with the word "schlecht" (bad). This term is serious, with a savor of Rumpelstiltskin's malevolent badness rather than merely the vain incompetent badness of the miller. These "bad" remarks worry over the relations of thoughts, words, and magic; is to summon a name to engage in a form of conjuring? "I think now that the right thing would be to begin my book with remarks about metaphysics as a kind of magic," Wittgenstein begins, but even as he sees through the conjuring of metaphysical thought, he recognizes a problem: "keeping magic out has itself the character of magic." The iconoclast is an idolater, after all.

The *Tractatus* had strained towards logically proper names and denied ordinary names, only allowing for "this" and "that" within the narrowed category of the logically atomist name. Hacker and Baker, in their neat summary of how names figure in Wittgenstein's thought, note that *Philosophical Investigations* then rethinks this earlier denial of ordinary names, but they also declare that the question of naming is not wildly interesting to the later Wittgenstein. Only §79 of the *Investigations* contributes to the lively philosophical conversation about names, they argue, and even then, this single remark about Moses is not really about names so much as vagueness.[37] It takes other readers, who have that more capacious concern with human symbol-making identified in Langer and Whitehead, to spot the other occasions where names are the subject in the *Investigations*. W. H. Auden, for example, an astute reader of Wittgenstein as well as the Brothers Grimm, notices §410 and reproduces it twice in his own writings about proper names.

> "I" is not the name of a person, nor "here" of a place, and "this" is not a name. But they are connected with names. Names are explained by means of them. It is characteristic of physics not to use these words.—Ludwig Wittgenstein.[38]

Attention more like Auden's opens up new possibilities for thinking about the sorts of questions about naming that Langer wants to ask after having read early Wittgenstein, with all their significance for how we relate feeling to form of words. For Wittgenstein's sensitivity to the ordinary proper name (particularly at the bookends of life, birth and death) is clearly more acute than a merely logical account recognizes.[39] Hacker and Baker, for all the flatness with which they treat the contribution to the philosophy of names made by the *Investigations*, do notice an enigmatic comment about the name as talisman, instrument, or ornament from Wittgenstein's enigmatic notes about Frazer and Freud.

> Why should it not be possible that a man's own name be sacred to him? Surely it is both the most important instrument given to him and also something like a piece of jewelry hung round his neck at birth.[40]

This is mentioned alongside their observation that, for later Wittgenstein, ordinary names, particularly proper names, can be a test for how far we can imagine forms of life. Wittgenstein shows us, they write, that "[a] community whose language contained no expressions comparable to proper names in their ramifying significance would have a way of life and thought very different from ours."[41] What would this near unimaginable community be like?

It would not quite be the world of fairy tale, for there are significant names and also, from time to time, plain, ordinary names. But it is the case that names are far from common here and, except in rare cases (such as the tale of Rumpelstiltskin), not momentous. Not only are the characters not significantly or often named, but the form itself has the stamp of anonymity; there is a "no name" quality to fairy tales.[42] Returning to this chapter's opening anecdote, then, what might Wittgenstein's wonder at Rumpelstiltskin's name-trickery suggest about his evolving sense of the name and the forms of life of the communities that speak it? When Wittgenstein exclaims about the imp's profundity, we might first discern a shadow of the contours of the Tractarian universe: what he is in awe of seems to be the ineluctable power of the withheld name, of what cannot be said. But also, in his deep appreciation for this tale, there is a suggestion of the deflationary anti-metaphysics of the *Investigations*. For the solution to the miller's daughter's problem is found through investigation of everyday behavior rather than through reaching into the realms of the ineffable. She merely sends out messengers, one of whom finds Rumpelstiltskin's cottage and overhears his name-revealing song. The quest is urgent but it is also ordinary. It may take a Wittgensteinian perspective to see this. Jack Zipes cannot, beginning an article about Rumpelstiltskin with these words: "the focus of folklorists, psychoanalysts, and literary critics has centered on Rumpelstiltskin's name and *his* role in the tale despite the fact that the name is meaningless. Indeed, it reveals nothing about Rumpelstiltskin's essence or identity. The naming simply banishes the threatening creature."[43] Zipes's tenacity about a correspondence theory of meaning (the syllables of Rumpelstiltskin must relate to the imp's "essence") blocks the philosophical force of the tale, which for Wittgenstein seems to perform the apprehension that a name (for all its temptations to the metaphysician) is something belonging to our verbal behavior, neither unspeakable nor a magical talisman. Names might be hung around our necks, at birth, like jewelry, and we might choose to hold them as of rare value, but they still operate according to some observable, if often complex, forms of behavior, susceptible to study, whether by the queen's messenger or the linguistic philosopher.

While it was Rumpelstiltskin that captivated Wittgenstein, other tales from the Grimms carry the same philosophical potential; "The Riddle," for example, where the princess is set the task of solving the verbal puzzle, "One killed none, but still killed twelve." She proceeds not by working it out as a problem of logic but rather by arranging for her maids to overhear the solution: again, the tale represents the triumph of observed verbal behavior over the analytical dissolution of a problem.[44] What these glimpses from the Grimms suggest is that, rather than a magical evasion of ethical and linguistic problems, the fairy tale world of charms, riddles, and enigmas is fertile ground for an ordinary language philosophical investigation. Fairy tales are among our techniques of language for carrying out the philosophical "battle against the bewitchment of our intelligence by means of our language."[45] A characterization of "the riddle" from Northrop Frye makes this clear:

> [T]he riddle ... represents the revolt of the intelligence against the hypnotic power of commanding words. In the riddle a verbal trap is set, but if one can "guess," that is, point to an outside object which the verbal construct can be related, the something outside destroys it as a charm, and we have sprung the trap without being caught in it.[46]

Riddles are a matter of configuration, of pointing to objects to which verbal constructs "can be related" but they are somewhat wary of isomorphism: once the relation is made, then the charm is torn in two, like Rumpelstiltskin. Hence the power of this tale for Wittgenstein at a time of transition from the metaphysics of the *Tractatus* and towards his later philosophical work of restoring names and symbols to the ordinary settings in which they are at home. A remark in *Culture & Value* makes the analogy between fairy tale process and ordinary language philosophical practice still more explicit:

> The solution of philosophical problems can be compared with a gift in a fairy tale: in the magic castle it appears enchanted and if you look at it outside in daylight it is nothing but an ordinary bit of iron or something of the sort.[47]

"Rumpelstiltskin" appears enchanted; a name so powerfully secret that it will claim the life of a baby. Looked at outside, in daylight, it is reduced to a name that has no magical force but is merely a piece of linguistic behavior to be pursued scrupulously. But the tearing force of the imp's self-destruction and the "unapproachable evil" he represents in the tale, still surround the name with the aura of power. Magic and metaphysics are not fully banished from the precincts of the tale and banishing would itself be a quasi-magical verbal act. For Wittgenstein, it is no wonder that this was a tale of awe.

In her own wonder tale, "The Cruise of the Little Dipper," Langer navigates this philosophical terrain through imaginative form. She chooses a diminutive name, so little that it is almost nothing, and the boy called Birdling is even said to have "no name." With this double gesture of quietly giving and yet also lightly withholding the name, Langer resists the grandeur of magical naming, but also of "keeping magic out," a gesture which, as we saw in the remarks from Wittgenstein quoted earlier in this chapter, "has itself the character of magic." Langer keeps carefully along this line of measured resistance to the bewitchment of intelligence by words. In her tale's version of the Rumpelstiltskin encounter, Birdling's duel of wits with Shag, Birdling is asked to find the most precious jewel in the King's vault, in order to be allowed to escape both with it and with his own life. He glances over "rubies and sapphires and emeralds, brooches and necklaces, pearl-set combs" but alights, correctly of course, on a locket "made of two oyster shells closed with a band of tin. There was nothing very precious about it."[48] Birdling's solution here is anti-metaphysical. He overlooks the shine of the jewels and the value of the hammered gold. He knows value to lie in use so sees that the act of positioning of the ordinary-looking locket in this treasure trove marks its value much more certainly than its mere materials. And so, he solves the enigma, rescues the fairy prince who is hidden within the locket, saves himself and his father, whistling blithely all the while. In such ways, this tale's oblique yet firm sense of how words relate to actions and to symbols lies close to the kind of philosophy of naming of middle and later Wittgenstein. Langer's Birdling sets us off on a journey that Wittgenstein, still somewhat in thrall to Rumpelstiltskin, takes later.

Langer's *märchen* can, then, be considered part of the expansion of symbolic forms that she sees as key to philosophy. The philosophy of language, as it proceeds along the lines of formal logic, is hampered by the paucity of its symbols, Langer argued from as early as *The Practice of Philosophy*, and her later works continue this quest for a richer symbolic repertoire, across architecture, art, literature, rituals, music, and dance. *Märchen* may, then, contribute not only to Langer's development of some areas in Wittgenstein's early thought, and to her reconfiguration of some areas of early to mid-twentieth-century aesthetic discussion, but also hint towards the underlying politics of a philosophy illustrated in this multifarious way. To return to Auden, symbolic poverty appeared to him to hold out particularly grave dangers in a climate of populist nationalism. "Half our troubles," Auden wrote in a 1944 review of a new edition of *Grimm's Fairy Tales*, "both individual neuroses and collective manias like nationalism, seem to me to be caused largely by our poverty of symbols, so that not only do we fail to relate one experience to another but also we have to entrust our whole emotional life to the few symbols we do have."[49] Philosophies of symbolic forms that can help us to configure experiences—as Langer argued that the *Tractatus* could, and as her own work bears witness—and that also refine the relation of thinking to "emotional life," are a prophylactic against pathologies, personal and global. If this first question, of configuration, is something Langer works through in her early writing, both creative and scholarly, then the best-known works of her middle period, *PNK* and *Feeling and Form*, refine questions of emotional life, in ways that show the traces of her earlier focus on wonder tales. *PNK*, for example, explores some genealogies for human symbolization, with Langer arguing that children are fantastical thinkers, natural denizens of *märchen*.

> Children mix dream and reality, fact and fiction, and make impossible combinations of ideas in their haste to capture everything, to conceive an overwhelming flood of experiences. Of course, the stock of their imagery is always too small for its purpose, so every symbol has to do metaphorical as well as literal duty. The result is a dreamlike, shifting picture, a faery "world."[50]

Rather than take the route from a stock of imagery "too small for its purpose" towards Auden's personal or collective manias, children live in a gestalt world, over-extending their few symbols.[51] Langer elaborates this sense of childhood symbolization and how it relates to "emotional life" in an important passage in *PNK*'s chapter 5, which describes childhood as "the great period of synesthesia," in which "over-active feelings" latch onto "flotsam material."

> Fear lives in pure Gestalten, warning or friendliness emanates from objects that have no faces and no voices, no heads or hands; for they all have "expression" for the child, though not—as adults often suppose—anthropomorphic form. One of my earliest recollections is that chairs and tables always kept the same look, in a way that people did not, and that I was awed by the sameness of that appearance. They symbolized

> such-and-such a mood; even as a little child I would not have judged that they felt it (if anyone had raised such a silly question). There was just such-and-such a look—dignity, indifference, or ominousness—about them. They continued to convey that silent message no matter what you did to them. A mind to which the stern character of an armchair is more immediately apparent than its use or its position in the room, is over-sensitive to expressive forms. It grasps analogies that a riper experience would reject as absurd.[52]

Children, Langer suggests, have a certain unripe absurdity to their imaginations, with over-active feelings and over-sensitivity to expressive forms. Their forms of wonder are not philosophically wise but they are also less "silly" than the hypothetical adult who assumes reductive versions of how feelings work. For these adults, feeling belongs to faces and voices, to human forms, missing the diffuse force of "mood" as it escapes anthropocentric limits. These adults want a correspondence theory, where a stimulus is related to an emotional response, which is legible in a movement or expression of the human body. Children, in their "faery world," know that inanimate objects do not have feelings but their moods can still be felt and judged. They are already beyond picture-theories of meaning. This childlike conception of feeling as something that relates to the object, without being isomorphic, has a strong presence in the more recent returns to Langer's philosophy within affect theories. Sianne Ngai, for example, picks up a related passage in her book, *Ugly Feelings* (2005), when she is explaining why her concept of "tone" owes little to formalists like the New Critics and much to Langer's conception of "significant form." Ngai selects an occasion from *Feeling and Form* where Langer is quoting one of the philosophers of gestalt who influenced her.

> We never think of regarding the landscape as a sentient being whose outward aspect "expresses" the mood that it contains subjectively. The landscape does not express the mood, but has it; the mood surrounds, fills, and permeates it ... the mood belongs to our total impression of the landscape.[53]

Fairy tales know well this common-sense point about the pervasiveness and non-expressive quality of mood. When Birdling takes up home in "a ground-sparrow's nest under the willow bushes, very near the fairy-bread flowers," everything is scaled to enclose and protect him, and the mood of the hillside is that of security, from which there can be bright song: the robin sings out to the sun and Birdling whistles with him. Later, at sea, there is peril in moods and in the elements as Shag's angry pursuit causes "a raging storm," with whitecaps "like ranks of horsemen," but when Birdling comes towards the Fairy Islands, at his whistling "the sea became calm and began to shimmer with a thousand lovely colors."[54] Simple as these mood impressions are, they represent the stage of thought that matters to Langer and Wittgenstein: thought not over-stretched by rational analogies and abstraction and so participating fully in the possibilities of symbolic forms. For, even if philosophers grow beyond child thought processes and their fairy world, its modes of configuration and experience linger: without reanimating something of this dream-like animism, it can be a struggle to see how Langer thinks symbols really work.

From her earliest publications, Langer points towards *märchen* as a "technique of our language" in a strong sense of that phrase: they are symbolic forms that refine our "emotional life" and negotiate the distinction between the sayable and unsayable, with all that this entails for the position of ethics in language. For Langer, who is not mystical, there is—even more than with Wittgenstein—a sense that these tales are part of life's ordinary symbolic forms, an expansion within experience rather than a stepping outside of it. Birdling brushes with magic, supernatural evil, and the limits of what can be said in the fairy world but still returns at the tale's end, "a life-sized boy once more" and as though from an ordinary journey, with souvenirs, "a gold thimble for his great-aunt and a little silver bell for Cat."[55] He experiences wonder, solves riddles, implies a stance on how words relate to the world and, with the fairy tale's true lightness of form, has no hard feelings. Had he read this wonder tale, Wittgenstein might not have reacted with quite the "hushed awe" solicited by Rumpelstiltskin and its lingering shadow of nominal magic. This only deepens the significance of Langer's gentle early strike against the poverty of philosophy's emotional symbols, and her important articulations of the need to develop the Tractarian vision away from the terminus of silence. It may be the case that "what we can't say we can't say, and we can't whistle it either"[56]—but can Birdling?

NOTES

1. Fania Pascal, "Wittgenstein: A Personal Memoir," in *Recollections of Wittgenstein*, ed. Rush Rhees (Oxford: Oxford University Press, 1984), 12–49.
2. Georg H. von Wright, "Intellectual Autobiography," in *The Philosophy of Georg Henrik Von Wright*, ed. Lewis E. Hahn and Paul A. Schilpp (La Salle, IL: Open Court, 1989), 3–58.
3. Cited by Joseph G.f. Rothhaupt, "Ludwig Wittgenstein and Wilhelm Busch: 'Humour is not a mood, but a Weltanschauung,'" In *Wittgenstein Reading*, ed. Sascha Bru, Wolfgang Huemer, and Daniel Steuer (Berlin: De Gruyter, 2013), 197–226, 198.
4. Marina Warner, *Once upon A Time: A Short History of Fairy Tale* (Oxford: Oxford University Press, 2014), 40.
5. Cora Diamond, "Ethics, Imagination, and the Method of Wittgenstein's *Tractatus*," in *The New Wittgenstein* (New York: Routledge, 2000), 149–73.
6. Diamond, "Ethics, Imagination, and the Method of Wittgenstein's *Tractatus*," 169–70. "The removal of thought and talk about the evil will from empirical talk is a technique of our language, just as stories about Rumpelstiltskin and the Fisherman's Wife are techniques of our language."
7. Alfred N. Whitehead, *Process and Reality: An Essay on Cosmology*, ed. David R. Griffin and Donald W. Sherburne (New York: The Free Press, 1978), 338. Glossing this, the Editors' Introduction to *The Lure of Whitehead*, ed. Nicholas Gaskill and Adam J. Nocek (Minneapolis: University of Minnesota Press, 2014), 11, notes: "Whitehead's inclusiveness is often astounding: A world in which the fairies dance is not a world familiar to those who pursue the powerful intellectual operations known as 'critique.'"
8. Louis Zukofsky, *Bottom: On Shakespeare* (Austin, TX: The Ark Press, 1963). Zukofsky's book is a collage of quotations, with many from Wittgenstein's *Tractatus Logico-Philosophicus*. Adrienne Dengerink Chaplin, *The Philosophy of Susanne Langer* (London: Bloomsbury, 2019), 133 points out that Wittgenstein is somewhat anomalous among Langer's influences: "In contrast to her minimal citing of those other thinkers, Langer quotes the early Wittgenstein both frequently and extensively."
9. As Colin Radford notes, this seems to be a reference to the Grimms's "The Gold-Youths," more often translated as "The Gold Children." Colin Radford, "Wittgenstein and Fairy Tales," *Merveilles & Contes* 2, no. 2 (1988): 106–10.
10. Susanne K. Langer, *An Introduction to Symbolic Logic* (New York: Dover, [1937] 1967), 118.

11. This is reiterated compellingly by Adrienne Dengerink Chaplin, *The Philosophy of Susanne Langer: Embodied Meaning in Logic, Art and Feeling* (London: Bloomsbury, 2019), ch. 8 and ch. 7, *passim*.
12. Susanne K. Langer, *The Practice of Philosophy* (New York: Henry Holt, 1930), 121.
13. See Langer's first article, "Confusion of Symbols and Confusion of Logical Types," *Mind*, New Series, 35, no. 138 (April 1926): 222–9.
14. See Dengerink Chaplin, *The Philosophy of Susanne Langer*, 8.
15. Richard Wollheim's review of Langer's *Feeling and Form* (1953), *Burlington Magazine* 97, no. 633 (1 December 1955): 400–1.
16. Garry Hagberg, "Art and the Unsayable: Langer's Tractarian Aesthetics," *British Journal of Aesthetics* 24, no. 4 (1984): 325–40.
17. Dengerink Chaplin, *The Philosophy of Susanne Langer*, 134.
18. See, for example, Joel Isaac, "Susanne K. Langer, *Philosophy in a New Key* (1942)," *Public Culture* 32, no. 2 (91) (May 1, 2020): 355–61. "Langer opened up the realm of meaning from the logicians' narrow treatment of propositions to the structured meanings implicit in modes of feeling and action." See also Dengerink Chaplin, *The Philosophy of Susanne Langer*, 69.
19. Susanne K. Langer, *Philosophy in a New Key: A Study in the Symbolism of Reason, Rite, and Art*, 3rd edn (Cambridge, MA: Harvard University Press, [1942] 1957), chapter 7, "Life-Symbols: The Roots of Myth."
20. Unpublished letter from the Susanne Langer Papers, MS Am 3110, Box 29, folder 2 (Harvard University, Houghton Library, July 20, 1942).
21. Dickens saw fairy tales as a mode of resistance to Utilitarianism and was angered by moralized retellings such as George Cruikshank's. "It is to be hoped," Auden wrote in a post-war introduction to a collection of fairy tales, "that the publication of the tales of Grimm and Andersen in one inexpensive volume will be a step in the campaign to restore to parents the right and the duty to educate their children, which, partly through their own fault, and partly through extraneous circumstances, they are in danger of losing for good." This remark comes in Auden's introduction to Wystan H. Auden, *Tales of Grimm and Andersen* (New York: The Modern Library, 1952), xiv.
22. Dengerink Chaplin notes in *The Philosophy of Susanne Langer* that Langer began her doctoral thesis, titled "A Logical Analysis of Meaning," with Whitehead the year he arrived in America, 1924.
23. Langer, *The Practice of Philosophy*, 115–16.
24. Ibid.
25. Ludwig Wittgenstein, *Wörterbuch für Volksschule*, ed. Paul Chan (New York: Badlands, 2020), 29.
26. Susanne K. Langer, *The Cruise of the Little Dipper, and Other Fairy Tales* (New York: Graphic Society, 1924), 3.
27. Cora Diamond makes this distinction key to her arguments about Rumpelstiltskin in "Ethics, Imagination, and the Method of Wittgenstein's *Tractatus*," *passim*.
28. Langer, *Philosophy in a New Key*, 175.
29. Philip Pullman, quoted in Warner, *Once upon a Time*, 116.
30. Ruth Leys, *The Ascent of Affect: Genealogy and Critique* (Chicago: University of Chicago Press, 2017), 156. Leys writes of Stanley Cavell's "brilliant thought" in his essay, "Aesthetic Problems of Modern Philosophy," "that the achievement of Wittgenstein's *Philosophical Investigations* was to de-psychologize psychology."
31. This is Sianne Ngai's phrase as she characterizes Langer's philosophy as instead emphasizing "feeling's synthesizing, epideictic, intellectual character" in *Ugly Feelings* (Cambridge, MA: Harvard University Press, 2005), 47.

32. “mullen” *[sic]*. Langer, *The Cruise of the Little Dipper*, 14.
33. Gaskill and Nocek, *The Lure of Whitehead*, 30.
34. Susanne K. Langer, “On Cassirer’s Theory of Language and Myth,” in *The Philosophy of Ernst Cassirer*, vol. 6, ed. Paul Arthur Schlipp (Evanston, IL: Library of Living Philosophers, 1949), 381–400.
35. Langer, “On Cassirer’s Theory of Language and Myth,” 390.
36. Ludwig Wittgenstein, *Remarks on Frazer’s Golden Bough*, ed. Rush Rhees (Harleston: Brynmill Press, 1979), 42e, 10e.
37. Ludwig Wittgenstein, *Philosophical Investigations*, 3rd edn, trans. Gertrude E. M. Anscombe (Oxford: Blackwell, 2001), 31e–32e.
38. Wystan H. Auden, *The Complete Works of W. H. Auden*, vol. 6: *Prose: 1969–1973*, ed. Edward Mendelson (Princeton, NJ: Princeton University Press, 2015), 204. On another occasion in his prose, Auden quotes the same words but with the error “It” for “I.” For Auden’s writing about fairy tales see, for example, “In Praise of the Brothers Grimm,” *New York Times Book Review*, November 12, 1944, which opens with the words “It is unlikely that there will be another event during the current publishing season as important as this ... it is hardly too much to say that these tales rank next to the Bible in importance.” Reprinted in Wystan H. Auden, “In Praise of the Brothers Grimm,” in *The Complete Works of W. H. Auden: Poems;* vol. 2: *1939–1948*, ed. Edward Mendelson (Princeton, NJ: Princeton University Press, 2002), 239–42.
39. For details about Wittgenstein’s use and erasure of his brother’s name, Paul, see Brian McGuinness ed., *Wittgenstein’s Family Letters* (London: Bloomsbury Academic, 2019), xl.
40. Wittgenstein, *Remarks on Frazer*, 5e.
41. Gordon P. Baker and P. M. S. Hacker, ch. XII, “Proper Names,” in *Wittgenstein: Understanding and Meaning*, vol.1, rev. edn (Oxford: Blackwell, 2008).
42. See James Merrill, quoted in Warner, *Once upon a Time*, 63.
43. Jack Zipes, “Spinning with Fate: Rumpelstiltskin and the Decline of Female Productivity,” *Western Folklore* 52, no. 1 (1993): 43.
44. In this tale, the princess sends her maids to listen in the prince’s bedroom in the hope he will sleep talk, revealing the answer to the verbal puzzle she is tasked with solving. Wise to this, the prince sets a trap to outwit the princess’s trap: he pretends to sleep talk so that she has her answer, but then steals her robe for material proof that the answer has been obtained by trickery not wisdom. The tale, it is true, does not fully approve the princess’s solution; it is determined that, because of her method, she fails the task, but the “punishment” still rewards the failure with a fair fairy-tale ending, marriage to the prince.
45. Wittgenstein, *Philosophical Investigations*, §109, 40e.
46. Northrop Frye, “Charms and Riddles,” in *Spiritus Mundi: Essays on Literature, Myth, and Society* (Bloomington: Indiana University Press, 1976), 137.
47. Ludwig Wittgenstein, *Culture and Value*, rev. edn, ed. Georg H. von Wright (Oxford: Blackwell, 1998), 11e.
48. Langer, *The Cruise of the Little Dipper*, 32.
49. Auden, “In Praise of the Brothers Grimm,” 242. Conversely, Marina Warner notes that, “[d]uring the process of de-Nazification, the Grimms were banned from schools and libraries; they had formed part of the propaganda machine for turning out fascists, and the Allies saw them as irredeemably tainted,” in Warner, *Once upon a Time*, 128.
50. Langer, *Philosophy in a New Key*, 148.

51. Adrienne Dengerink Chaplin notes that, during the first year of her marriage, 1921–2, spent in Vienna, Susanne Langer attended lectures with Karl Buhler, a leading philosopher and gestalt psychologist. Dengerink Chaplin also argues that Buhler was a friend of the Wittgenstein family. Dengerink Chaplin, *The Philosophy of Susanne Langer*, 135.
52. Langer, *Philosophy in a New Key*, 123.
53. Langer is quoting Otto Baensch. See Ngai, *Ugly Feelings*, 43–4.
54. Langer, *The Cruise of the Little Dipper*, 12, 36–8.
55. Ibid.
56. Frank P. Ramsey's quip about the end of the *Tractatus*. See "General Propositions and Causality," in *F. P. Ramsey: The Foundations of Mathematics*, ed. Richard B. Braithwaite (London: Routledge and Kegan Paul, 1931), 238.

BIBLIOGRAPHY

Auden, Wystan H. *The Complete Works of W. H. Auden,* vol. 6: *Prose: 1969–1973*. Edited by Edward Mendelson. Princeton, NJ: Princeton University Press, 2015.

Auden, Wystan H. "Introduction." In *Tales of Grimm and Andersen*, edited by Frederick Jacobi, xiii–xxi. New York: The Modern Library, 1952.

Auden, Wystan H. "In Praise of the Brothers Grimm." In *The Complete Works of W. H. Auden: Poems,* vol. 2: *1939–1948*, edited by Edward Mendelson, 239–42. Princeton, NJ: Princeton University Press, 2002.

Baker, Gordon P., and P. M. S. Hacker. "Proper Names." In *Wittgenstein: Understanding and Meaning*, vol. 1, rev. edn, 227–49. Oxford: Blackwell, 2008.

Dengerink Chaplin, Adrienne. *The Philosophy of Susanne Langer*. London: Bloomsbury, 2019.

Diamond, Cora. "Ethics, Imagination, and the Method of Wittgenstein's Tractatus." Reprinted in *The New Wittgenstein*, edited by Alice Crary and Rupert Read, 149–73. New York: Routledge, 2000.

Frye, Northrop. "Charms and Riddles." In *Spiritus Mundi: Essays on Literature, Myth, and Society*, 123–47. Bloomington: Indiana University Press, 1976.

Gaskill, Nicholas, and A. J. Nocek, eds. *The Lure of Whitehead*. Minneapolis: University of Minnesota Press, 2014.

Hagberg, Garry. "Art and the Unsayable: Langer's Tractarian Aesthetics." *British Journal of Aesthetics* 24, no. 4 (1984): 325–40. https://doi.org/10.1093/bjaesthetics/24.4.325.

Isaac, Joel. "Susanne K. Langer, Philosophy in a New Key (1942)." *Public Culture* 32, no. 2 (91) (May 1, 2020): 355–61. https://doi.org/10.1215/08992363-8090117.

Langer, Susanne K. "Confusion of Symbols and Confusion of Logical Types." *Mind*, New Series, 35, no. 138 (April 1926): 222–9.

Langer, Susanne K. *The Cruise of the Little Dipper, and Other Fairy Tales*. New York: Graphic Society, 1924.

Langer, Susanne K. *An Introduction to Symbolic Logic*. 1937. Rev. edn. New York: Dover, 1967.

Langer, Susanne K. "On Cassirer's Theory of Language and Myth." In *The Philosophy of Ernst Cassirer*, vol. 6, edited by Paul Arthur Schilpp, 381–400. Evanston, IL: The Library of Living Philosophers, 1949.

Langer, Susanne K. *Philosophy in a New Key: A Study in the Symbolism of Reason, Rite, and Art*. 1942. 3rd edn. Cambridge, MA: Harvard University Press, 1957.

Langer, Susanne K. *The Practice of Philosophy*. New York: Henry Holt, 1930.

Leys, Ruth. *The Ascent of Affect: Genealogy and Critique*. Chicago: University of Chicago Press, 2017.
McGuinness, Brian, ed. *Wittgenstein's Family Letters*. Translated by Peter Winslow. London: Bloomsbury Academic, 2019.
Ngai, Sianne. *Ugly Feelings*. Cambridge, MA: Harvard University Press, 2005.
Pascal, Fania. "Wittgenstein: A Personal Memoir." In *Recollections of Wittgenstein*, edited by Rush Rhees, 12–49. Oxford: Oxford University Press, 1984.
Radford, Colin. "Wittgenstein and Fairy Tales," *Merveilles & Contes* 2, no. 2 (1988): 106–10.
Ramsey, Frank P. *F. P. Ramsey: The Foundations of Mathematics*. Edited by Richard B. Braithwaite. London: Routledge and Kegan Paul, 1931.
Rothhaupt, Joseph G. F. "Ludwig Wittgenstein and Wilhelm Busch: 'Humour is not a mood, but a Weltanschauung.'" In *Wittgenstein Reading*, edited by Sascha Bru, Wolfgang Huemer, and Daniel Steuer, 197–226. Berlin: De Gruyter, 2013.
Susanne Langer Papers, MS Am 3110, Box 29, folder 2. Harvard University, Houghton Library, July 20, 1942.
von Wright, Georg H. "Intellectual Autobiography." In *The Philosophy of Georg Henrik Von Wright*, edited by Lewis E. Hahn and Paul A. Schilpp, 3–58. La Salle, IL: Open Court, 1989.
Warner, Marina. *Once upon a Time: A Short History of Fairy Tale*. Oxford: Oxford University Press, 2014.
Whitehead, Alfred North. *Process and Reality: An Essay on Cosmology*. Corrected edn. Edited by David R. Griffin and Donald W. Sherburne. New York: The Free Press, 1978.
Wittgenstein, Ludwig. *Culture and Value*. Rev. edn. Edited by Georg H. von Wright. Oxford: Blackwell, 1998.
Wittgenstein, Ludwig. *Remarks on Frazer's Golden Bough*. Translated by A. C. Miles. Edited by Rush Rhees. Harleston: Brynmill Press, 1979.
Wittgenstein, Ludwig. *Wörterbuch für Volksschule*. Translated by Bettina Funcke. Edited by Paul Chan. New York: Badlands, 2020.
Wollheim, Richard. "Review: Feeling and Form by Susanne K. Langer." *Burlington Magazine* 97, no. 633 (December 1, 1955): 400–1.
Zipes, Jack. "Spinning with Fate: Rumpelstiltskin and the Decline of Female Productivity." *Western Folklore* 52, no. 1 (1993): 43–60.
Zukofsky, Louis. *Bottom: On Shakespeare*. Austin, TX: The Ark Press, 1963.

CHAPTER FOUR

Susanne K. Langer on Logic as the Study of Forms and Patterns of Any Sort

GIULIA FELAPPI

In *An Introduction to Symbolic Logic*, Langer maintains that logic, as the science of forms and patterns, "is to the philosopher what the telescope is to the astronomer: an instrument of vision."[1] In fact, for Langer logic is an "indispensable tool"[2] for philosophy, and not just because philosophy needs correct reasoning and logic is indeed "an inestimable aid in reasoning."[3] While it is well known that Langer was a professed enthusiast of logical analysis and the analytic method in philosophy, her point is more general. Langer stresses that "[a]ll knowledge, all sciences and arts,"[4] philosophy being no exception, have their beginning in the recognition of structures and patterns, which can help us systematize and understand our "rapidly changing, shifting, surprising world."[5] Hence, philosophy requires a certain ability "to conceive of things in general, to appreciate formal relations,"[6] and logic is a means for philosophers to "see the world in its clear light."[7] One example, Langer stresses, is that, thanks to the developments of mathematical logic, "infinity has ceased to be a magic word."[8]

But in the 1920s and 1930s logic for Langer is not just a means but also "the most elementary, restricted and definite philosophical science."[9] Hence, logic is a subject of study, which she pursued while claiming that she was presenting philosophical questions "with hesitation, with the discomfort which a mere logician quite properly feels in the presence of philosophical problems."[10] As a philosophical science, logic is, moreover, for Langer itself a domain of philosophical investigation, as there are "philosophical problems, which arise directly from logical considerations."[11]

By being conversant with different logical traditions, Langer's reflections in logic, and on the philosophical problems logic gives rise to, famously led her to endorse two claims: first, logic should be concerned not only with propositions and propositional forms, as it was then orthodox, but rather with forms for anything that follows a pattern of any sort; second, there is nothing like *the* logical form of any thing, as any matter can be analyzed as exemplifying radically different forms.

The aim of this paper is to unfold Langer's main reasons toward these two claims and to show how they stem from considering logic both as a tool for philosophy and as itself a subject of study and philosophical investigation.

BEYOND THE PROPOSITIONAL

Langer stresses that "Bertrand Russell, in one of his facetious moments, defined mathematics as 'the science in which we never know what we are talking about.'"[12] The same is true of what logic is for her, as it is the "*analysis of systems*, disregarding entirely the nature of their elements."[13] For Langer, logic "*is the science of forms as such, the study of patterns*,"[14] the tracing and description of the forms exemplified by systems of elements of "experience (or Reality, or what-not)"[15] and of their relations. The notion of *form* or *pattern* is then for Langer central to logic, and she characterizes it by combining ideas coming from quite different logical traditions.[16] While she maintains that for the notions of *system* and *pattern* she is indebted to Sheffer,[17] she considers Whitehead and Russell's *Principia Mathematica*[18] to be an "elaborate masterpiece"[19] and Russell to be "blessed with both candor and acumen."[20] She is then happy to characterize her central notion of *form* explicitly in line with Russell's "excellent account,"[21] with his "admirably lucid exposition of logical forms,"[22] in the following way: "[t]he logical form of a thing is the way that thing is *constructed*, the way it is put together."[23] Explicitly in line with Russell,[24] but also with Frege[25] and Moore,[26] she endorses anti-psychologism in logic, as for Langer forms are not a "subjective ingredient; forms are found in experience, not added to it."[27]

While Langer is happy to align with Russell in her characterization of form, at the same time she rejects the then "well-established"[28] claim, which she says was endorsed by him,[29] but also by Frege,[30] and many others, that "the study of propositions and of the relations which obtain between propositions is the only legitimate claimant to the title of 'logistic', and is, in fact, formal logic itself."[31] She maintains that logic "certainly … includes the *forms of propositions*,"[32] but should not be confined to those forms. When it comes to what "the material of logic"[33] is, Langer suggests following Royce:[34]

> Josiah Royce defined logic as *the study of types of order*. This is essentially the point of view I wish to advocate, that *logic is the study of forms as such*, regardless of content ("forms" is a somewhat less restricted term than "order").
>
> "Orderliness and system," says Royce … are much the same in their general characters, whether they appear in a Platonic dialogue, or in a modern textbook of botany, or in the commercial conduct of a business firm, or in the arrangement and discipline of an army, or in a legal code, or in a work of art, or even in a dance or in the planning of a dinner. Order is order. System is system. Amidst all the variations of systems and of orders, certain general types and characteristic relations can be traced.[35]

So, this is the first of Langer's famous claims coming from putting together different logical traditions: while Russell characterized the notion of logical form correctly, on what the material of logic is, we should rather follow Royce and maintain that "anything may be said to have form that follows a pattern of any sort, exhibits order, internal connection."[36]

Why maintain this? Let us start from the reasons Langer provides that stem from considering logic as a philosophical tool. Logic can help our endeavors to see the world in its clear light in two equally important and related ways. First, logic, in involving "the analysis of systems *qua* systems, the discovery of relations which hold for all possible worlds … [the] analysis of all possible things, given in abstract terms,"[37] "liberates the human mind from the finitude of actuality and opens to it the endless reaches of *potentiality*,"[38] by also providing us with the boundary between what patterns, forms, and structures are possible, and which ones are instead impossible. Second, by the recognition of forms "we find *analogies*, and come to understand one thing in terms of another":[39] logic improves our ability to represent something we are trying to understand, as a thing exemplifying a certain logical form "may be *represented* by another which has the same structure,"[40] and will aid our understanding of it, as "any essential configuration"[41] in one system will "find its analogue"[42] in the other system, "just as the lines and proportions of a suit are analogous to those of its paper prototype."[43]

Now, if propositional forms can be taken to be "not peculiar to propositions,"[44] that is, can be taken to be logical forms also of things that are not propositions, those very forms or something very close to them can also be a means to understand those other domains. In 1929, Langer thinks that "[a] good case in point is the structure of music."[45] While offering her set of postulates for the logical structure of music "to delimit the field in which any musical configuration whatever must necessarily lie,"[46] she notes that such logic "looks enough like Boolean algebra."[47] Since for Boolean algebra there surely is a "propositional interpretation,"[48] if in logic we do not confine ourselves to propositions, propositional forms or forms very close to the propositional ones can then help us also understand music. Moreover, if there are philosophical disciplines that concern some matter that exemplifies forms very different from those propositions can be taken to exemplify, as there seem to be indeed in the domain of "emotional and aesthetic experience: the recognition of intrinsic values,"[49] then a logic that goes beyond propositions and then beyond propositional forms can be a tool also in those philosophical endeavors.

Langer also hints:

> There is a further point of interest in this attempt to discern the purely logical structure of the musical universe—a matter of such philosophical import, howbeit of such unsubstantiated, visionary character, that I offer it as the merest suggestion: is it possible that music is not the only interpretation for this algebra, but that some logician versed in the arts, especially in arts other than music, might trace similar structures in some other form of aesthetic expression? The implication of such a hypothesis for the philosophy of art is obvious and vital. Psychology and metaphysics alike have failed so far to put aesthetics on any better basis than a purely empirical one; is it conceivable that logic might bridge the gap between those two disciplines and discover truly fundamental principles whereon to build a rational science of aesthetics? I have added this speculative paragraph with hesitation ... but add it I must, even as a fantastic hypothesis, the timid, scientific version of Schopenhauer's bold poetic dictum, "*die Baukunst ist erstarrte Musik*."[50]

So, a logic that is not concerned only with propositions is a better tool for philosophy for two related reasons, as exemplified by the case of music. First, going beyond propositions allows us to employ logical forms to understand also other subjects of inquiry, such as music. Second, if it were the case that those forms that music exemplifies were typified by all forms of aesthetic expression, a logic that goes beyond propositions could lead to something vital and of such philosophical import, that is, a rational science of aesthetics. Similarly, for Langer, in 1930 there is room to think that "ethically interested logicians will probably be the founders of scientific ethics."[51] Since many philosophical domains were at that time logically unexplored, in 1930 for Langer logic then has "a significance for philosophy"[52] that is "incalculable."[53]

For Langer, logic should go beyond propositions and propositional forms not just to become a better philosophical instrument of vision, but also for reasons that stem from considering logic as itself a subject of philosophical inquiry. For a logic that goes beyond propositions and propositional forms can account for some notions, such as the notion of meaning, which are central to it, instead of having to leave them logically undefined.

Let us see her argument concerning the notion of meaning.[54] First, Langer claims that meaning "in its profoundest sense is one of the fundamental notions of logic."[55] The reason is the following. Langer stresses that "[t]he only way we can express logical facts is through the employment of symbols,"[56] that is, logic is not "strictly a *mere* 'string of marks,'"[57] logic is *symbolic*. While a logical symbol "is not assumed to have any specific meaning … it exemplifies things which are true of many systems."[58] Since logic is symbolic, then meaning is crucial to it, as meaning is "that in virtue of which we have a *symbolism* at all,"[59] and that in virtue of which we can distinguish "between a mark on paper which is a *symbol* and one which is due to spilled ink or the murder of a mosquito."[60] Langer urges that the question as to what object a word refers to is not a logical question: in logic we do not establish "the actual forms in which [meaning] does occur,"[61] as in this way logic would be trafficking with "real poets, lovers, unicorns"[62] and would lose its formal character. But still there is a logical question and, Langer maintains while quoting Russell, it is this: "What relation must one fact (such as a sentence) have to another in order to be *capable* of being a symbol for that other?"[63] So there is a question about meaning that is a logical question, and it is the one that concerns the "logical prerequisites for meaning,"[64] "the logical situations in which meaning is possible,"[65] "the formal possibility of meaning,"[66] the "structure of symbolism."[67]

Second, for Langer the account of the formal possibility of meaning put forward by Wittgenstein in his *Tractatus Logico-Philosophicus* (1921), which she describes as an "extraordinary prophetic gospel,"[68] but also by Whitehead, in *Symbolism, Its Meaning and Effect* (1927), is "on the whole, a true account."[69] According to that account, Langer stresses, at the logical basis of meaning there is a "common element of formal structure,"[70] there is "a system related to another system."[71] For the propositional system, for example, meaning, at least in most of the cases, "must lie somewhere else than in the formal properties of propositions":[72] the relation should be toward "something *outside* the system which is the proposition,"[73] as most pieces of language do not mean pieces of language but something extra-linguistic.

But Langer then urges that meaning is therefore "definitely outside the scope of *Principia Mathematica*,"[74] and of any logic according to which the material of logic is exclusively propositional. If we take logic to be concerned only with propositional forms and the system of propositions, and we take the basis of meaning to be a relation to *another* system, "[s]uch logical problems as the nature of meaning … remain perfectly insoluble"[75] and we cannot but follow Wittgenstein's dictum, "There is indeed the inexpressible. This shows itself; it is the mystical,"[76] and we cannot but end "in perfect alogicism,"[77] so that meaning "lives in the underworld (or superworld?) of Mysticism."[78]

Langer then remarks that it is "rather unfortunate that logic should be characterized by certain arbitrary alogical elements,"[79] and should be governed by some "deus ex machina,"[80] as this is a "metaphysical dead-stop,"[81] adding that "Mysticism has ever been the graveyard for logical doctrines."[82] A logic able to bring the notion of meaning into its scope is then to be preferred, and this, Langer maintains, is precisely what can be done if we follow Royce on what the material of logic is. For:

> if we allow our logical interest to cover forms of every sort, merely as forms, we shall find that there are innumerable systems, or patterns, in the world, of which the propositional system is merely a special one; that these patterns may be compared, and the systems which exemplify them may be brought into relation with one another, and the traditional "alogical" notions may be brought into the scope of logic as we include not only the relations of elements within one system, but the relations of certain systems to each other (relations such as similarity, analogy, etc.) ... If we treat the system of propositional forms as merely one formal system which may be compared with other logical structures, I think we shall ... find perfectly definite relations between propositional structures and other structures ... Every thing, situation, idea, or what not, has a logical pattern; propositions follow such a pattern, and, as Royce has pointed out ... all other things, from dialogues to dinners, have patterns of their own.[83]

The Roycean logic can then account for the possibility of "the sign-function of the barometer"[84] and the possibility that "the system of physics 'describes' the world we know through sense,"[85] that is, the possibility that "its formulae *mean* our world."[86]

So, Langer concludes, the correct account, put forward by Wittgenstein and by Whitehead, of the logical basis of meaning "really presupposes the less restricted view of logic."[87] Wittgenstein should have combined differently his account of the logical basis of meaning and his own proposition:

> 4.014 The gramophone record, the musical thought, the score, the waves of sound, all stand to each other in that pictorial internal relation, which holds between language and the world. To all of them the logical structure is common.[88]

For in a less restricted view of logic, like the Roycean one, whose material includes the gramophone record, the musical thought, the score, and the waves of sound, the logical prerequisite of meaning, as correspondence between systems, is not outside the scope of logic.

As shown by her argument about the notion of meaning,[89] Langer then finds a motivation for a less restricted view of logic also by considering logic as a source of philosophical problems and then in the fact that a less restricted view of logic, as she suggests, "promises to save some important logical relations from their present metaphysical limbo."[90]

RADICALLY DIVERSE FORMS

The Roycean move of going beyond propositional forms is not the only famous claim Langer put forward concerning the Russellian notion of logical form. Again, by putting together different logical traditions, from Whitehead[91] and Sheffer[92] she also inherited the idea that "no structure is absolute, no relation peculiar to the material in hand, no analysis of fact the only true one ... Living experience may come to us in most various forms."[93] For Langer "*the* form of an object, if taken to denote a single absolute notion, suffers from ... non-significance,"[94] as any thing might be taken to exemplify "radically different forms."[95] For example,

> [i]f we take points as our basic terms we will have other postulates and theorems than if we started with volumes, or still more, if we started with notions such as "space-time events" or with Leibnitzian "monads." Yet our various systems of geometry, of physics, and of metaphysics are all designed to describe the actual world, that is

> to say, they are all supposed to present a pattern which is to be found in the actual world.[96]

So, let us go back to Langer's claim above of how knowledge of forms allows us to understand one thing in terms of another. It should not be understood as stating that knowledge of forms will allow us to see *the* logical form of the matter under consideration. As Langer urges "[i]ngenuity in thinking—whether in practical, scientific, or philosophical thinking—is primarily the art of … discovering new ways in which a familiar thing may be treated so as to reveal some hitherto unknown relation."[97]

Also in the case of this claim of hers, according to which "there is no such things as *the form*,"[98] "'form' always means *a* form,"[99] Langer thinks that it is motivated both by considering logic as a tool and by an investigation into logic as a philosophical science. Let us start again from Langer's points stemming from considering logic as the philosopher's telescope. In a review she published in 1930, Langer marks the difference between *philosophical logic* and *logical philosophy*, which, she stresses, "have nothing in common except words."[100] Philosophical logic, she explains, "begins with a metaphysical attitude, and employs alleged logical principles for its defense"[101] and, in doing this, it is not a "legitimate way in which the recent advances of logic can influence metaphysics."[102] In order for logic to genuinely be a tool when the metaphysical and philosophical endeavors will take place, it cannot presuppose any metaphysical claim, it cannot rely on any metaphysical assumption. Any "metaphysical notion," as she highlights, "must be an unwelcome stranger in the logical field,"[103] and "metaphysical gods"[104] are "not to be worshipped openly in the realm of logic."[105] For example, logic should not tell us the answers to questions such as the question, typical of Langer's time, as to "whether what is 'given' in our experience is a property or a relation."[106] If it did, it could not be used as a metaphysical tool, as the "ante chamber"[107] of metaphysics, as it could tell us that only if it was already making some metaphysical assumptions. Hence,

> "[t]he only legitimate way in which the recent advances of logic can influence metaphysics is by giving rise to a *logical philosophy*, such as Professor Whitehead … represents … logical philosophy begins with a single-minded and rigorous devotion to logic, from which, by long acquaintance, a certain new metaphysical outlook is born."[108]

A logic that is confined to one form as *the* logical form for some thing is already making some metaphysical assumptions concerning that thing and then in assuming that there is *the* logical form, such a logic has moved from logical philosophy to philosophical logic. Only if we reject the claim that there is *the* logical form, can logic be a genuine tool for philosophers and can it indeed be the case that "every advance in logic is a gain in metaphysical insight."[109]

Also in the case of her rejection of the claim that there is anything like *the* logical form for some thing, Langer's reasons do not stem only from considering logic as a tool. In her review of Langer's *An Introduction to Symbolic Logic*, Susanne L. Stebbing observes:

> Controversial issues are avoided, which is all to the good from the point of view of the elementary student. Here and elsewhere, as for example in her discussion of "logical form," Dr. Langer may give the student a misleading impression that there are no real difficulties to be overcome. Whilst it is desirable that these difficulties should not be discussed in an elementary introduction to the subject, it would have been better to hint that they exist.[110]

While it is true that in *An Introduction to Symbolic Logic* Langer does not tackle the issues that the notion of logical form gives rise to, in that very book, and among the philosophical problems, which arise directly from logical considerations, she mentions "the relativity of language, logical patterns,"[111] and "the problem of form and content."[112] Moreover, already in her dissertation,[113] Langer proves to be well aware of the philosophical problems that spring from the notion of form and in particular the problem of relating the form of anything to its content, which seems indeed to lead to a logical paradox. For Langer, as shown by her paper "Form and Content: A Study in Paradox" (1926), the rejection of the notion of *the* logical form also finds its motivation in the ability of such a rejection in aiding us to avoid that logical paradox. Here is the way in which she presents the paradox:

> At first sight it appears obvious that there can be such a relation; but if there is, then it can be expressed symbolically, as *R(f, c)*; and thereby we have transformed our empirical content into a term of the formal structure, *i.e.*, we have formalized it, and are no longer dealing with the non-logical content. Thus it seems there can be no such thing as the relation between the form of a thing and the content of that form, since this relation would entail a true paradox.[114]

Clearly, if we reject the notion of *the* form, there is nothing like the relation between *the* form of a thing and its content, as for that relation we would need there to be its relata, but one of them, *the* form, is just not there.

It should be said, though, that, as Langer knew, this is not the end of a solution to the problem. For, as Langer stresses, one might hold that there is *the* form, as "the class of all possible forms under which the object in question can be conceived."[115] But, Langer maintains, there is nothing like the class of all possible forms either. In order to have such a class, we would need "a single system wherein all these forms are conceived,"[116] we would need a language in which we could have all these forms together. But even though "there are types of logical language, which yield various types of system,"[117] for Langer each "[l]anguage ... determines by its structure just the sorts of ... forms, which can be expressed in it. And whatever object we are talking about, we are limited to some definite language, with its idiosyncrasies of structure, and consequently we are limited as to the things we can say."[118] So, in order for there to be the class of all forms we would need the ability to have a language in which we could speak about *all* those forms together. But, Langer maintains, languages all have themselves structures, have themselves forms, which will inevitably make each language unable to speak about some of the forms. Hence, there is nothing like the class of all forms. So, for Langer, there is nothing like the relation between *the* form of an object and its content as a *relatum* is missing, since there is neither *the* one form nor *the* class of all possible forms. Without the relation, the paradox that relation would lead to is then avoided.

For Langer, some other relations are there, though, but for them the paradox does not arise. There are the relations between *a* form of an object, the form it takes in a particular logical structure, and *a* content, which is "that which is not given as part of *this* logical structure."[119] But Langer says it is clear that each of these relations "abstruse and complex as it may be, exhibits no true paradox."[120] Langer does not explain this point in detail, but it might be taken to be the following. The paradox, concerning these relations, if there were one, would be: on the one hand, if there were these relations, it would be possible to formalize them; on the other, if we formalize them, we have formalized content, and then it is not content anymore. But then there is no paradox for these relations because there is no reason to think that it should be possible to formalize the relation between *the form that an object takes in a structure* and *that which is not given as part of that logical structure*. There is no reason to think that it can be formalized in the very structure, as there is no reason to think that in that structure; we can formalize *that which is not given as part of that logical structure*. Moreover, there is no reason to think that it can be formalized in any of the other structures, as there is no reason to think that in them we can formalize *the form that an object takes in the original structure*.

It can be disputed whether Langer genuinely avoided the paradox that the relation between the form of a thing and its content seems to lead to, because it can be disputed that she genuinely proved that there is nothing like the class of all forms by applying an observation she herself makes in *The Practice of Philosophy* (1930). While claiming that "[i]f now we would describe the location of any place, we must use one geometry or another,"[121] she adds in a footnote: "We could, of course, assert propositions *about the systems* and *involve* propositions from both of them, but we could not *use* them."[122] So, one might hold that Langer has not genuinely proved that there can be no language in which all forms are *involved*, even though she might be perfectly correct that there is no language in which all forms are expressible, that is, in which the propositions that can be *used* can together exemplify all forms. Maybe, one can urge, there is a language, with a particular structure for sure, in which we can nonetheless name all forms, in which we could then speak about all forms together, and the class of all forms would follow. Still, if *the* form cannot be taken to be "the class of all possible forms under which the object in question can be conceived," either because the former is not tantamount to the latter or because the latter is to be rejected for one reason or another, Langer did genuinely solve the paradox thanks to her rejection of *the* form. Be that as it may,[123] Langer's reflections on the paradox surely show how her motivations for going beyond *the* logical form also stem from considering logic as itself a source of controversial issues, of whose existence she was indeed well aware.

CONCLUSION

Langer maintained that logic "becomes useful and important to the philosopher"[124] only after she has "really grasped its technique,"[125] and so "we must understand its power and difficulties thoroughly before we can use it,"[126] and "we must work with a genuine interest in our restricted, abstract subject."[127] In the 1920s and 1930s, Langer did exactly what she then thought a philosopher must do and, as her reflections on the notion of form and pattern show, she showed all her genuine interest in logic. From then on, she aimed at increasing "our understanding of any forms or facts which are hidden in the kaleidoscope of experience."[128]

NOTES

1. Susanne K. Langer, *An Introduction to Symbolic Logic* (New York: Allen and Unwin, 1937), 41.
2. Ibid., 40.
3. Ibid., 41.
4. Ibid., 21.
5. Ibid., 22.
6. Susanne K. Langer, "Algebra and the Development of Reason," *The Mathematics Teacher* 24, no. 5 (1931): 288.

7. Langer, *An Introduction to Symbolic Logic*, 41.
8. Susanne K. Langer, "A Logical Analysis of Meaning," (PhD diss., Radcliffe College, Cambridge, MA, 1926), 11.
9. Susanne K. Langer, *The Practice of Philosophy* (New York: Henry Holt, 1930), x.
10. Susanne K. Langer, "A Set of Postulates for the Logical Structure of Music," *The Monist* 39, no. 4 (1929): 570.
11. Langer, *An Introduction to Symbolic Logic*, 334.
12. Langer, "Algebra and the Development of Reason," 292–3.
13. Langer, "A Logical Analysis of Meaning," 2.
14. Langer, *The Practice of Philosophy*, 83.
15. Langer, "A Logical Analysis of Meaning," 30.
16. As Juliet Floyd stressed: "because she was fluent in German, Langer was able to access the works of Husserl and Frege, as well as Wittgenstein and Russell ... Her work forms a bridge between the American Idealist tradition in which the status of logic, intentionality, and the categories are central (Royce, Peirce, Sheffer, and C. I. Lewis), the British tradition of Russell and Whitehead (Whitehead was her advisor), the German phenomenological tradition of Husserl, and the neo-Kantian tradition of Cassirer that investigates meaning through analysis of symbolic forms" (Juliet Floyd, "Recent Themes in the History of Early Analytic Philosophy." *Journal of the History of Philosophy* 47, no. 2 (2009): 199). See also Sander Verhaegh's Chapter 1 in this volume, "Susanne K. Langer and the Harvard School of Analysis".
17. Langer, "A Logical Analysis of Meaning," iii, 27, 48–52, 74. She refers to Henry M. Sheffer, *The General Theory of Notational Relativity* (manuscript, 1921, Harvard Widener Library); Langer, *The Practice of Philosophy*, 97f.5. Langer also claims that she had the "unusual opportunity, to discuss the question of logical systems with Prof. Whitehead" (Langer, "A Logical Analysis of Meaning," iv).
18. Alfred N. Whitehead and Bertrand Russell, *Principia Mathematica*, 2nd edn (Cambridge: Cambridge University Press, 1925/7).
19. Langer, "A Logical Analysis of Meaning," 57.
20. Susanne K. Langer, "Form and Content: A Study in Paradox," *Journal of Philosophy* 23, no. 16 (1926): 436.
21. Langer, *An Introduction to Symbolic Logic*, 31, where she refers to Bertrand Russell, "Logic as the Essence of Philosophy," in *Our Knowledge of the External World* (London: The Open Court Publishing Company, 1914), 43.
22. Langer, *The Practice of Philosophy*, 91. Also here, she refers to Russell, "Logic as the Essence of Philosophy," 42 ff.
23. Langer, *An Introduction to Symbolic Logic*, 24. It is worth stressing that this characterization is of what a form of a thing is with respect to that thing. But out of this characterization, we cannot evince anything about the metaphysics of forms. For example, we are completely in the dark concerning what kind of thing a way is. For a discussion of the metaphysics of forms in relation to what Langer says about them, see Kris McDaniel, "Ontology and Philosophical Methodology in the Early Susanne Langer," in *Innovations in the History of Analytical Philosophy* (London: Palgrave Macmillan, 2017), 280–3.
24. Langer refers to Bertrand Russell's "Mathematics and the Metaphysicians," 74, in *Mysticism and Logic* (London: Allen and Unwin, 1917), and §37 of *The Principles of Mathematics* (London: Allen and Unwin, 1903) already in her dissertation "A Logical Analysis of Meaning," 16f.7, and 66f.27.

25. Langer regards Gottlob Frege to be the first to state that logical forms and actual judgments are "separable." See Langer, "A Logical Analysis of Meaning," 16. She refers to the preface of his *Grundgesetze der Arithmetik* (1893/1903).
26. Langer refers to George E. Moore's "The Nature of Judgment" (*Mind* 8, no. 30 (1899): 179) in "A Logical Analysis of Meaning," 38–9.
27. Langer, *The Practice of Philosophy*, 143.
28. Ibid., 57. In favor of this point, she refers to Clarence I. Lewis, *A Survey of Symbolic Logic* (Berkeley: University of California Press, 1918), 354.
29. Langer, "A Logical Analysis of Meaning," 57. She refers to Whitehead and Russell's *Principia Mathematica*, vol. 1: Introduction; Langer, "A Logical Study of Verbs," *Journal of Philosophy* 24, no. 5 (1927): 122–3.
30. Langer, "A Logical Analysis of Meaning," 54, 57, 61f.17.
31. Ibid., 57.
32. Langer, *The Practice of Philosophy*, 85.
33. Langer, "A Logical Study of Verbs," 122.
34. Langer, "A Logical Analysis of Meaning," 25, 70; Langer, "A Logical Study of Verbs," 122; Langer, *An Introduction to Symbolic Logic*, 39. In all, Langer refers to Josiah Royce, "The Principles of Logic," in *Encyclopaedia of the Philosophical Sciences*, vol. 1: Logic (London: Palgrave Macmillan, 1913), 73.
35. Langer, "A Logical Study of Verbs," 123.
36. Langer, *An Introduction to Symbolic Logic*, 23.
37. Langer, "A Logical Analysis of Meaning," 25–8.
38. Langer, *The Practice of Philosophy*, 50–1.
39. Ibid., 88.
40. Langer, *An Introduction to Symbolic Logic*, 42.
41. Langer, *The Practice of Philosophy*, 99.
42. Ibid.
43. Ibid.
44. Langer, "A Logical Analysis of Meaning," 68–9.
45. Langer, "A Set of Postulates for the Logical Structure of Music," 562. The *structure* of music is to be distinguished from the *meaning* of music, with which Langer deals as early as in her dissertation, see Langer, "A Logical Analysis of Meaning," Appendix D. On Langer's reflections on music and meaning, see Lona Gaikis, "Music as the DNA of Feeling, and some Speculations on Whitehead's Influence on Susanne K. Langer's Philosophy," Chapter 11 in this volume.
46. Langer, "A Set of Postulates for the Logical Structure of Music," 562.
47. Ibid., 563.
48. Langer, "A Logical Analysis of Meaning," 57. In the dissertation, Langer suggests also another interpretation for the Boolean algebra: "It seems to me at present that another interpretation is possible; that the Algebra is capable of expressing, for instance, the system of "ideas"" (ibid., 147) of "empiricist psychology" (ibid., 159).
49. Langer, *The Practice of Philosophy*, 153.
50. Langer, "A Set of Postulates for the Logical Structure of Music," 570.
51. Langer, *The Practice of Philosophy*, 213.
52. Ibid, 83.
53. Ibid.
54. For Langer's reflections on the notion of meaning after the 1930s, see Adrienne Dengerink Chaplin, *The Philosophy of Susanne Langer: Embodied Meaning in Logic, Art and Feeling* (London/Oxford: Bloomsbury Academic, 2020), ch. 8.

55. Langer, "A Logical Analysis of Meaning," 6.
56. Ibid.
57. Ibid., 87.
58. Ibid. She takes her claim that logical symbols can mean different systems to be aligned with what Frege maintains, and refers to Frege, *Grundgesetze der Arithmetik*, vol. 2, 100–1.
59. Langer, "A Logical Study of Verbs," 128.
60. Langer, "A Logical Analysis of Meaning," 6.
61. Ibid., 126. She also stresses: "Here I wish to guard against a misunderstanding which is almost inevitable ... this is the error of supposing that I consider meaning essentially as a logical relation, and hope by logical analysis to exhaust all its constituent factors. I do not believe it to be ever a purely logical affair, any more than judgment, or empirical existence" (ibid., 9).
62. Ibid., 126. Here Langer is in fact criticizing Russell, who she maintains claims that "logic is built up on atomic propositions; therefore logic is built up on things that cannot occur in logic" (ibid.). According to Langer, this is in opposition to his claim that "[c]onstants do not occur in logic; that is to say, the *a*, *b*, *c* which we have been supposing constant are to be regarded as obtained by an extra-logical assignment of values to variables" (Whitehead and Russell, *Principia Mathematica*, xxx), which Langer instead takes as correct (Langer, "A Logical Analysis of Meaning," 30).
63. Bertrand Russell, "Introduction," in *Tractatus Logico-Philosophicus* (London: Kegan Paul, 1922), 7, quoted in Langer, "A Logical Analysis of Meaning," 10f.; 28.
64. Langer, *The Practice of Philosophy*, 118.
65. Langer, "A Logical Analysis of Meaning," 126.
66. Ibid., 5.
67. Ibid., 84.
68. Langer, *The Practice of Philosophy*, 108.
69. Langer, "A Logical Study of Verbs," 124. On Langer's understanding and interpretation of Wittgenstein's supposed picture theory, see Adrienne Dengerink Chaplin, "Scientific Models and Artistic Images: Susanne K. Langer and the Early Wittgenstein," Chapter 2 in this volume.
70. Langer, "A Logical Study of Verbs," 124.
71. Ibid., 127. Langer, ibid., 124 also states that "Mr. Russell in large measure subscribes" to Wittgenstein's account. See also Langer, "A Logical Analysis of Meaning," 7–8, 19, where Langer quotes and refers to Russell's Introduction to the *Tractatus*. As a further proof that Langer worked on different logical traditions, it is worth mentioning that she also maintains that the claim that there is a relation between "a symbol and its object" at the logical basis of meaning, was also endorsed by Husserl, in Investigation 1, §9 of *Logische Untersuchungen* (1900/1). See Langer, "A Logical Analysis of Meaning," 119–20.
72. Langer, "A Logical Study of Verbs," 122.
73. Ibid., 125.
74. Ibid., 122.
75. Ibid., 123.
76. *Wittgenstein's Tractatus* 6.522, quoted in Langer, "A Logical Analysis of Meaning," 8f.25.
77. Ibid., 68.
78. Langer, "A Logical Study of Verbs," 124.
79. Langer, "A Logical Analysis of Meaning," 68.
80. Ibid., 72.
81. Ibid.
82. Langer, "A Logical Study of Verbs," 124.

83. Ibid., 123–4.
84. Langer, "A Logical Analysis of Meaning," 107.
85. Ibid., 93.
86. Ibid.
87. Langer, "A Logical Study of Verbs," 124.
88. Wittgenstein quoted in Langer, "A Logical Analysis of Meaning," 136f.7 and in Langer, *The Practice of Philosophy*, 120.
89. In Langer, "A Logical Study of Verbs," she also aims at showing how endorsing the Roycean stance on what the material of logic is, allows us to characterize the Fregean assertion-sign and the Russellian notion of logical assertion. For further discussion, see Giulia Felappi, "Saving Logic from a Metaphysical Limbo: Susanne Langer on Logical Assertion," manuscript.
90. Langer, "A Logical Study of Verbs," 129.
91. Langer, "Form and Content: A Study in Paradox," 437. Langer quotes and refers here to Whitehead's *An Enquiry Concerning the Principles of Natural Knowledge* (Cambridge: Cambridge University Press, 1919), 59–60.
92. Susanne K. Langer, "Facts: The Logical Perspectives of the World," *Journal of the Philosophy* 30, no. 7 (1933): 181–2. Langer refers to Henry M. Sheffer's *The General Theory of Notational Relativity* and his paper "Notational Relativity," in the *Proceedings of the Sixth International Congress of Philosophy* (1927).
93. Langer, "Facts: The Logical Perspectives of the World," 182–3.
94. Langer, "Form and Content: A Study in Paradox," 437.
95. Ibid.
96. Langer, *The Practice of Philosophy*, 135. The selection of one form rather than another depends for Langer on the "intellectual purpose" (Langer, "Facts: The Logical Perspectives of the World," 183) of our enquiry. For example, "the analysis of a line into an infinity of points is sufficient, and therefore correct, for the purpose of establishing its relations to other parts of space, but not for establishing the line itself in the Euclidean system" (ibid.) But while for Langer the purpose might dictate which form to consider in each of the various endeavors to see the world in a clear light, still for her purposes are in no way the business of logic, as shown by her discussion of James's and Dewey's pragmatist take on logic (Langer, *The Practice of Philosophy*, 80–1). In it, she notes that with them "we arrive at a logical theory whose interest is the proposition in almost any aspect except the formal one. The conception, plausibility, content, purpose of propositions; their usefulness, their truth; anything but their form and their place in a closed abstract system" (ibid., 80). In line with her anti-psychologism, she then laments that since what pragmatists are interested in is "psychological principles such as belief and interest" (ibid., 81), in following them "logic is not expanded, but simply abandoned, superseded by … psychology" (ibid.).
97. Langer, "Algebra and the Development of Reason," 295.
98. Langer, *The Practice of Philosophy*, 135.
99. Langer, "Form and Content: A Study in Paradox," 438.
100. Susanne K. Langer, "Book Review: *The Logic of Events: An Introduction to a Philosophy of Time* by Andrew P. Uchenko," *Journal of Philosophy* 27, no. 13 (1930): 362.
101. Ibid.
102. Ibid.
103. Langer, "A Logical Study of Verbs," 122.
104. Ibid., 127.

105. Ibid.
106. Susanne K. Langer, "Book Review: *Non-Aristotelian Logic and the Crisis in Science* by Oliver L. Reiser," *Journal of Symbolic Logic* 2, no. 2 (1937): 89.
107. Langer, "Algebra and the Development of Reason," 297.
108. Langer, "Book Review: *The Logic of Events: An Introduction to a Philosophy of Time* by Andrew P. Uchenko," 362.
109. Langer, *The Practice of Philosophy*, 101.
110. Susan L. Stebbing, "Book Review: *An Introduction to Symbolic Logic* by Susanne K. Langer," *Philosophy* 13, no. 52 (1938): 482.
111. Langer, *An Introduction to Symbolic Logic*, 334. For our purposes, we can take this to be the problem, but it should be said that for Langer the problem is "the relativity of language, logical patterns, and 'facts'," and facts are the main topic of Langer's paper, "Facts: The Logical Perspectives of the World." On Langer's notion of fact, see Giulia Felappi, "Susanne Langer and the Woeful World of Facts," *Journal for the History of Analytical Philosophy* 5, no. 2 (2017): 38–50.
112. Langer, *An Introduction to Symbolic Logic*, 334.
113. Langer, "A Logical Analysis of Meaning," 43, 96–7.
114. Ibid., 436.
115. Ibid., 437.
116. Ibid.
117. Langer, "Facts: The Logical Perspectives of the World," 182.
118. Langer, "Form and Content: A Study in Paradox," 437.
119. Ibid., 438.
120. Ibid.
121. Langer, *The Practice of Philosophy*, 137.
122. Ibid., f.3.
123. While not observing explicitly that her claim that there is nothing like *the* logical form is in conflict with Russell, Langer maintains that her solution to the paradox the relation between form and content gives rise to relies on Mr. Russell's "fallacy of 'illegitimate totalities'" (Langer, "Form and Content: A Study in Paradox," 436, where she refers to both Russell, *The Principles of Mathematics*, ch. 10 and Appendix B, and Whitehead and Russell's *Principia Mathematica*, in particular to the Introduction of vol. 1 and ch. 2). But in her point about why the alleged totality of forms would be illegitimate she does not rely, at least not explicitly, on the Russellian criteria for a totality to be illegitimate. Langer, moreover, sees some similarities between her solution to the problem of relating the abstract form of anything to its specific content, and Russell's discussion of Wittgenstein's mystical point seen above that the logical basis of meaning, that is, the structure, common to a proposition and a fact it is a picture of, cannot be put into words, cannot be part of logic. Russell suggests that the only way to escape Wittgenstein's mysticism would be, for the totality of languages, to "deny that there is any such totality" (Russell, "Introduction," *Tractatus Logico-Philosophicus*, 19), to maintain that such totality is "not merely logically inexpressible, but a fiction, a mere delusion" (ibid.). Langer urges explicitly: "[s]ince the logical forms we can designate are determined by the language, or medium of designation, what holds for language holds also for logical forms; if, as Mr. Russell maintains, there could be no totality of them, then … there yet could be no totality of [forms]" (Langer, "Form and Content: A Study in Paradox," 438). But it should be said that Russell does not seem to *maintain* what she takes him to be maintaining, as concerning his suggestion he calls it a "possibility" (Russell,

"Introduction,"*Tractatus Logico-Philosophicus*, 19) and adds: "Such an hypothesis is very difficult, and I can see objections to it which at the moment I do not know how to answer" (ibid.).

124. Langer, *An Introduction to Symbolic Logic*, 41.
125. Ibid.
126. Ibid.
127. Ibid.
128. Ibid., 126.

BIBLIOGRAPHY

Dengerink Chaplin, Adrienne. *The Philosophy of Susanne Langer: Embodied Meaning in Logic, Art and Feeling*. London/Oxford: Bloomsbury Academic, 2020.

Felappi, Giulia. "Saving Logic from a Metaphysical Limbo: Susanne Langer on Logical Assertion." Manuscript, 2022.

Felappi, Giulia. "Susanne Langer and the Woeful World of Facts." *Journal for the History of Analytical Philosophy* 5, no. 2 (2017): 38–50. Available online: https://doi.org/10.15173/jhap.v5i2.2812.

Floyd, Juliet. "Recent Themes in the History of Early Analytic Philosophy." *Journal of the History of Philosophy* 47, no. 2 (2009): 157–200. Available online: https://muse.jhu.edu/article/262927 (accessed September 13, 2022).

Frege, Gottlob. *Grundgesetze der Arithmetik*. Jena: Verlag Hermann Pohle, 1893/1903.

Husserl, Edmund. *Logische Untersuchungen*. Halle: Max Niemeyer, 1900/1.

Langer, Susanne K. "Algebra and the Development of Reason." *The Mathematics Teacher* 24, no. 5 (1931): 285–97. Available online: https://www.jstor.org/stable/27951356.

Langer, Susanne K. "Book Review: *Non-Aristotelian Logic and the Crisis in Science* by Oliver L. Reiser." *Journal of Symbolic Logic* 2, no. 2 (1937): 88–9. Available online: https://www.jstor.org/stable/2267382 (accessed September 13, 2022).

Langer, Susanne K. "Book Review: *The Logic of Events: An Introduction to a Philosophy of Time* by Andrew P. Uchenko." *Journal of Philosophy* 27, no. 13 (1930): 361–3. Available online: https://www.jstor.org/stable/2015257 (accessed September 13, 2022).

Langer, Susanne K. "Facts: The Logical Perspectives of the World." *Journal of the Philosophy* 30, no. 7 (1933): 178–87. Available online: https://www.jstor.org/stable/2015751 (accessed September 13, 2022).

Langer, Susanne K. "Form and Content: A Study in Paradox." *Journal of Philosophy* 23, no. 16 (1926): 435–8. Available online: https://www.jstor.org/stable/2014823 (accessed September 13, 2022).

Langer, Susanne K. *An Introduction to Symbolic Logic*. New York: Allen and Unwin, 1937.

Langer, Susanne K. "A Logical Analysis of Meaning." PhD diss., Radcliffe College, Cambridge, MA, 1926. Available online: https://langercircle.sites.uu.nl/wp-content/uploads/sites/662/2021/02/Langer-Susanne-K.-1926-A-Logical-Analysis-of-Meaning-doctoral-dissertation-scanned-PDF-from-photocopy.pdf (accessed September 13, 2022).

Langer, Susanne K. "A Logical Study of Verbs." *Journal of Philosophy* 24, no. 5 (1927): 120–9. Available online: http://www.jstor.org/stable/2015082 (accessed September 13, 2022).

Langer, Susanne K. *The Practice of Philosophy*. New York: Henry Holt, 1930.

Langer, Susanne K. "A Set of Postulates for the Logical Structure of Music." *The Monist* 39, no. 4 (1929): 561–70. Available online: http://www.jstor.org/stable/27901211 (accessed September 13, 2022).

Lewis, Clarence I. *A Survey of Symbolic Logic*. Berkeley: University of California Press, 1918.
McDaniel, Kris. "Ontology and Philosophical Methodology in the Early Susanne Langer." In *Innovations in the History of Analytical Philosophy*, edited by Sandra Lapointe and Christopher Pincock, 265–98. London: Palgrave Macmillan, 2017. Available online: https://link.springer.com/chapter/10.1057/978-1-137-40808-2_9 (accessed September 13, 2022).
Moore, George E. "The Nature of Judgment." *Mind* 8, no. 30 (1899): 176–93. Available online: https://www.jstor.org/stable/2247657 (accessed September 13, 2022).
Royce, Josiah. "The Principles of Logic." In Arnold Ruge, Wilhelm Windelband, Josiah Royce, Louis Couturat, Benedetto Croce, Federigo Enriques, and Nicolaj Losskij, *Encyclopaedia of the Philosophical Sciences*, vol. 1: Logic, translated by B. Ethel Meyer, 67–135. London: Palgrave Macmillan, 1913.
Russell, Bertrand. "Introduction." In Ludwig Wittgenstein, *Tractatus Logico-Philosophicus*, edited by Charles K. Ogden, translated by Charles K. Ogden and Frank P. Ramsey, 7–19. London: Kegan Paul, 1922.
Russell, Bertrand. "Logic as the Essence of Philosophy." In *Our Knowledge of the External World*, Lecture II. London: The Open Court Publishing Company, 1914.
Russell, Bertrand. "Mathematics and the Metaphysicians." In *Mysticism and Logic*, ch. 5. London: Allen and Unwin, 1917.
Russell, Bertrand. *The Principles of Mathematics*. London: Allen and Unwin, 1903.
Sheffer, Henry M. *The General Theory of Notational Relativity*, manuscript, 1921. Harvard Widener Library.
Sheffer, Henry M. "Notational Relativity." *Proceedings of the Sixth International Congress of Philosophy* (1927): 348–51.
Stebbing, L. Susan. "Book Review: *An Introduction to Symbolic Logic* by Susanne K. Langer." *Philosophy* 13, no. 52 (1938): 482–3. Available online: https://www.jstor.org/stable/3746393 (accessed September 13, 2022).
Verhaegh, Sander. "Susanne Langer and the American Development of Analytic Philosophy." In *Women in the History of Analytic Philosophy*, edited by Jeanne Peijnenburg and Sander Verhaegh. Cham: Springer, 2022.
Whitehead, Alfred N. *An Enquiry Concerning the Principles of Natural Knowledge*. Cambridge: Cambridge University Press, 1919.
Whitehead, Alfred N. *Symbolism, Its Meaning and Effect*. New York: Macmillan, 1927.
Whitehead, Alfred N., and Bertrand Russell. *Principia Mathematica*. 2nd edn. Cambridge: Cambridge University Press, 1925/7.
Wittgenstein, Ludwig. "Tractatus Logico-Philosophicus." *Annalen der Naturphilosophie* 14 (1921): 195–262.

[illegible] Berkeley: University of California Press, 1993). [illegible]

CHAPTER FIVE

The Horizontal, Vertical, and Transversal Mechanics of Susanne K. Langer's Card-Index System

IRIS VAN DER TUIN

Susanne K. Langer (1895–1985) studied Philosophy with a specialization in Logic at Radcliffe College, the so-called women's annex of Harvard University in Cambridge, Massachusetts, for her BA (1916–20), MA (1922–4), and PhD degrees (1924–6). During all these years of studying and for the entire duration of her lifelong career as a professional philosopher, Langer, an American woman born in New York City from German immigrants, kept a card-index system for her personal use that ended up consisting of thirty-seven drawers and approximately 25,000 cards in total. With its drawers and cards, the system resembles a technology characteristic of Western European and Northern American library and office spaces in the period from the early 1920s to the late 1960s.[1] My archival research on Langer's card-index system conducted in January 2020 in the Houghton Library at Harvard, where the system is currently being preserved, revealed that Langer did her research in company of the work of at least 345 women and that these were female professionals from across academia and the arts.[2] In the system I encountered women of many different nationalities, generations, disciplines, and professions. The indexed women that I was enabled to list and count were accompanied by more, and more widely known, male figures. Langer has stored the work of both women and men with a single reference, with a meticulous summary, or in opinionated manner. Their work was either commented on in isolation or put in connection with the work of others. Langer herself denied any impact on her career of the fact that she was a woman.[3] Nor did she push a feminist agenda with her work on "feeling."[4] She did, however, work in what can be poetically called "a universe of women," whose scholarship she discovered, studied, evaluated, and integrated.

This chapter on Langer's card-index system has been written with the aim of doing theoretical justice to what is, in more than one way, a hybrid system with many layers of storage architecture and mechanics. Historically or perhaps I should say *biographically*, we find written and typed cards in the archived system, from both the hand of Langer herself and from those of the several assistants upon whom Langer had come to rely already during her active career and, in particular, with whom she worked closely later in her solitary life as a philosopher on a research and writing grant from the Edgar J. Kaufmann Charitable Trust of Pittsburgh, Pennsylvania that she was lucky to acquire in 1956. Langer worked on and used her card-index system from 1916 until she had to give up philosophy entirely for reasons of increasing blindness and, eventually, old age. Some index cards are part of a series and numbered I, II, III [...] and some have been written on both sides. In terms of its structure and contents, the system has elements of a traditional card-index system in which identically sized and shaped index cards on similar topics are stored back-to-front and upright on their edges, divided by manila tabs, in several drawers and of a more modern vertical "filing cabinet" containing differently sized papers, in this case handwritten or typed index cards stored alongside hand-cut and/or folded paper clippings, often glued to standardly sized and shaped cards. Researching the card-index system back in 2020, I also found a dried flower attached to the back of one of the many cards I diligently and eagerly fingered through. Some of them were damaged by use, water, mold, red sealing wax, or even fire. During my research, I came to understand the idiosyncratic card-index system as embodying an Alfred Lord Tennyson (1809–92) quote inked down by Langer on one of the many cards that I reviewed: "I am part of all that I have known."[5]

In his profile piece on Langer in *The New Yorker* of December 3, 1960, music critic Winthrop Sargeant uses "card-index file" and "card-indexing system" interchangeably.

FIGURE 5.1 "Photograph of Dr. Susanne K. Langer at her desk in Old Lyme Connecticut" by permission of the Estate of Susanne K. Langer. Photographer: James Lord (1968). From the article "A Lady Seeking Answers," *The New York Times Book Review*, May 26, 1968. Reprinted with permission of Harold Ober Associates. Copyright © 1968 by James Lord.

Sargeant describes the system as "a sort of hand-made mechanical memory that she has kept ever since her undergraduate days."[6] The long and ubiquitous presence of the system of index cards in Langer's life is confirmed by the artist and botanist Wesley Wehr, who met Langer in Seattle, Washington, in the first half of the 1950s, stayed in touch with her for several decades, and wrote his posthumous profile piece on Langer from memory. He remembers:

> "That's very interesting! I may need to remember that!" Susanne exclaimed, taking a small brown manila folder out of her purse. It was filled with file cards. She recorded [the poet] Richard [Selig]'s remarks, and then read it back to him, wanting to assure herself that she had quoted him correctly. I often saw her write down a remark that particularly interested her. These duly recorded remarks, she explained, might eventually fit somewhere into her work.[7]

So, indeed, we can say that Langer's set of thirty-seven drawers applies to the logic unearthed by the American media historian Craig Robertson in his monograph *The Filing Cabinet: A Vertical History of Information*, being that "the arrival of the typewriter mechanized the act of writing, and the introduction of the filing cabinet mechanized the act of remembering."[8] Mechanization, here, refers to automatic ordering, that is, to a practice in which the system (not the user, whether office clerk or otherwise) wills *and remembers* the location of certain papers for easy and fast retrieval.[9] Yet, in *The New Yorker*, Sargeant claims on the basis of observation during house visits and personal interview data that Langer "uses her card index *not only* as a storehouse of reference but also as *a stimulation to thought*" (emphasis added).[10] And Sargeant goes on to explain: "Many of her ideas have arisen suddenly from the fortuitous congruence of notions she has come upon while leafing through it."[11] Here, a transversal practice of leafing through the card-index system is added to what Robertson has convincingly elaborated (and what I will shortly explain) as a vertical paradigm of information management that came into being with the invention of the stacked-up filing cabinet. The addition points to the fact that whereas the concept of "verticality" is useful, more is needed in order to truly understand and theorize what it meant for a scholar like Langer to work not just with but also *within* her card-index system.[12] This is to say that the oppositional subject-object relation of the philosopher and the ideas, knowledge, and insights stored on cards shifts once efficiency, ease, and speed are exchanged for the more liberal practice of "leafing," perhaps at leisure or with a bit of academic anxiety or even pure angst. Langer *speaks back* to the contents of the cards and to their systematization, just as well as the very contents of the cards *speak back* to the system and to Langer herself.[13] This is not a dialectical process as per one of Langer's cards headed "Note—dialectic" that reads: "Could the problematical dialectic of subj. + objectification be handled on the model of metabolism rather than discourse?"[14] Indeed, the German media theorist Markus Krajewski, who speaks less about the filing cabinet and more about traditional card-index systems, argues that "[t]he architecture of the idiosyncratic scholar's machine [i.e., the personal card-index system] requires no mediation for, or access by, others. In dialogue with the machine, an intimate communication is permitted."[15] He expresses beautifully what this systematized machine does: "It *sorts addresses* [of published scholarship and, in the case of Langer, unpublished remarks] so as to *address thoughts*."[16] Later we will see that whereas Langer criticizes the implied "mechanicism" of the machine (and computer) metaphor, she also strongly agrees with the intimacy and perhaps even the *serendipity* of Krajewski's practice of dialoguing.

The dialectically oppositional subject-object relation of the vertical paradigm underlies the mechanical storehouse metaphor, alluded to also by Sargeant, that allowed for the contents on the cards to have become paradigmatically reduced to "information" after the 1890s invention of the filing cabinet. However, it does *not* underlay the practice of working with free-floating and active ideas, knowledge, and insights. Robertson writes:

> When mechanized, the association of ideas was no longer a mental connection [as per a preceding horizontal paradigm], a recollection that linked a memory and an object. With the object being information, not ideas, it became necessary to make it coordinate, to place information in a proper position relative to the other information.[17]

Robertson, in his book on office spaces in the first half of the twentieth century in North America, does however acknowledge the *aspiration* of certain filing systems to also produce knowledge and insights and to stimulate the formation of ideas, thus invoking Sargeant's "stimulation to thought" and Krajewski's "intimate [and dialogical] communication," when the media historian of the filing cabinet states that "cross-referencing systems created a secondary memory that could produce new knowledge."[18] Here Robertson refers to the early-modern period and to cross-referencing via keywords, calling such horizontal and networked systems "open,"[19] as if a transversal connection between horizontality and verticality may not have been an option either in the early-modern period or afterwards. This transversality, both historically and conceptually a distortion of linear progression, is in fact what Krajewski dug up in his study across the 1800s and 1900s in which we can read, among other things and in slightly anachronistic parlance, that there were to be found in the early-modern period,

> [C]lassification systems using both software, meaning the question of what principles can order scientific and library data, and hardware, meaning long-term storage devices: (1) the book ([Konrad] Gessner); (2) the nearly immobile, heavy piece of furniture, as yet unnamed, but ... clearly ... a kind of card index cabinet ([Vincentius] Placcius); and (3) the loosely sorted pile of papers on a table, at times filed in envelopes ([Joachim] Jungius).[20]

These systems, *and their next iterations* into the modern and postmodern periods and into contemporary times,[21] culminated in algorithmic techniques consisting of sets of fixed instructions for book placement and human movement through library stacks.[22]

Conceptually then, horizontal, vertical, and transversal paradigms and practices in knowledge, information, and data management can each be found in Langer's intricate, yet under-studied card-index system.[23] These paradigms and practices reveal in their combination how we may want to theorize the systematicity of the card-index system as well as they reveal what happens within and between drawers, on and behind tabs, among sets of cards in manila folders or tied together with elastic bands, and on the very cards themselves in frozen-in-time sections that were clearly prepared by Langer and her assistants for use during the writing of the trilogy *Mind: An Essay on Human Feeling*.[24] It is significant that some of the cards that I found in the preserved system at the Houghton are used explicitly as "x-reference cards," that is, for cross-referencing purposes. Some of these cards transgress a list—or web-like presentation yet they do refer explicitly to other cards and, significantly, to a writing process ("from this, make the transition to ...") thus suggesting that the latter cards have in fact been on Langer's desk while she was engrossed in the writing process itself and that they were not just used to "will and remember" locations and/or for their informational content. The word

FIGURE 5.2 "Photograph of Dr. Susanne K. Langer's Desk/Study" by permission of the Estate of Susanne K. Langer. Contributed by Donald Dryden.

"cross-references" has been penciled or penned on certain tabs, and it can be deduced from the architecture of the card-index system that other tabs not explicitly mentioning the practice of cross-referencing were in fact used as such. This is important because, as Krajewski makes explicit, "[o]nly through this skill does the index card box grow from a mere filing instrument to an author's assistant, or even ... into a regular communication partner during textual production."[25] Here we find Krajewski moving beyond Robertson's vertical paradigm, indeed, and toward some sort of hybridization.

Let me continue the conceptual journey of this chapter by citing verbatim one of Langer's index cards:

> Note—machine + mental operations [↵ Return] There is something wrong with the machine model of the brain as it influences conception just at present. The machine works entirely with <u>units</u>, "stored away" and "desired"* and "put together" in the processes that simulate thinking. But in the brain ideas are <u>formed</u>, more the way frost flowers and prism effects are formed; perhaps even more the way forms are made by erosion or by the action of moving waters. They are carved out of chaotic activity, or minted suddenly by catalytic transforming agents. They are activities that many things can start, not "products" in which the elements keep their identity and can be stored away again after use, like the elements in a machine when it is to be set for another

FIGURE 5.3 "Photograph of Dr. Susanne K. Langer's Study with Card Index System" by permission of the Estate of Susanne K. Langer. From the Susanne K. Langer Papers, MS Am 3110, Box 28, Houghton Library, Harvard University.

run. [↵ Return] The nearest familiar analogy to idea-formation is artistic composition. Notice that the activity of thought or fantasy is rhythmic, not repetitious like the "circulating messages" in elaborate machines. All artistic composition involves living mechanisms, but the activity is probably very different from any machine. A reacting organism may exemplify the same forms as complex inorganic units, but the relation is intimate on the chemical level only, + quite distant on the mechanical.—Is <u>rhythm</u> peculiar to organism?*[*] It is not (I think) the same as periodicity.

*Nielsen, Agnosia, Apraxia + Aphasia, p. 28: "When an idea is to be executed (by motor act) an impulse travels from somewhere in the brain to the precentral gyrus where the proper group of association cells is stimulated to effect utilization of the desired engrams." Who desires, + plans to execute the idea?

*[*]no.[26]

What Langer does on this card is critiquing the storehouse metaphor for human thinking, whether supported by rudimentary or advanced technological devices or not. She acknowledges a certain "willing" (here: desiring) on the part of the machine, but only to quickly affirm that mechanicism by necessity works from units outward (horizontal paradigm) or upward (vertical paradigm), whereas the process of thinking is a practice that is less linear, less predictable (it is a transversal practice). Langer's very statement about ideas as being (becoming) *formed* versus being (yes, *being*) mechanically combined can also be found in the first volume of *Mind* for which this card has presumably been used.[27] In *Mind*, Langer critiques what she calls "computerism"[28] in all possible directions: (1) conceptually, the brain is not like a computer;[29] (2) the computer should, in turn, not be seen as a "mechanical brain";[30] and (3) computers do not think.[31] For our purposes in this chapter, the first appearance of a critique of computers in *Mind I* is perhaps the most relevant. Here, Langer deconstructs the conceptualization of the brain as a computer, whilst also touching upon the other aspects of computerism, by arguing the following:

> The principles of logic are exhibited both by the "mechanical brains" of systems engineering and by human thought. ... But there is much more to rational thinking than the highly general form which may be projected in written symbols or in the functional design of a machine. Thinking employs almost every intuitive process, semantic and formal (logical), and passes from insight to insight not only by the recognized processes, but as often as not by short cuts and personal, incommunicable means. The measure of its validity is the possibility of arriving at the same results by the orthodox methods of demonstrating formal connections. But a measure of validity is not a ground of validity. Logic is one thing, and thinking is another; thought may be logical, but logic itself is not a way of thinking—logic is an abstract conceptual form, exemplified less perfectly in our cerebral acts than in the working of computers which can outdo the best brains a thousandfold in speed, with unshakable accuracy.[32]

This compares, in fact, to one of Krajewski's cautious statements that "[e]ven if it is clear that a card catalog does not perfectly resemble the digital calculator or computer ... the card catalog is *one* precursor of computing."[33] And, importantly, Langer pushes this entire historical and conceptual argument to the extreme by claiming, as per her characteristic philosophy of art and life, that idea-formation is not analogous to mechanical combination but to artistic composition instead, and that a much better metaphor for human thinking may be *chemical*. The latter hypothesis has, in fact, also been worked out in *Mind I* when Langer states in a footnote:

> To any one who has ever worked with living matter *in vitro* or under the microscope, the synthetic production of a chemical particle that metabolized for a brief period would be a much more impressive approach to the creation of a brain than the invention of Eniac and all its successors. A machine, however powerful and versatile, is an entirely different mechanism from a cell, a multicellular organ or a complex organism controlled by its own brain.[34]

In conclusion, what can be said about the paradigms and practices that we have encountered and evaluated in this chapter with the aim of delivering a theoretically justified account of Langer's hybridized card-index system and its use? Let me sum up my findings by discussing consecutively the horizontal, vertical, and transversal aspects of the system and their characteristics as if they were a series of distinct "types." First, there is what I call "horizontality 1" or the systematicity of a card-index system that functions in Euclidean space in 2D (length, width) or chronologically, as per Robertson's historical review of flat filing on the office floor and in the drawer. Chronological ordering is a fixed and linear ordering. In Krajewski, we find mainly alphabetical, alphanumeric, or, later, decimally systematized and therefore *symbolical* orderings of books in a library as well as the early, not exclusively horizontal technique of writing book titles and sometimes ideas of reader-writers on small and rearrangeable "paper slips." The decimal system was fundamentally a mobile system predicated on the logic of horizontally networked openness. Discrete informational cards, however, whilst allowing for internal mobility, albeit in a purely machinically *mechanical* manner and driven from the outside by a human hand or a pair of human hands, is something entirely different that does not exhaust a theory of Langer's card-index system and its use in the thinking process of this unique twentieth-century female philosopher.[35]

The second aspect that I want to highlight is the *verticality* that arrived on the office and library scenes, in both public institutions and in the homes of intellectuals, around the 1890s. Verticality functions in a Euclidean/Cartesian space in 3D, a space established by the length, width, and height of a container organized as a grid or as per thematic subject-headings such as the ones replacing chronology as an organizing principle. Put in some early twentieth-century quotes taken from Krajewski's monograph: "The card index overcomes the book. Its proper characteristic is vertical order." And: "Card indexes are books broken up into their components."[36] The tricky issue here, in offices, is the impossible move from units (specific tasks) to wholes (one person, the larger assignment).[37] Behind the tabs cutting up the intellectual flowing of ideas or the rhythmic movement of work life or life itself, however, connections are being revealed immediately. The system of the index or the file provides the context for understanding its contents with the steel drawers or the steel case in its entirety keeping everything together. Here we see that in the very use of an index or a file, there is no escape from knowledge: whereas information may be seen as impersonal and transparent, not as knowledge seen as connected to a knower,[38] it *must* be argued that a user like Langer is a knower. So, it can *not* be argued that the hierarchical direction of the vertical paradigm of understanding something like indexing and filing is the only or the right direction. Perhaps we should, as per Robertson, make the differentiation between archive (knowledge, power, control) to library (retrieval and circulation according to a system of classification).[39] And, indeed, there is a lot that resembles or evokes the archive when we look at Langer's home library and her personal card-index system from which, or *with which*, her knowing was being done in seeming solitude.

The third aspect (or "type") that I want to highlight for the purpose of grasping theoretically Langer's card-index system is "horizontality 2," which arrived on the scene in the 1950s.[40] I prefer to call this aspect or perhaps this "mode" *transversality*. This mode functions in Riemannian/Einsteinian spacetime in 4D, which allows us to transcend mobility and move to "flow."[41] In the words of Krajewski:

> [The idiosyncratic card-index system for personal use] not only reliably reproduces everything the scholar gradually invested in it, recalling the extended present back to the time when each entry was made. Provided that the scholar knew how to tie new material together with the existing stock of excerpts, and marked connections and associations to similar texts and themes, the scholar's machine *as a text generator* delivers these very connections by branching out into forgotten memories as virtually new, served up as well as unexpected connections. The apparently insignificant, but regularly marked cross-reference yields rich profits when its recombinatory linkages enrich the power of the excerpts with chains of references.[42]

The cross-referencing here described implies a "diffractive" technique,[43] i.e., a technique of "weaving" text and/or textual fragments through one another.[44] I wrote about this technique in an earlier piece on Langer, claiming that Langer's 1953 monograph *Feeling and Form: A Theory of Art Developed from Philosophy in a New Key* showed diffractive patterning in its dealing with the work of other philosophers and a wide variety of thinkers, thus transgressing both historical (i.e., chronological) and systematic (i.e., thematically subject-driven) ways of doing philosophy.[45] All this is to say that transversality also has to do with the linearity of one book, the multidimensionality of the card-index system, and the ontological condition of the generative *multiplicity* of many books or excerpts.[46] Langer's practice speaks directly to Krajewski's conjecture that "the apparatus returns infinitely more than the user feeds into it. As soon as one regularly cross-references new input with older material, the index database blazes associative trails that may serve as clarifying creative prompts for different connections and unexpected arguments."[47] Where Langer says about her writing process that "[q]uotations could be multiplied almost indefinitely,"[48] Krajewski argues, with reference to the German social theorist Niklas Luhmann, who famously kept his own card-index system, that the connections and arguments are made "almost autonomously."[49] Sargeant writes: "The cards on which [Langer's] own and other thinkers' ideas are preserved are methodically cross-indexed in a separate file, so that she can instantly lay her hands on everything pertaining to a given subject."[50] What needs to be added to this discussion of endlessness, autonomy (not automation!), and instantaneousness is *modularity*. Riemannian/Einsteinian spacetime implies that the drawers are not vertically or horizontally fixated in space but flexible instead through the possibility of recombination, thus allowing for the cards themselves to afford "not only the organization of information, but also mobility, portability, flexibility, modularity, representativity, transitivity, manageability, updatability, legibility, and combinability."[51] The profile piece by Wehr provides the biographical details that accompany these concepts:

> Wherever Susanne traveled, two things invariably accompanied her: her card catalog file and her cello. She always had to be within easy reach of her card files, which contained hundreds of her carefully notated file cards. These cards were filled with copious notes from her far-ranging reading in philosophy, biology, anthropology, art, and a long list of other such subjects, with her personal observations, and with remarks made to her by friends, remarks that stimulated her reflective imagination.[52]

There is movement in many directions and at many directions at once in the user-history of Langer and her card-index system. We may want to imagine Langer surrounded by the modular system, a modularity and its corollary movement that manifested itself within the drawers as well as between them. Her task was: how to arrange the drawers internally (where to "cut" the flows of art and life by using tabs) and in relation to one another (which drawers to put on one's desk and how to arrange them vis-à-vis one another). After all, she was a part of all that she had known.

NOTES

1. Markus Krajewski, *Paper Machines: About Cards & Catalogs, 1548–1929* (Cambridge, MA: The MIT Press, 2011); Craig Robertson, *The Filing Cabinet: A Vertical History of Information* (Minneapolis: University of Minnesota Press, 2021).
2. Susanne K. Langer Papers, 1895–1985, MS Am 3110, Houghton Library, Harvard University.
3. Arabella Lyon, "Susanne K. Langer: Mother and Midwife at the Rebirth of Rhetoric," in *Reclaiming Rhetorica: Women in the Rhetorical Tradition*, ed. Andrea A. Lunsford (Pittsburgh, PA: University of Pittsburgh Press, 1995), 266.

4. Adrienne Dengerink Chaplin, *The Philosophy of Susanne Langer: Embodied Meaning in Logic, Art and Feeling* (London: Bloomsbury Academic, 2020), 56–7; cf. Jean Barr and Lynda Birke, *Common Science? Women, Science, and Knowledge* (Bloomington/Indianapolis: Indiana University Press, 1998), 91.
5. Box 25 of the Langer Papers.
6. Winthrop Sargeant, "Profiles: Philosopher in a New Key," *The New Yorker*, December 3, 1960, 75.
7. Wesley Wehr, "Susanne K. Langer: Philosopher of Art & Science," in *The Accidental Collector: Art, Fossils, Friendships* (Seattle: University of Washington Press, 2004), 110. Wehr goes on to state, however, that Langer once said in a conversation about memory: "'My memory is like flypaper. Everything sticks to it. I don't have to try to remember things. I just remember them,' she answered" (Wehr, "Susanne K. Langer," 113). This, in turn, stands at sharp contrast to Langer's habit of index-card writing that Sargeant comes up with. He notes, "Behind the scattering of papers is a large card-index file, in which she has recorded, for many years, references to philosophical, anthropological, and psychological items she has discovered in books, and also ideas of her own, jotted down in moments of reflection. The moments are apt to occur at almost any time. She often thinks of a theory in the middle of the night, and has developed an efficient technique for writing in the dark. She has also been known to stop her station wagon on the road to record an idea before it escapes her mind, and she remembers once doing this while she was on her way to the dentist with a toothache" (Sargeant, "Profiles," 75).
8. Robertson, *The Filing Cabinet*, 25.
9. Ibid., 163–73.
10. Sargeant, "Profiles," 75–6.
11. Ibid., 76.
12. Cf. Krajewski, *Paper Machines*, 52 and 64 for exactly this dual point, the contents of which I will represent at the end of this chapter.
13. I borrow the language of "objects speaking back" from Dutch cultural analyst Mieke Bal; Bal, "Imaging Madness: Inter-Ships," *InPrint* 2, no. 1 (2013). Bal, in turn, borrowed the language from French art theorist Hubert Damisch; Yve-Alain Bois, Rosalind Krauss, and Hubert Damisch, "A Conversation with Hubert Damisch," *October* 85 (1998). See the chapter "Theoretical Object, Knowledge Object," in Iris van der Tuin and Nanna Verhoeff, *Critical Concepts for the Creative Humanities* (Lanham, MD: Rowman & Littlefield, 2022), 193–5.
14. The underlining is original. This card can also be found in Box 25 of the Langer Papers. On another card, from the same box, Langer writes: "Note [↵ Return] The reason I shy at computor language is that the 'on—off' language is appropriate only at the high level of human thinking where the negative comes into play."
15. Krajewski, *Paper Machines*, 50–1.
16. Ibid., 52; emphasis in original.
17. Robertson, *The Filing Cabinet*, 169.
18. Ibid., 174.
19. Ibid.
20. Krajewski, *Paper Machines*, 20.
21. See Jean-François Lyotard, *The Postmodern Condition: A Report on Knowledge* (Manchester: Manchester University Press, 1984); Felicity Colman, Vera Bühlmann, Aislinn O'Donnell, and Iris van der Tuin, *Ethics of Coding: A Report on the Algorithmic Condition* (Brussels: European Commission, 2018).

22. Krajewski, *Paper Machines*, 38–41.
23. An example of a well-studied card-index system is Niklas Luhmann's. See Markus Krajewski, "Paper as Passion: Niklas Luhmann and His Card Index," in *"Raw Data" Is an Oxymoron*, ed. Lisa Gitelman (Cambridge, MA: The MIT Press, 2013); Johannes F. K. Schmidt, "Niklas Luhmann's Card Index: The Fabrication of Serendipity," *Sociologica* 12, no. 1 (2018).
24. Susanne K. Langer, *Mind: An Essay on Human Feeling*, 3 vols. (Baltimore: Johns Hopkins University Press, 1970, 1974, 1984).
25. Krajewski, *Paper Machines*, 63.
26. The underlining and the use of asterisks are, again, original and this card can also be found in Box 25 of the Langer Papers. Langer's published critique of the "machine model of the brain" can be found throughout Langer, *Mind*, vol. 1 and will be commented on in this chapter.
27. In the preserved system at the Houghton Library, the card is stored for use during the writing of chapters 14 and 15 of *Mind*, vol. 2. The system as it has been handed down to us demonstrates a re-ordering done in the 1960s, 1970s, and 1980s by Langer and her assistants while writing *Mind*. We see that the tabs correspond to earlier and published versions of *Mind* and that the cards stored are both handwritten and typed up, so they offer a journey through time. The version history of the *Mind* trilogy is currently (from October 2021 onward) being studied in the archive of Connecticut College, where Langer landed a full professorship in Philosophy as late as in 1954, by Tereza Hadravová from the Department of Aesthetics of Charles University, Prague.
28. Langer, *Mind*, vol. 1, 276 n. 40.
29. Ibid., 148.
30. Ibid., 272 n. 32, 304–5.
31. Ibid, 320–1 n. 26.
32. Ibid., 148–9.
33. Krajewski, *Paper Machines*, 8; original emphasis.
34. Langer, *Mind*, vol. 1, 272 n. 32.
35. Krajewski, *Paper Machines*, 3, 7, 23, 30.
36. These are quotes from Wilhelm Dux resp. Elsa Herrmann in Krajewski, *Paper Machines*, 127.
37. Here, the intersection of Krajewski's and Robertson's work with theories and practices of "rationalization" comes to the front here. Cf. Jan Overwijk, "Rationalization: Paradoxes of Closure and Openness," PhD diss., University of Amsterdam, 2021.
38. Craig Robertson, "File." In *Uncertain Archives: Critical Keywords for Big Data*, ed. Nanna Bonde Thylstrup, Daniela Agostinho, Annie Ring, Catherine D'Ignazio, and Kristin Veel (Cambridge, MA: The MIT Press, 2021), 245.
39. Ibid.
40. Robertson, *The Filing Cabinet*, 35.
41. Cf. Krajewski, *Paper Machines*, 61.
42. Ibid., 52; emphasis in original.
43. Iris van der Tuin, "Diffraction as a Methodology for Feminist Onto-Epistemology: On Encountering Chantal Chawaf and Posthuman Interpellation," *Parallax* 20, no. 3 (2014).
44. Krajewski, *Paper Machines*, 61. See also Krajewski, Markus. "Card." In *The Oxford Handbook of Media, Technology, and Organization Studies*, ed. Robin Holt, Timon Beyes, and Claus Pias (Oxford: Oxford University Press, 2020), 72. Krajewski argues that the weaving technique formed the very basis of mobile classification by card-index systems, tracing the invention of technique and system back to the Swiss bibliographer Konrad Gessner (1516–65).

45. Iris van der Tuin, "Bergson before Bergsonism: Traversing 'Bergson's Failing' in Susanne K. Langer's Philosophy of Art," *Journal of French and Francophone Philosophy* 24, no. 2 (2016).
46. Krajewski, *Paper Machines*, 137.
47. Ibid., 63.
48. Susanne K. Langer, *Philosophy in a New Key: A Study in the Symbolism of Reason, Rite and Art* (Cambridge, MA: Harvard University Press, 1957), 27.
49. Krajewski, *Paper Machines*, 65.
50. Sargeant, "Profiles," 75.
51. Krajewski, "Card," 70.
52. Wehr, "Susanne K. Langer," 109. A significant number of cards in Langer's personal system refer, not only to author-year-title-publisher of books or journal articles but also to university-library and museum card-index systems (or rather, filing cabinets) for collective use, and these "call numbers," penciled on the cards by Langer, can be used to trace historically whereabouts Langer was based while reading and processing a certain text. Examples (that are also an indication of the interdisciplinary breadth of Langer's scholarship) are: "Harvard College Library," "Philos. Lib," "Wid.," "Brown Univ.," "Columbia, Deutsches Haus," "Barnard," "Mus. of Nat. Hist., N.Y.," "N.Y. Acad. of Medicine," "Yale," "Sterling," "Yale Med.," "(Y) Art Library," "(J. Hopkins)," "L. of Congr.," "Dartmouth," "Vassar L.," "Wesleyan Univ.," and, of course, "CC." Some cards mention that a text was *not* available at, especially, Connecticut College where the writing of the *Mind* trilogy was being done. Whereas the great majority of cards in Langer's personal system are standard-sized and shaped index cards bought commercially, some cards are in fact smaller than the bulk of them and from a library, e.g., from Harvard College Library's circulation desk.

BIBLIOGRAPHY

Bal, Mieke. "Imaging Madness: Inter-Ships." *InPrint* 2, no. 1 (2013): 51–70.

Barr, Jean, and Lynda Birke. *Common Science? Women, Science, and Knowledge*. Bloomington/Indianapolis: Indiana University Press, 1998.

Bois, Yve-Alain, Rosalind Krauss, and Hubert Damisch. "A Conversation with Hubert Damisch." *October* 85 (1998): 3–17.

Colman, Felicity, Vera Bühlmann, Aislinn O'Donnell, and Iris van der Tuin. *Ethics of Coding: A Report on the Algorithmic Condition*. Brussels: European Commission, 2018.

Dengerink Chaplin, Adrienne. *The Philosophy of Susanne Langer: Embodied Meaning in Logic, Art and Feeling*. London: Bloomsbury Academic, 2020.

Foucault, Michel. *Discipline and Punish: The Birth of the Prison*. Translated by Alan Sheridan, London: Penguin Books, [1975] 1991.

Krajewski, Markus. *Paper Machines: About Cards & Catalogs, 1548–1929*. Translated by Peter Krapp. Cambridge, MA: The MIT Press, 2011.

Krajewski, Markus. "Paper as Passion: Niklas Luhmann and His Card Index." In *"Raw Data" Is an Oxymoron*, edited by Lisa Gitelman, 103–20. Cambridge, MA: The MIT Press, 2013.

Krajewski, Markus. "Card." In *The Oxford Handbook of Media, Technology, and Organization Studies*, edited by Robin Holt, Timon Beyes, and Claus Pias, 68–83. Oxford: Oxford University Press, 2020.

Langer, Susanne K. *Feeling and Form: A Theory of Art Developed from Philosophy in a New Key*. London: Routledge & Kegan Paul, 1953.

Langer, Susanne K. *Mind: An Essay on Human Feeling*, vols. 1–3. 1967–82. Baltimore: Johns Hopkins University Press, 1970.

Langer, Susanne K. *Philosophy in a New Key: A Study in the Symbolism of Reason, Rite and Art*. 3rd edn. Cambridge, MA: Harvard University Press, [1942] 1957.

Langer, Susanne K. "Photograph of Dr. Susanne K. Langer's Desk/Study," "Photograph of Dr. Susanne K. Langer's Study with Card Index System." MS Am 3110, Box 28, Houghton Library, Harvard University.

Lord, James. "A Lady Seeking Answers." *The New York Times Book Review*, May 26, 1968, 4–5, 34.

Lyon, Arabella. "Susanne K. Langer: Mother and Midwife at the Rebirth of Rhetoric." In *Reclaiming Rhetorica: Women in the Rhetorical Tradition*, edited by Andrea A. Lunsford, 265–84. Pittsburgh, PA: University of Pittsburgh Press, 1995.

Lyotard, Jean-François. *The Postmodern Condition: A Report on Knowledge*. 1979. Translated by Geoff Bennington and Brian Massumi. Manchester: Manchester University Press, 1984.

Overwijk, Jan. "Rationalization: Paradoxes of Closure and Openness." PhD diss., University of Amsterdam, 2021.

Robertson, Craig. "File." In *Uncertain Archives: Critical Keywords for Big Data*, edited by Nanna Bonde Thylstrup, Daniela Agostinho, Annie Ring, Catherine D'Ignazio, and Kristin Veel, 241–7. Cambridge, MA: The MIT Press, 2021.

Robertson, Craig. *The Filing Cabinet: A Vertical History of Information*. Minneapolis: University of Minnesota Press, 2021.

Sargeant, Winthrop. "Profiles: Philosopher in a New Key." *The New Yorker*, December 3, 1960, 66–100.

Schmidt, Johannes F. K. "Niklas Luhmann's Card Index: The Fabrication of Serendipity." *Sociologica* 12, no. 1 (2018): 53–60.

van der Tuin, Iris. "Bergson before Bergsonism: Traversing 'Bergson's Failing' in Susanne K. Langer's Philosophy of Art." *Journal of French and Francophone Philosophy* 24, no. 2 (2016): 176–202.

van der Tuin, Iris. "Diffraction as a Methodology for Feminist Onto-Epistemology: On Encountering Chantal Chawaf and Posthuman Interpellation." *Parallax* 20, no. 3 (2014): 231–44.

van der Tuin, Iris, and Nanna Verhoeff. *Critical Concepts for the Creative Humanities*. Lanham, MD: Rowman & Littlefield, 2022.

Wehr, Wesley. "Susanne K. Langer: Philosopher of Art & Science." In *The Accidental Collector: Art, Fossils, Friendships*, 106–43. Seattle: University of Washington Press, 2004.

CHAPTER SIX

Susanne K. Langer's Foray into Art as a "Phenomenology of Feeling"

ROLF LACHMANN

TRANSLATION BY MÝ HUÊ MCGOWRAN

INTRODUCTION

This chapter departs from Susanne K. Langer's almost casual formulation of art as a "phenomenology of feeling" in *Feeling and Form* (1953). In reconstructing where this reference comes from, and why there is no further note on the source of its sudden appearance, Langer's relationship to phenomenology will be examined in further detail. This analysis will take into account the various direct and indirect references to phenomenology scattered throughout Langer's writings. A close look at Langer's major work *Mind: An Essay on Human Feeling* (1967–82) will outline the importance of how art articulates the phenomenal character of feeling. Its discussion will show that Langer had seen a more suitable alternative in conceptualizing artistic expression as a symbolic form than to follow a phenomenological method, and why she did so. It will conclude with a critical reflection of the scope and usefulness of Langer's approach.

TRACING LANGER'S LINKS TO PHENOMENOLOGY

In a passage from her main work on the philosophy of art, *Feeling and Form*, published in 1953, Susanne K. Langer casually uses the phrase "phenomenology of feeling," which appears entirely extemporaneously and is also not further discussed. The exact passage reads: "... instead of seeking for elements of feeling among the sensuous contents, or qualia, literally contained in the art object, we are led straight to the problem of created form (which is not always sensuous) and its significance, the phenomenology of feeling."[1] As regards content, this formulation is understandable to the extent that, according to Langer, artistic forms are "symbols for the articulation of feeling."[2] However, the term "phenomenology" is surprising. Neither had Langer used the formulation "phenomenology

I would sincerely like to thank Lona Gaikis for inviting me to contribute to this book project and for her valuable comments and suggestions on my contribution.

of feeling" before, nor was she to use it again in further publications. That she quite consciously used it, however, is apparent in that she cites "phenomenology" in the index of the book. But there Langer does not direct to the passage quoted above; she alludes to it as a footnote in a later context in which she explicitly relates philosophical phenomenology with what she sees as the vivid consciousness of time articulated in music. Specifically—by mentioning an essay by Philip Merlan[3]—Langer critically engages the phenomenology of time consciousness in Husserl and Heidegger: "Phenomenology attempts to describe in discursive terms this complex experience; and it tries to do so in terms of momentary impressions and actual feelings. The result is a tremendous complication of 'states' in which the sense of passage is entirely lost in the parade of 'moments.'"[4] It thus becomes evident that Langer's use of "phenomenology" is not incidental and that we must consider the references she makes to philosophical phenomenology in her reflections. While she identifies the genuine issue of articulation in art as the "phenomenology of feeling," she criticizes philosophical phenomenology by referencing Philip Merlan's essay as being deficient because it articulates only states by means of linguistic discourse, and the specific experience of the time in passage is lost. These two references to phenomenology describe the field of tension that are to be illuminated more closely.

How are these mentions of phenomenology to be understood? This question is of particular interest because the role of phenomenological (re-)presentification is centrally important in the late phase of her thinking, which revolves around the initiation of a new approach to scientific research of the human mind. This becomes very clear in the vehemence with which Langer, in the first chapters of her late work *Mind: An Essay on Human Feeling*, objects to the contemporary paradigms in psychology for not making any effort at all to (re-)presentify their subject matter independently. This was precisely the cardinal error responsible for the often trivial results of contemporary behavioral psychology. This error had to be corrected. Accordingly, the very first task in human sciences, including psychology, was to develop descriptions that were oriented on the phenomena of human subjectivity and ultimately also concepts.

Thus, sensitized to this topic, it should firstly be recalled that her citing of philosophical phenomenology is not surprising because Langer was an outstanding expert on German-language philosophy. As the daughter of wealthy parents from Leipzig and Chemnitz who had emigrated to New York, she grew up in a distinctly German-educated, middle-class atmosphere. At home, only German was spoken. Her study visit to Vienna from 1921 to 1922 deepened her knowledge of German culture and philosophy. As is well known, she adopted Ernst Cassirer very early on and paved the way for him into the English-speaking discourse also through her English translation of his essay "Sprache und Mythos" (*Language and Myth*, New York: Dover, 1946). However, while she regularly alludes to Cassirer and characterizes his philosophy as pioneering, she makes only sparing and at first rather critical references to Edmund Husserl.

Indeed, Langer already deals with Husserl's *Logische Untersuchungen* in her attempt to define a general concept of "meaning" in her dissertation, "A Logical Analysis of Meaning," written in 1926. Langer's tenor is, however, critical: Husserl did not succeed in defining a general definition of "meaning" because the phenomenological method he used was inductive and he was therefore unable to develop any general principles.[5] She repeats this criticism in *The Practice of Philosophy* and in *Philosophy in a New Key*, in each of which she criticizes Husserl's approach as inadequate with the same allusion only once.[6] Husserl (similarly to Charles Sanders Peirce and Lady Welby) had wanted to analyze the characteristics of "meaning" by finding the common essence in the different types in a comparative view. "Meaning," however, was not a quality but a relation. Based on looking into Langer's publications up until the end of the 1940s, one therefore initially comes to a negative conclusion and could assume that Langer's formulation "phenomenology of feeling" does not distinguish a deeper systematic meaning.

That this is a rash conclusion is supported by the fact that Susanne K. Langer and her husband, the Harvard historian William L. Langer, visited Edmund Husserl on August 28, 1933, in Schluchsee during their stay in Germany.[7] Even afterwards, Husserl and Langer stayed in touch through letters.[8]

Langer likely refers to this visit to Husserl in her essay "The Deepening Mind: A Half-Century of American Philosophy" (written in collaboration with Eugene T. Gadol) published in 1950. As the title indicates, the essay is a localization of American philosophy in the first half of the twentieth century. Starting with the works of William James, Josiah Royce, and Charles Sanders Peirce, which were decisive for contemporary American philosophy, Langer outlines five significant European influences on American thought. She names mathematical logic (Gottlob Frege, Giuseppe Peano, Bertrand Russell, and Alfred North Whitehead), the Vienna Circle around Moritz Schlick, the philosophy of language and meaning that emerged in the initial decades (Lady Welby, Ludwig Wittgenstein, and Rudolf Carnap), the gestalt psychology of Max Wertheimer, Kurt Koffka, and Wolfgang Köhler, and finally Edmund Husserl's phenomenology. In the last pages of the article, there is a closer exposition and interpretation of Husserl's philosophy. It is probably the most detailed discussion of Husserl in Langer's publications. Langer presents Husserl's phenomenology as the modern successor to idealism because, in contrast to the empiricist philosophical tradition, "consciousness" is the central starting point. This became clear in that Husserl took a transcendental philosophical position and asked how a transcendental consciousness "constituted" the world of experience. In Husserl's reconstructing of external reality as a result of the "constituting" activities of consciousness, he opted for a certain interpretation of what "constitution" meant. Many of his students and successors would have held a less strong view. The method of "phenomenological reduction" said nothing about the constitution of the world but only about the origin of the phenomena. Accordingly, phenomenological reduction for the systematic analysis of the conditions of consciousness was also possible without having to make far-reaching metaphysical assumptions, and, indeed, this path was preferred by Husserl's American successors: it was about the discovery of nature and the limits of knowledge. In the fact that empiricist thought developed towards a semiotic theory and phenomenology focused on the phenomenon of "sensemaking," Langer recognizes a convergence that revolved around the concepts of symbol and sense, language and meaning, form and fact that was so significant that it marked a turning point in the history of philosophy. Langer expresses her astonishment that, despite its technical and systematic rigor, phenomenology did not procure more detailed results but stopped at determining laws of the general world constitution. Perhaps, however, the implicit goal of phenomenology was in fact what Husserl told her in a conversation; she quotes him by saying, "The natural outcome of phenomenology, I believe, will be a powerful, autonomous science of psychology."[9]

Langer notes that she finds Husserl's view an interesting notion, because logical empiricism is essentially oriented towards the theorizing of physics. However, this was of little help in researching social, historical, and value-related topics. In contrast, psychology was "for all its experimental techniques ... closer to the humanities than to the so-called exact sciences."[10] The positivist and behaviorist methods did not uncover new phenomena or lead to new factual connections. "The birth of a 'powerful, autonomous psychology,' and perhaps of ethics and esthetics too, is probably awaiting the full articulation of philosophical ideas that are rapidly in the making just now. Phenomenology may be the vanguard, though I do not think so, for reasons that are much too technical to be stated here."[11]

Langer regards the studies in the theory of meaning undertaken by thinkers as diverse as Rudolf Carnap, Alfred North Whitehead, and Ernst Cassirer as the starting point for new approaches to perception and reason in one direction and to emotions, feeling, and deeper biological foundations in the other. Langer sees herself supported in this assessment by the remarkable fact that the most fruitful (but perhaps also most problematic) contemporary psychological conceptions, namely gestalt psychology and Sigmund Freud's psychoanalysis, made no attempt at all to adopt the languages or methods of the natural sciences.

LANGER'S LOGICAL INVESTIGATION OF "FEELING" IN ART

It is clear, then, that phenomenology was important for her late work, which was—to use her reference to Husserl once again—aimed precisely at the initiation of a "powerful, autonomous science of psychology" that she aspired to. Incidentally, this is also clear from the fact that in her preparatory work for *Mind: An Essay on Human Feeling* Langer dealt intensively with the book by Herbert Spiegelberg *The Phenomenological Movement*,[12] published in 1960, as the notes preserved in her estate prove. In her distinction between empathy and sympathy as well as the ethics of values outlined in volume 3, one can now also assume an incorporation of Schelerian concepts. In *Mind: An Essay on Human Feeling*, Langer mentions Max Scheler, among others, but she accuses him of hastily attributing the perception of expression to the subject of understanding foreign states of mind.

Against this background, one is sensitized to the fact that the term "phenomenon" is used very frequently, both in her book *Philosophical Sketches*, published in 1962, and *Mind: An Essay on Human Feeling*—especially in the preface. In these contexts, however, "phenomenon" certainly does not have the strict terminological meaning of a methodically secured and final *Wesensschau*, but, rather, means the effort to take into account the diversity of concrete realities, forms of appearance and findings which must be acknowledged without bias and serve as a touchstone for the formation of theory and concepts. Langer sees a challenge facing the human sciences. In her sense:

> only an image can hold us to a conception of a total phenomenon, against which we can measure the adequacy of the scientific terms wherewith we describe it. We are actually suffering today from the lack of suitable images of the phenomena that are currently receiving our most ardent scientific attention, the objects of biology and psychology. This lack is blocking the progress of scientifically oriented thought toward systematic insight into the nature of life and especially of mind: the lack of any image of the phenomenon under investigation, whereby to measure the adequacy of theories made on the basis of physical models.[13]

The point is that the conceptualization of psychology and the other human sciences must be oriented towards images in which the relevant phenomena regarding the entire scope of human feeling and thinking have been manifested objectively. According to Langer, it was precisely this effort of phenomenological (re-)presentification that the positivist and behaviorist understandings of science that prevailed until well into the 1960s, as well as the cognitive sciences oriented towards information theory, skipped, which is why they remained unfruitful. Langer notes,

> Our first acquaintance with the material of any research has to be negotiated by images which organize and present the phenomena as such, for it is always phenomena that we ultimately wish to explain, and this requires detailed empirical knowledge ... If we want to study the phenomena of emotion, sensibility, ideation and especially the integral mental life in which they are all what Husserl's English-speaking disciples call 'moments' (Momente), the most promising method is to study art.[14]

The decisive turn that Langer takes in contrast to phenomenology becomes clear in this passage. It is art which, according to Langer, has an extraordinary power to express the phenomenon of feeling in its complex and subtle dynamics, and which can therefore guide the development of appropriate concepts and models for the human sciences. This brings us one step closer to answering the question of what Langer's objections to phenomenology are, as indicated in "The Deepening Mind: A Half-Century of American Philosophy."

In order to determine the similarities and differences between philosophical phenomenology and the artistic "phenomenology of feeling" more precisely, we should first recall the basic idea of the phenomenological *Wesensschau*. For Langer's critique of thoughtless and uncritically implemented terminology applied to the human sciences connects her approach with philosophical phenomenology.

Phenomenology also denies the direct knowledge from structures of our consciousness and attempts to make the phenomena, which we mostly find obscured by conventional opinions, prejudices, scientific methods of analysis and traditional philosophical conceptualization, accessible through a particular theoretical effort. As a fundamental prerequisite for researching the interconnections of our consciousness, Edmund Husserl mentions the elimination of our natural, practical, but mundane attitudes. Only through an act of "phenomenological reduction" can we see its laws by adopting the attitude of a "disinterested onlooker," with the "sole remaining interest being to see and to describe adequately what he [or she] sees."[15]

Langer shares this reluctance with philosophical phenomenology towards everyday language, the conceptual prejudice in the sciences, and philosophical traditions. The basic methodological movement in art is the same as in phenomenological reduction, for the elementary abstraction of artistic observation also exists in overriding our everyday practical attitude in order to turn to an object being freed from it. Langer writes,

> The function of artistic illusion is ... disengagement from belief—the contemplation of sensory qualities without their usual meanings of 'Here's the chair,' 'That's my telephone,' 'These figures ought to add up to the bank's statement,' etc. The knowledge that what is before us has no practical significance in the world is what enables us to give attention to its appearance as such.[16]

Nothing other than this elementary process—which Langer calls "abstraction"—creates the illusory, unreal character of the work of art, through which it dissociates from practical terms but can function as a symbol in the context of imagination. An important difference to phenomenological reduction, however, is that in Langer's view, the expression of a work of art (and thus the articulation and phenomenological (re-)presentification that arises from it) does not develop through an "aesthetic attitude," i.e., a consciously adopted basic attitude. Rather, a work of art itself emerges from its environment, "It detaches itself from its actual setting and acquires a different context."[17]

COUNTERING ART'S REDUCTION TO LANGUAGE: THE ACT MODEL

The thesis is thus that, for Langer, art as a "phenomenology of feeling" takes the place of phenomenological reduction, and that her objections to philosophical phenomenology as the base of the human sciences, which she asserts but does not elaborate in her essay "The Deepening Mind: A Half-Century of American Philosophy," have to do with the fact that she sees art as the superior approach. In view of the objections that can be found in scattered passages of Langer's text, at least four advantages to art can be identified.

Firstly, Langer emphasizes the inadequacy of language to adequately reflect the dynamic complexity and processuality of feeling. Already in *Philosophy in a New Key*, Langer writes,

> Everybody knows that language is a very poor medium for expressing our emotional nature. It merely names certain vaguely and crudely conceived states, but fails miserably in any attempt to convey the ever-moving patterns, the ambivalences and intricacies of inner experience, the interplay of feelings with thoughts and impressions, memories and echoes of memories, transient fantasy, or its mere runic traces, all turned into nameless, emotional stuff.[18]

For this reason, a language-bound description, as is maintained in phenomenology, reaches its limits: "The unfolding of even so simple a phenomenon as an emotion defies any verbal account, as I think the brave efforts of the phenomenologists show."[19]

Secondly, works of art are able to objectify and secure the quality that characterizes emotional dynamics, and thus represent a permanent point of reference to formulate a theory. "The great value of a permanent image is that one can resort to it to recover an elusive idea, and reorient one's intellectual progress, when enticing simplifications and reductions have turned it away from its long course into shorter alleys that do not really lead to the same goal."[20] Due to the fleeting nature of current feelings, one gains an "image" through artistic articulation, "which can be held for contemplation."[21]

Thirdly, Langer is convinced that artistic symbolization goes much "deeper" and even articulates the vital and biological foundations and structural conditions that we do not feel but on which the dynamics of feeling are built. Langer exemplifies the phenomena of growth or "organic memory," which we cannot feel directly but which can be expressed in works of art.[22]

Fourthly, the articulation of far deeper, living relationships of form that take place in artworks, and from which sensation ("feeling") emerges, makes it possible to develop a "model," i.e., a conception of the dynamics in action, and from which life and feeling emerge and function. In this way, organic and psychic phenomena can ultimately be linked with a "single system of facts."[23] In this sense, the approach that emerges in a note found in Langer's preparatory work to *Mind: An Essay on Human Feeling*, "Note—import of phenomenology," is probably to be understood as follows,

> The main importance of phenomenology is that it emphasizes the complexity and individuality of psychological material. It is a protest against the self-assured simplifications of psychology. In practical psychology, the variability of the "clinical

> picture" has somewhat the same virtue. The trouble with phenomenology is that a method—and even more so, a methodological principle—is not an organizing principle, i.e., not a fecund idea. Husserl had no central idea.[24]

This highlights the central divergence between Langer's conception and philosophical phenomenology. Firstly, art is a far more fruitful starting point for articulating the structures of our feelings. Secondly, Langer's aim is not to develop a method but a model that identifies the structural conditions of life and feeling and relates them to each other in order to give impetus to concrete scientific research. This is precisely what the concept of the *act*—this "fecund idea"—and the multitude of principles that specify this model (individuation, involvement, tolerance, motivation, entrainment, etc.) are meant to do in *Mind: An Essay on Human Feeling*.

"For an image may—indeed, must—be dropped when it has done its work."[25] Langer closes the second part of *Mind: An Essay on Human Feeling* with these words in order to develop a fruitful conceptual framework with the "act model" for a scientific study of human mind that grounds in an artistic "phenomenology of feeling."

On closer inspection, however, it is noticeable that only the act model is based on a phenomeology of art. The concepts and hypotheses developed in the subsequent parts of the work are no longer referred to the "phenomenology of feeling" in a systematic and detailed form. This is astonishing insofar as, according to Langer, not only the dynamics of our feelings but the entire scope of higher intellectual achievements and experiences can find expression in works of art: "that image [created by art] seems to be capable of encompassing the whole mind of man, including its highest rational activities."[26] But none of the concepts and distinctions introduced in the later parts of *Mind: An Essay on Human Feeling*, such as the concept of value, the distinctions between communication and communion, suggestion and imitation, empathy and sympathy, or the use of symbols that constitute human thinking, are introduced on the basis of the phenomena articulated in works of art. All in all, there are only very isolated allusions to art.

One can explain this finding by the fact that Langer's own philosophical development ran from art to living form. Identifying the dynamics of the living, which is conceptualized in the act model, is the actual new step beyond her previous writings, which is why it receives such detailed treatment.

But there is also a systematic reason. For obviously the artistic symbolization of even the highest intellectual activities is limited to articulating their living dynamics. Langer writes: "The prime function of art is to make the felt tensions of life, from the diffused somatic tonus of vital sense to the highest intensities of mental and emotional experience, 'stand still to be looked at,' as Bosanquet said, 'and, in principle, to be looked at by everybody.'"[27] So even the highest emotional and intellectual experiences are "merely" objectified in terms of their "felt tensions of life." Thus, works of art are limited to the articulation—formulated in the act model—of the living dynamics and structural relations of feeling and thinking. This interpretation is supported by the fact that in the two formulas that have become "canonical" in her late writings, which denote the expressivity of art, namely "felt life" and "the life of feeling,"[28] are about "life." The formulations "life of the mind" or "mental life," which Langer uses several times, and obviously terminologically, also prove this.[29]

Therefore—in conclusion—an independent art-phenomenological justification for the derivation of the specifically human intellectual abilities of communication, sympathy, intuition, imitation, and symbolization, etc., is not necessary, because it already exists in the form of the act model. Instead, human thinking would have to be proven as a specific individuation of pre-human precursors such as communion, suggestion, and practical understanding of signs on the basis of the act model, which Langer, however, did not carry out.

CONCLUSION

How are Langer's positions to be evaluated and what lasting significance do they have for the human sciences? The following theses will answer this question and attempt to draw a conclusion.

Firstly, both Langer's and philosophical phenomenology's skepticism about the scientific concepts and understandings adopted unreflectively from everyday language, the philosophical tradition, and also from the scientific tradition is of lasting relevance. Scientific research into human feeling and understanding must be accompanied by an effort to ensure that the explanations of research can always be shown to be appropriate on the basis of independent phenomenological (re-)presentifications of our feeling.

Secondly, works of art can play an important role in this continuous task, because they represent an independent and—due to their symbolic form—particularly appropriate objectification of our feelings. Langer's objections to phenomenology are plausible and it seems promising to use the phenomena objectified in works of art for psychological and neuroscientific research.

Here, the difficulty must be noted that the "meaning" of artistic articulations is inherently problematic insofar as works of art cannot be definitively referenced. Langer has taken this characteristic of art, which distinguishes it from linguistic symbolizations, as an opportunity to speak of "unconsummated symbols"[30] in the case of artistic symbolizations. Unlike words, which have a concrete and definable "meaning," works of art have an "import" that cannot be discursively determined.

However, there is no reason to pit language-bound phenomenological descriptions (which secure a relation to the object) against the qualities of feeling/sensations objectified in artworks (which are particularly well suited to articulating the dynamic nature of feeling). Rather, these two methods of phenomenological (re-)presentification can be combined, thus eliminating their respective deficits and achieving both a detailed description close to the phenomena and ensuring their attribution through the language-bound description. This, incidentally, is precisely Langer's approach in her chapter on art theory in *Mind: An Essay on Human Feeling*, in which she "brings up" the structural properties that are decisive for the expression of works of art—here also alluding to art-theoretical literature. It is possibly precisely this "dual method" of phenomenological (re-)presentification that yields approaches for describing the structures of our experience, whereby these can be brought into a fruitful connection with neuroscientific findings. As a bridge for the connection of the descriptions gained through this dual method with the neurobiological processes and structures, the process—or system-theoretical[31] act model can provide a comprehensive and differentiable framework of understanding.

Thirdly, without diminishing the importance of Langer's process-theoretical model of life, these considerations also lead us to ask whether a research method can also be derived from her positions. This idea arises directly from Langer's work, but it has not been considered, or at least not made explicit by Langer. In a sense, this perspective amounts to examining Langer's own method, which she follows in *Mind: An Essay on Human Feeling*, to see if it can be established as a general research method. In doing so,

Langer's own approach would certainly need to be specified. Langer herself described the challenge facing her project thus: "This means, of course, that to make art illuminate a field of science one has to be intellectually at home in both realms."[32] As an independent research method, it cannot—unlike Langer's late work—be about art illuminating an entire field of science and introducing new basic concepts in a singular mega-project. More than fifty years after the publication of the first volume of *Mind: An Essay on Human Feeling*, this task seems to me to no longer be necessary, not least because the process—and system-theoretical understanding of life conceptualized in the act model can, in any case, be regarded as a widely established research paradigm. This opens up space for the idea that, in the highly specialized research fields of neuroscience and psychology, it can and must nevertheless be a matter of systematically including the range of description and articulations of phenomena that come from other, and especially artistic, articulations of our subjectivity in a more "modest" project.[33] In this sense, Langer's notion of art as a "phenomenology of feeling" can become a systematic component of psychological or neuroscientific investigations, whereby the biological processes and explanations are also related to our experience and thereby also gain an independent source of confirmation as to their adequacy.

NOTES

1. Susanne K. Langer, *Feeling and Form* (New York: Charles Scribner's Sons, 1953), 57.
2. Ibid., 52.
3. Philip Merlan, "Time Consciousness in Husserl and Heidegger," *Journal of Phenomenology* 8, no. 1 (1947): 23–53.
4. Langer, *Feeling and Form*, 113.
5. See pages 3 and 126 in Susanne K. Langer, "A Logical Analysis of Meaning," PhD diss., Radcliffe College, Cambridge, MA, 1926. Here, Langer's formulation of the "phenomenology of meaning" simply pertains to the representation (or the '(re-) presentification') of the various contexts in which "meaning" is brought forth.
6. Susanne K. Langer, *The Practice of Philosophy* (New York: Henry Holt, 1930), 106, and, later on, in *Philosophy in a New Key: A Study in the Symbolism of Reason, Rite and Art* (Cambridge, MA: Harvard University Press, 1957), 54.
7. A short account can be found in a letter from M. Husserl to E. Rosenberg dated September 7, 1933. In *Edmund Husserl. Briefwechsel*, ed. Karl Schuhmann (Dordrecht: Kluver, 1994), 430.
8. According to the Husserl Archive in Freiburg and Leuven, there exists no correspondence between Langer and Husserl. The author has a copy of a letter (provided by Donald Dryden) from Husserl to Susanne K. Langer dated July 5, 1935. Here, Husserl introduces Langer to the eminent medical doctor Professor Thannhauser and his family to initiate mutual contact. Due to his Jewish confession, Thannhauser was dismissed from his job as director of the Medical Clinic at the University of Freiburg by the National Socialists in 1934 and emigrated to the United States.
9. Susanne K. Langer and Eugene T. Gadol, "The Deepening Mind: A Half-Century of American Philosophy," *American Quarterly* 2, no. 2 (1950): 130.
10. Ibid., 131.
11. Ibid., 131.
12. Herbert Spiegelberg, *The Phenomenological Movement: A Historical Introduction* (The Hague: Martinus Nijhoff, 1960).

13. Susanne K. Langer, *Mind: An Essay on Human Feeling*, vol. 1 (Baltimore: Johns Hopkins University Press, 1967), xviii.
14. Ibid., 68.
15. Edmund Husserl, *Cartesian Meditations: An Introduction to Phenomenology*, trans. Dorion Cairns (The Hague: Martinus Nijhoff, 1995), 35.
16. Langer, *Feeling and Form*, 49.
17. Ibid., 45.
18. Langer, *Philosophy in a New Key*, 100–1.
19. Susanne K. Langer, "'Expressive Language' and the Expressive Function of Poetry," in *On Expressive Language*, ed. Heinz Werner (Worcester, MA: Clark University Press, 1955), 6.
20. Langer, *Mind*, vol. 1, xx.
21. Ibid., 67.
22. Ibid., 199.
23. Ibid., 65.
24. This is found in Susanne K. Langer "Finished Chapters 1–5," Susanne Langer Papers, MS Storage 199 (Cambridge, MA: Harvard University, Houghton Library, n.d.).
25. Langer, *Mind*, vol. 1, 243.
26. Ibid., 150.
27. Ibid., 115.
28. Ibid., 112. Here, both formulations can be found in the same context.
29. Cf. Rolf Lachmann, "'Das Leben des Geistes' als zweite Lebensrevolution," in *Naturalisierung des Geistes und Symbolisierung des Fühlens. Susanne K. Langer im Gespräch der Forschung*, ed. Cornelia Richter and Petra Bahr (Tectum Verlag: Marburg, 2008), 129–46.
30. Langer, *Philosophy in a New Key*, 240.
31. On the interpretation of the "act-model" as a "system theory," see Rolf Lachmann, *Susanne K. Langer. Die lebendige Form menschlichen Fühlens und Verstehens* (Munich: Wilhelm Fink Verlag, 2000), 146–7.
32. Ibid., xi–xx.
33. This perspective is thus connected with earlier aspirations of the field of so-called "neurophenomenology." See Francisco Varela and Jonathan Shear, "First-Person Methodologies: What, Why, How?" *Journal of Consciousness Studies* 6, no. 2–3 (1999): 1–14.

BIBLIOGRAPHY

Husserl, Edmund. *Cartesian Meditations: An Introduction to Phenomenology*. 1960. Translated by Dorion Cairns. The Hague: Martinus Nijhoff, 1982.

Lachmann, Rolf. *Susanne K. Langer. Die lebendige Form menschlichen Fühlens und Verstehens*. Munich: Fink Verlag, 2000. Available online: http://digi20.digitalesammlungen.de/de/fs1/object/display/bsb00043407_00001.html (accessed March 2, 2022).

Lachmann, Rolf. "'Das Leben des Geistes' als zweite Lebensrevolution." In *Naturalisierung des Geistes und Symbolisierung des Fühlens. Susanne K. Langer im Gespräch der Forschung*, edited by Cornelia Richter and Petra Bahr, 129–46. Marburg: Tectum Verlag, 2008.

Langer, Susanne K. "'Expressive Language' and the Expressive Function of Poetry." In *On Expressive Language*, edited by Heinz Werner, 3–9. Worchester MA: Clark University Press, 1955.

Langer, Susanne K. *Feeling and Form: A Theory of Art Developed from Philosophy in a New Key*. New York: Charles Scribner's Sons, 1953.

Langer, Susanne K. "Finished Chapters 1–5." Cambridge, MA: Harvard University. Houghton Library. Susanne Langer Papers, MS Storage 199, n.d.

Langer, Susanne K. "A Logical Analysis of Meaning." PhD diss., Radcliffe College, Cambridge, MA, 1926.Available online: https://langercircle.sites.uu.nl/wp-content/uploads/sites/662/2021/02/Langer-Susanne-K.-1926-A-Logical-Analysis-of-Meaning-doctoral-dissertation-scanned-PDF-from-photocopy.pdf (accessed July 19, 2023).

Langer, Susanne K. *Mind: An Essay on Human Feeling*, vol. 1. Baltimore: Johns Hopkins University Press, 1967.

Langer, Susanne K. *Philosophy in a New Key: A Study in the Symbolism of Reason, Rite and Art*. 1941. Cambridge, MA: Harvard University Press, 1957.

Langer, Susanne K. *The Practice of Philosophy*. New York: Henry Holt, 1930.

Langer, Susanne K., and Eugene T. Gadol. "The Deepening Mind: A Half-Century of American Philosophy," *American Quarterly* 2, no. 2 (1950): 118–32. https://doi.org/10.2307/3031449.

Merlan, Philip. "Time Consciousness in Husserl and Heidegger." *Philosophy and Phenomenological Research* 8, no. 1 (1947): 23–54. https://doi.org/10.2307/2102915.

Schuhmann, Karl, ed. *Husserliana: Documents 3/1–1: Edmund Husserl: Briefwechsel*. Dordrecht, NL: Kluwer, 1994.

Spiegelberg, Herbert. *The Phenomenological Movement: A Historical Introduction*. The Hague: Nijhoff, 1960.

Varela, Francisco J., and Jonathan Shear. "First-Person Methodologies: What, Why, How?" *Journal of Consciousness Studies* 6, no. 2–3 (1999): 1–14.

CHAPTER SEVEN

Susanne K. Langer's Theory of Self-Liberation through Culture

ANNE POLLOK

INTRODUCTION

Following the exodus of neo-Kantian philosophers such as Ernst Cassirer from Germany to the United States during the 1940s, we witness a strengthening of so-called continental philosophy in the United States. One rather atypical representative of this trend is Susanne K. Langer, whom I would call one of Ernst Cassirer's most important students.[1] Both attempted to expand philosophy beyond the boundaries of reason. Yet, in distinction from the existentialists or the new mythology à la Heidegger, they found this in the human modes of expression in culture. Contrary to many other approaches to Langer that concentrate on her naturalistic outlook,[2] I focus on her uptake on the rather idealistic claim by Cassirer that the process of symbolic formation enables human self-liberation.

Langer makes it clear on several occasions that Cassirer's philosophy deeply influenced her thinking.[3] "It was Cassirer," she closes the study she dedicated to him, *Feeling and Form* (1953), "who hewed the keystone of the structure [of a theory of the art symbol], in his broad and disinterested study of symbolic forms."[4] Langer sees her work as a comprehensive development of these keystones. While her take on meaning is without doubt also influenced by Whitehead, Wittgenstein, Peirce, and Dewey, among others,[5] it is Cassirer's oeuvre she constantly comes back to and adapts most closely. Like Cassirer, she sees its most fruitful seed in the idea that human beings create their worlds by meaningful symbols. Cassirer discusses the most fundamental of those meaning-creating functions that cover different areas of human life and thought in his treatment of myth, language, and science (among some more cursory reflections on history, art, and technology) in his main oeuvre, *The Philosophy of Symbolic Forms,* vols. 1–3 (1923, 1925, 1929), as well as in numerous related articles. Its most mature form, with the strongest push towards a new philosophy of culture, we find in the studies during his exile.[6] These later discussions are of particular importance in our given context as they reflect Cassirer's concern regarding

I thank Lona Gaikis not only for inviting me to contribute to this wonderful volume but also for offering many thoughtful comments and constructive criticism, and for her patience. I also thank Samantha Matherne for commenting on an earlier version of this paper.

the human tendency to react to difficult situations—such as the challenges of modernity—with a re-mythification of their world.[7] And as such, these later works take a decisively more "political" perspective on the philosophy of symbolic forms,[8] arguing that our capacity to establish and change a shared culture ultimately enables human freedom. On the flipside, this same capacity also bears the danger of a return to pre-rational thinking; a development Cassirer diagnoses as being in full force in Nazi Germany.

In this chapter, I will clarify Langer's affinity to the basic tenets of Cassirer's philosophy of symbolic formation in culture, paying particular attention to their supposed relation to human freedom. I will explicate why Langer's philosophy of art is of central importance for the project of self-liberation as it showcases the non-discursive aspect of symbolic formation and its implications on human culture. Finally, I will discuss how this non-discursive aspect plays an important role in Langer's take on civilization versus culture that she develops in the last part of *Mind* (and some related writings). Similar to Cassirer, Langer was well aware of the danger of a re-mythification of our modern world, as was obvious in the rise of oppressive, cult-like ideologies such as the Third Reich's Teutonic-Barbaric fantasies. Langer argues instead for a new integration of modern myths into the concept of a free society, and sets out to formulate the conditions for a philosophy adequate for this task.

THE FREEDOM OF SYMBOLIC FORMATION

By "freedom" Cassirer first of all means the ability to form a world we can truly inhabit by the full use of our capacities. He combines the Kantian requirement to align one's will to the moral law with the concept of rational agents who are also cognizant of their emotive and sensible dimension as humans in a shared culture: the "ideal" is our self-realization in and through not just one, but a complex of symbolic formations.[9] The mere ability to symbolize is thus the basis of human freedom. Additionally, any such symbolic act has to be reflective of our basic ability to give form to our thoughts and experiences: we are not free while completely immersed into our world, but when we become aware of our own activity of giving form. This is Cassirer's modern version of Kant's categorical imperative in its expanded usage for all human expressions. This "freedom" is a form of inner orientation and truthfulness: in the tradition of the Enlightenment, we can only be free if we know of the basic tenets of our functioning as meaning-creating beings (who are still subject to events in the world that are beyond our control). In its ethical dimension, freedom amounts to a unity of character in light of our ability to subsume the particular, situational demands to act under universal norms.[10] The ultimate freedom in the Kantian sense is the ability to determine oneself in accordance with one's insight into moral duty. With Hermann Cohen, Cassirer understands this not merely as a simple reality, but a *task* that we strive to fulfill.[11] More specifically, Cassirer now understands this task as one we fulfill *from within* the various forms of culture: freedom is only possible through association with others, and within an environment. This means that freedom can be expressed and practiced only in a world that we share with others, by insights and actions that we can and must reflect upon, make our own, and communicate about. This enables us to bring our particular interests into harmony with more general structures, not just aligning ourselves to them but also altering them in turn.

Cassirer mentions "self-liberation" as the goal of humanity in *Essay on Man* (1944), and elaborates on the nature of human freedom in *Axel Hägerström. Eine Studie zur schwedischen Philosophie der Gegenwart* (Gothenburg, 1939) as well as *The Myth of the*

State (New York, 1946), but we also find earlier discussions on freedom in all volumes of his *Philosophy of Symbolic Forms*. This process of liberation understood as an awareness of us forming and using symbols shows itself in our growing ability to realize the difference between what something *is* and what something *means*.[12] Once we understand that we are indeed agents within a symbolic universe, we can determine ourselves and this universe in light of the symbols we are surrounded with, and which are also to a significant extent of our own making.

Before we come to a discussion of the possibility of freedom through the symbolic, we need to look at the act of symbolization proper. The key insight in both Cassirer's and Langer's philosophy is this: symbolization as the basic human function of world-formation does not keep us away from reality, but is the key to having a reality at all.[13] Symbols are not arbitrarily applied onto a ready-made reality, but are the means by which we establish a reality. Cassirer summarizes this very elegantly in the first volume of *Philosophy of Symbolic Forms*:

> Every authentic function of the human spirit has this decisive characteristic in common with cognition: it does not merely copy but rather embodies an original, formative power ... This is as true of art as it is of cognition; it is as true of myth as of religion. All live in particular image-worlds, which do not merely reflect the empirically given, but which rather produce it in accordance with an independent principle. Each of these functions creates its own symbolic forms which, if not similar to the intellectual symbols, enjoy equal rank as products of the human spirit. None of these forms can simply be reduced to, or derived from, the others; each of them designates a particular approach, in which and through which it constitutes its own aspect of "reality."[14]

Langer's and Cassirer's central thought is that the deepest need of humans is to render their experience, emotions, and thoughts in symbolic forms. A symbolic form, roughly, is a systematic pattern of symbols concerning an aspect of our life world, such as art, myth, science, or law.[15] Our world comes in certain shapes[16] that express our particular experience of it, our thoughts, and feelings. Such a shape makes them recognizable, and then either testable or at least integrable into a more complex set of meanings, i.e., a symbolic form. This is also a fundamental function of *self*-formation: only through symbolic formation can we reflect on these forms of life, and either incorporate them into our personality and community, or reject them. Here we also see how important our consciousness of the symbolic *as* something symbolic (and not "reality itself," the ever-lasting "myth of the given") still is, since only if we are aware of our forms of symbolization can we determine the direction of our self-determination among those symbolic systems. We can, however, never leave the world of the symbolic for something "more immediate."

Some human affairs cannot be grasped by articulate, concept-based thought, as their meaning cannot fully deciphered by a rational analysis. In particular in her seminal work *Philosophy in a New Key* (1942), as well as all subsequent works on the art symbol, Langer argues that such non-conceptual, "presentational symbols" are not mere "blind, inconceivable, mystical affairs," but have their own "symbolistic schemata." Langer is even convinced that non-discursive symbolic systems are the foundation of discursive ones. This clearly follows from Cassirer's insights which he voiced, for instance, in "Language and Myth,"[17] where he calls myth and language the most foundational forms of the objectification of sense impressions and feeling.[18] A proper knowledge about the ways in which these non-discursive symbolic formations have developed and how they interact with discursive formations is key to understand the power of the human mind to establish its world. Langer discusses such non-discursive forms of "feelings" in *Feeling and Form* (1953),[19] but she never quite settles on a final nomenclature—here, I use the term "presentational symbol." First and foremost, in these symbols *meaning* is intuited and established *as a whole*, not through the addition and combination of independent parts, i.e., as *Gestalten*. Hence, such symbols are not translatable into other forms without losing an essential part of their meaning. This applies, in particular, to art.

What interests us now is how the art symbol in particular manifests a movement towards greater freedom of mind, and in what sense this progress is endangered by a return to mythological patterns of thinking and feeling.

THE ART SYMBOL

Essential for Langer's concept of the art symbol are the categories of virtuality, semblance, and significance. Art, in general, is a formation of feeling—a way in which the artist encapsulated their stance, a particular sense of being in the world cognizant of this feeling. Such a symbol offers us a means to look at this feeling without being tethered to the actual emotion. Art is "about" emotions, but is in itself not an emotion, nor is it an abstract tool of dealing with, or even getting rid of emotions. Art unfolds these feelings, offering us a semblance of the world,[20] a vivid impression that makes feelings familiar, yet utterly "other" at the same time. In Langer's words, art offers us a "vital import."[21]

It has been, however, notoriously hard to pin down what this "vital import" means, other than an actually felt meaning that covers more than abstract information. I will try to capture this "import" by the notion of "meaning." In his encompassing study on Cassirer's and Langer's concept of myth, Schultz claims that the meaning of art becomes apparent in "the unity of the human act of its production."[22] But how are we to understand such a unity of the human act of symbolization? For Schultz, such a unity depends on "the unity of our human culture."[23] Still, the "human act in [artistic] production" seems to be much more limited than the much grander unity of "human culture." It also remains unclear why and how both "unities" should become easily available through the art symbol, as it also is not easy to grasp why we should even seek any "unity" in human culture at all (other than in its function to be expressive of the symbolic forms we developed). Still, maybe Schultz is onto something here that Langer herself also struggled to formulate, because he could very well mean—in good Cassirerian fashion—that this unity is a *functional* unity. I thus propose to understand this equivalence between an individual act and a culture in functionalistic terms. Both the unity expressed in the act of artistic production and the unity of culture are built on a connection between an individual expression and a shared form, or pattern. What the individual expresses is not a merely personal feeling, and what people share in a culture is more than a means to collectively survive. Let us first look at the unity in artistic formation.

According to Langer, when an artist formulates a feeling, they do not merely express some subjective state of mind. Rather, they have discovered something more general, law-like about the form and function of a feeling. In "The Cultural Importance of Art" (1958), Langer describes the process of symbolic transformation as being both a subjectification and objectification—it does not just turn a subjective "feeling" into a form, but also makes this form subjectively significant. Cassirer, in turn, speaks of a "intensification" and "condensation" of feeling in symbols.[24] Concerning subjectification, Langer stresses that the artistic form gives the object a dimension of inner articulation, so that it becomes the expression of an inner feeling. Regarding objectification, said formulated feeling is not a mere personal cry, but it is given a visible (or audible etc.) form that allows others to spell out this feeling as well. To be accessible to others, however, the artwork has to follow a shared pattern of expression, or one that has not been used conventionally so far, but that is still part of the audience's emotional lexicon. Langer assumes here that an unconsciously shared memory of feelings can be stimulated by a form that is *somehow* akin to them.[25] This interplay of a subjective and objective side in our appreciation of art also gives us the possibility of bringing to consciousness our feelings in and judgments about art. This, in turn, allows us to gain insight into how not only *art* works but also how *we* work. In this sense, art offers the indispensable cognitive value of "self-knowledge, insight into all phases of life and mind,"[26] and it offers an emotive value, in that it heightens and enlivens our individual and social life, imbuing "mere reality" with a deeper meaning and value.[27]

Such an aesthetic "education of feeling"[28] enables us to give feelings a shape and hence make them perceivable, discernable, thus allowing us to take a step back from them, look at them, experience them aesthetically but also critically. What does this mean? An aesthetic experience in the neo-Kantian tradition (and I wish to claim here that Langer is part of this tradition) is marked by its disinterestedness, which entails that we are emotionally engaged, but do not react to an artistic presentation as we would to the real event. In Langer's terminology, we are not "deluded,"[29] but capable of a certain critical or at least reflective distance. This also implies that we are never fully and completely carried away without any capacity for critical thought. And this is why "art is a public possession, because the formulation of 'felt life' is the heart of any culture, and molds the objective world for the people. It is their school of feeling, and their defense against outer and inner chaos."[30] What is more, by means of making worlds of feeling accessible, it allows for an intuitive way of connection, a "shared communication in non-discursive expressions,"[31] a deep bond of a people. The vast repository of presentational symbols in the arts, but also in other regions of our emotional, spiritual, and communal life is thus part of a common undertaking that becomes visible in rituals, traditions, and cultures. For art in particular, these patterns are not just mere repetitions of the very same, but each part of the pattern in itself offers something "significant": it either represents an insight or a feeling that we deem noteworthy, or it is the formation of something original, so far unseen, that highlights a previously unnoticed aspect of our reality. Langer stresses the dynamic nature of these patterns. They thus exhibit an "open potentiality"[32] that seems worthwhile to be explored in contemplation; their being "open" indicates that they are accessible for us, their potentiality offering us a way to understand a new facet of reality we might have overlooked before, thus changing our previous perception of reality. We find the roots for Langer's stance on the dynamics of symbolic forms in Cassirer as well. In *Freiheit und Form*, as well as in *The Philosophy of Symbolic Forms*, vol. 2: *Mythical Thought* (1925), he traces the changes of mythical and artistic patterns, insisting that these are the embodiment or expression of a "continuously effective, ideational fundamental force of the intelligible event itself."[33] The symbolic systems such as myth, art, and language are constantly changing, building new aspects while demolishing old ones. In his treatment on myth, he calls this the specific "progress" of symbolic forms.[34]

Langer's account on human symbol formation in art thus captures what is individually *felt* in patterns that are inter-subjectively shared. We can articulate our unique experience using forms of symbolic presentation, which in turn changes and reorients these patterns (sometimes only just slightly so, sometimes with more disruptive effects). Let us now look at Langer's conception of these patterns in terms of culture and civilization, highlighting both the advantages and dangers of their influence on our self-understanding.

THE CRISIS OF THE SYMBOL: CULTURE AND CIVILIZATION

The general direction of Langer's opus magnum, *Mind*, seems to be the explanation of the development of human culture out of the evolutionary role of the human need to formulate experience symbolically. The development of culture here seems to be an issue within evolutionary biology, not a philosophy of culture. But how can this naturalistic understanding of human culture go together with Cassirer's neo-Kantian claim that a deeper knowledge of our symbolic forms leads to our self-liberation? In the last chapter of Innis's comprehensive account on Langer's notion of the "symbolic mind," which attempts to place her philosophy in the general philosophical landscape, he claims her "central thesis" to be

> that Mind, in the human sense, is a functional matrix of acts, not a substance, and is defined by diverse and stratified powers of symbolization and sign use ... The mind becomes the "locus" or "place" for the generation and interpretation of meaning(s). Minding is, at bottom, semiosis, the creation and interpretation of signs, symbols, and meanings. And this process encompasses the total range of human "feeling."[35]

The later parts of *Mind* in particular discuss this act of "minding" in relation to culture formation, starting with chapter 23, "The Breaking," which delineates "the rise of civilization."[36] Here, Langer marks the turn from myth-centered cultures to our modern civilization. Main aspects of this development are the discovery of the notion of the individual and the breaking up of tribal structures,[37] a rediscovery of the ancient category of causation for science,[38] and even the development of a historical consciousness as a new and "highly interesting cognitive construct."[39] Also very instructive on this issue are Langer's various essays collected in *Philosophical Sketches* from 1962. It is important to note that neither Langer nor Cassirer claims that Western society reached a higher "level" than others. Rather, she points out the differences in these cultures' approach to their shared reality, and considers the potential dangers when only parts of the modern "civilized" world view are integrated into foreign cultures. As Cassirer considered the re-mythification of Nazi Germany, so Langer considers the West's colonial efforts in other countries that far too often resulted (and still result) in the annihilation of cultural plurality.

Like Cassirer, Langer tries to conceptualize the failings of civilization that we came to witness in the first half of the twentieth century. If we did not have the tendency to counteract present developments in a civilization, how could we ever claim any sort of "self-liberation?" In her essay "Scientific Civilization and Cultural Crisis" (1961), Langer clarifies her focus on modern society through a systematic and widely critical distinction between *civilization* and *culture*.

A *culture* is "the symbolic expression of developed habitual ways of feeling."[40] Through culture we have a symbolic way of understanding, expressing, and channeling our "feelings" in established "patterns."[41] We go beyond mere pragmatic exchange and express ourselves in "gestures" and "style"[42] that all articulate our humanity as well as our individual situatedness within this particular group, forming a "personal expression [of] our social heritage."[43] Culture is based on an interplay of conservation and change. On the one hand, if expressions had never stayed the same on a basic level, we would not recognize them as being made "by *our* people." On the other, a living culture requires dissent, conflict, and departure to develop adequately in light of new challenges (for instance, new technologies, but also the ever-more obvious presence of *other* cultures, claiming the same validity as "ours"). It is easy to see that this mode of conversation and change is the same for the ever-changing lexicon of the arts as discussed above: we appreciate patterns, but also original changes to their old structures. In other words, we can "read" an artwork as it makes use of common techniques and modes of expression, but we only find these artworks interesting if they also change these common patterns into something new.

Civilization, in contrast, deals with "the practical implementation of life."[44] Whereas culture thrives on close contact of its people with each other and their environment, civilization organizes the artificial and short-timed clustering of people in the city; it goes away from tradition and "decorum" to foster the "venturesome, personalistic pole"[45]—at the same time allowing a more complex society consisting of various "cultures" in close contact, in which even the mobility across cultures can grow and give room for different accounts on individual activity.[46] What Langer identifies as the true threat of civilization is that its fruits can be uprooted and replanted in all kinds of soil, even a soil that it does not really fit into. On the surface, this can have rather beneficial outcomes, such as the ability to provide electricity, health care, and farming equipment to secluded parts of the world. But what Langer sees here is the danger that these developments do not take seriously the present tribal structures of self-understanding and self-worth, thus pressing the population into forms of work and a life that is incompatible with their culture, ultimately leaving them in shiny new shackles. There are numerous examples in older cultures/civilizations that developed in scientific and technical areas without a sufficiently strong corresponding artistic and religious-philosophical stratum,[47] expanded into areas that were not fit for these innovations, and hence faltered, taking the whole culture/civilization with them.

Westerners cannot claim to be exempt from such a version of a cultural downfall, as we can see in the hordes storming the US Capitol in 2021, or in the more than abundant anti-vaxxers and conspiracy theorists all over the Western world. These parallel cultures can very well take on the means of civilization and blow other cultures off the grid. If we lack the means to symbolically—and consciously—express our relation to technology and science, we will develop this dangerous tendency to ignore the capacities of said sciences, or we might even increasingly engage in a battle against their outcomes without any good measure nor alternative (other than to fall back to a pre-modern state, or by turning dystopian novels such as Margaret Atwood's *The Handmaid's Tale* into an upsetting reality).

We thus need a culture which gives form to the realities of life, a culture that allows us to articulate our experience and make its unique aspects comprehensible by aligning them to stable patterns that can be recognized by others. Only by a continual balancing act of realigning individual and society—symbol-monger and symbol-beholder—we form cultures and ultimately civilizations which last amidst change through inventions, wars, or environmental changes. As one inter-subjective tool to align culture and civilization Langer offers the concept of ritual. Already in mythical thinking a ritual is—as an overt act—fundamentally "public."[48] Out of the basic tenets of life, human beings developed the basic "life-symbols" (for birth and death, for instance) that we attempt to conquer[49] in myth, and that are expressed (and made manageable) in ritual. Ritual enables us to face changing situations with an arsenal of stable mental responses. Ritual, in Langer's view, is a means to connect to the "divine forces" present in our daily lives. In some native cultures, the rain dance does not "produce" rain, nor facilitate its probability; rather, it expresses the awareness of the changes of the seasons, and sets the practitioners of such a dance into an active relation to those forces. Rituals may have emerged within native cultures, but are very much still present in any kind of society. Rather than purging a society from it (as civilization might attempt for the sake of progress), we ought to reform ritual as an artistic part of our culture (to offer a means to deal with said progress). Such rituals should be a means of stability and empowerment. In Langer's words, they can facilitate the transition from symbols of power to *life symbols*.[50] Myth became less an instrument for symbols of power by securing a divine order, but for life symbols by offering ways to instantiate the idea that, for instance, death will not be the end for us,[51] thus allowing for a more constructive course of action in society. When art takes up this function of mythical thinking, it offers us the ability to become cognizant of the form of life as a means to counteract the claim for power by science or economics. The ritualistic element is contained in the traditions in which we came to appreciate any art form; the form of behavior we expect in an exhibition, a classical symphony, a popular stadium concert. Such behavior goes beyond mere decorum, but is the condition of the possibility of the imaginative space that the art form offers.

It is important to note here, however, that this form of reliance on ritual is not meant as a defiance of autonomy and enlightenment. Langer rather argues for rituals as an artistic means of reflection that also allows for self-transparency. This can become a tool for self-liberation from the shackles of work, as well as the daunting of the unknown, in that these rituals give the individual the tool to rely on its own[52] expressions, share them, and make them thus accessible to (but also changeable by) others. The ritualistic aspect offers a form of a self-given limitation parallel to the moral law in the area of human agency. "Any miscarriage of the symbolic process is an abrogation of our human freedom ... A life that does not incorporate some degree of ritual, of gesture and attitude, has no *mental anchorage*" (emphasis added).[53] In these rituals, we formulate and express our attitude toward our world. These rituals are at best open, narrational, and interpersonal instead of forbidding, stiff, and restrictive; the creation of meaning is a dynamic and interpersonal act, not something we shall instantiate as fixed forms of behavior.

However, the incorporation of cultures in a civilization is more complex, in that many of its structures do not develop in an organic way, but an artificial one. Langer attempts to offer a solution to this conundrum: philosophers can help by reflecting on (already existing) myths. With this, we could, first, make everyone conscious of a possible imbalance in a civilization that might be highly advanced in science, but lacks a sufficiently vital backdrop in its culture(s). Secondly, critical reflection should help to avoid falling back into inadequate, dated myths that bar individual agency. To be effective, this merely reflective stance, however, is in dire need of input from the arts, which, as I developed in "The Art Symbol," are a more self-transparent means to make visible the emotive and emotional side of culture, but also its common patterns. Already in *Philosophy in a New Key* Langer first and foremost advocates a philosophy that does not merely contrast logical, scientific thinking with its supposed only alternative: "folly"[54]—instead, she argues for a new way of philosophizing that considers "symbolic function" and "the morphology of significance"[55] present not just in scienc but also in art and culture at large. We can only freely and accurately think about the grand ideas of modernity such as "Race, Unity, Manifest Destiny, Humanity"[56] if we understand that although these concepts cannot be adequately captured by logic, they are also not *averse* to it. We need to consider the full range of the power of mind, "the impulse toward symbolic formulation, expression, and understanding of experience"[57] to capture the depth of those concepts, and establish new narratives to realize them.

CONCLUSION

Philosophy needs to be able to at least make us understand the function of the human mind fully and recognize that "mind" is not tantamount to scientific rationality. This, of course, must not mean that we have to indulge each and any wild and new "mythology." But what science and reason need to take into consideration is the wealth of "mind," and its ability to reach further into non-conceptual territory; the territory of myth, and, more importantly, of art, to quench the justified human need for "self-assertion [and] self-justification."[58] In *Philosophy in a New Key*, Langer closes her discussion with reference to the conditions of the possibility of human freedom. If we continue to be ignorant of the specifics of our current situation, if we are deprived of "food for thought, or imagination to envisage our problems clearly and negotiably,"[59] we will witness the next fall of a civilization—as Langer herself might have feared to do in light of the events of the Second World War. "[D]isorientation, the failure or destruction of life-symbols and loss or repression of votive acts"[60] would be the greatest threat to a functioning civilization. For Langer, we might be able to avoid this by allowing the "life symbols" to return into our daily lives. A life without art or ritual would be "prosaic to the point of total indifference, purely casual, devoid of that structure of intellect and feeling, which we call 'personality.'"[61] In order to counteract the destructive effects of civilization, Langer holds, we had better remember the benefits of culture.

Langer, with Cassirer, values rituals "to restore the balance between the mind's individuation and the earthbound hold of its roots in animal nature, the enormous potential of mental life and its tiny allowance of time for realization."[62] A first break with native civilization was necessary, as she argues further, for the advancement of science.[63] In light of the catastrophic events in the twentieth century which are in the background of *Philosophy in a New Key* and *Mind*—and we should also consider more recent developments that Langer could not quite foresee, but that were dawning in the development of mass culture and only gained traction through the expansion of virtual reality, finally leading people to mistrust science and the painfully obvious again—it is understandable why Langer thought that we are "now" in an era of a necessary renewal of culture(s) as a fundamental function of human self-formation.

NOTES

1. Stefan Morawski notes Langer's closeness to Cassirer in her style of argumentation that draws from analytic as well as continental philosophy: "Langer's stance, i.e., her reiterated polemics with empiricism, naturalism, and with neo-positivist philosophers of science, her insistent cutting herself off from metaphysics and psychology—all these bear on the Marburg version of neo-Kantianism which characterizes Cassirer's work." In Stefan Morawski, "Art as Semblance," *The Journal of Philosophy* 81, no. 11 (1984): 655.
2. Robert E. Innis, "Art, Symbol, and Consciousness: A Polanyi Gloss on Susan [*sic*] Langer and Nelson Goodman," *The International Philosophical Quarterly* 17, no. 4 (1977): 455–76; Cornelia Richter, "The Body of Susanne Langer's *Mind*," *Embodiment in Cognition and Culture*, ed. John Michael Krois et al. (Amsterdam: John Benjamins, 2007), 107–25.
3. Another interesting instance of this influence is Cassirer's usage of "living form" to designate the art symbol. See Ernst Cassirer, *The Myth of the State. Gesammelte Werke* (ECW 25), 26 vols, ed. Birgit Recki et al. (Hamburg: Meiner, 1995–2006), 39. In the following, I will reference Cassirer's *Gesammelte Werke* by ECW vol., page. The best translation of Cassirer's *Philosophy of Symbolic Forms* was done by Steve G. Lofts for Routledge, 3 vols, 2021.

4. Langer, *Feeling and Form*, 410.
5. For an encompassing discussion of the influence of Whitehead, (the early) Wittgenstein, and Sheffer, see Adrienne Dengerink Chaplin, *The Philosophy of Susanne Langer* (London: Bloomsbury, 2020), and Robert E. Innis, *Susanne Langer in Focus* (Bloomington: Indiana University Press, 2009), as well as Chapters 2, 3, and 10 in this volume. Innis emphasizes Whitehead's influence when considering the non-reductivist nature of her naturalism, as well as her stress on the *dynamics* of symbolic formation. As he notes, "[h]er world, like Whitehead's, is a processual world, a world of events and relations, not of stable 'things' embodying 'essences,'" ibid., 5. We should also consider, however, that her theory of presentational symbols is in sharp contrast to the basic tenets of analytic philosophy. Cf. ibid., 43.
6. Notably during his exile in Sweden we see Cassirer's explicit turn to practical issues in philosophy, in particular his discussion of the Uppsala school and his critique of naturalism in Ernst Cassirer, *Axel Hägerström: eine Studie zur schwedischen Philosophie der Gegenwart* (Gothenburg: Elanders Boktryckeri Aktiebolag, 1939). The later works starting in the late 1930s represent the focus of Cassirer's ideas during the time of his interaction with Langer.
7. Cassirer wrote *The Myth of the State* in the US during the winter of 1944/5, but did not live to see its publication in 1946. However, as the editor notes in his preface, they finished the revision of parts I and II before Cassirer's death, so that the final edition should give a clear view of Cassirer's thoughts. See Ernst Cassirer, *The Myth of the State* (ECW 25), 4–5. The shorter essay "The Myth of the State" published in 1946 in *Fortune* reflects the issues Cassirer discussed in chapter 18 of the final version of *The Myth of the State*. The ideas presented here form the decisive part of Cassirer's argument against the volatile combination of myth and technology in modern states, namely the Nazi regime.
8. See Samantha Matherne, *Ernst Cassirer*. The Routledge Philosophers (London: Routledge, 2021), ch. 7, and Edward Skidelsky, *Ernst Cassirer: The Last Philosopher of Culture* (Princeton, NJ: University Press, 2008), ch. 9. Even though Skidelsky argues that Cassirer starts engaging with contemporaneous debates in his *Philosophy of Symbolic Forms*, those offered indeed a rather "contemporary slant." In Skidelsky, *Ernst Cassirer*, 221.
9. See Cassirer, *Essay on Man* (ECW 23), 228.
10. See Cassirer, *Axel Hägerström. Eine Studie zur schwedischen Philosophie der Gegenwart*, in ECW 21, 64, 75–6, 79. Further, Cassirer on freedom, and its promise, 100–4.
11. Cassirer identified the re-emergence of mythical thinking as the reversal of such a task, exemplified in the myth of folk and race in the Nazi regime. Such a reversal into myth as a demolition of personal will and responsibility cannot be attacked by philosophy directly, but only by unveiling the hidden mechanisms in newly emerging myths. Cassirer, *The Myth of the State* (ECW 25), 290–1. While Cassirer referenced the Nazi regime, we can still see such mechanisms at work in contemporary Western societies. That Cassirer's stance indeed reached much further is argued in Chiara Bottici, "Who is Afraid of the Myth of the State? Remarks on Cassirer's Forgotten Manuscript," *Social Imaginaries* 3, no. 2 (2017): 213–27.
12. As Cassirer discusses, for instance, in the "dialectics of mythical consciousness" in *Philosophy of Symbolic Forms*, vol. 2 (ECW 12), 280.
13. See *Sprache und Mythos* (ECW 16), 233; *Philosophy of Symbolic Forms*, vol. 1 (ECW 11), 16; *Philosophy of Symbolic Forms*, vol. 3 (ECW 12), 4; *Die Bedeutung des Sprachproblems für die Entstehung der neueren Philosophie* (ECW 17), 3–11, 4. Langer's concept of symbolization summarizes this most succinctly: "a symbol is a form *through which the object is being perceived*." In Dengerink Chaplin, *The Philosophy of Susanne Langer*, 190.

See also Langer, *Philosophy in a New* Key (Cambridge, MA: Harvard University Press, 1957), 218.

14. Ernst Cassirer, *Philosophy of Symbolic Forms*, vol. 1 (ECW 11), 7.
15. Samantha Matherne translates the German *Recht* in Cassirer as "right" to stress the wider circumference of the term, which covers not only legal issues, but also more basic issues of right and wrong. See Matherne, *Ernst Cassirer*, ch. 7.
16. To be sure, Langer seeks to replace the term "symbol," in particular for art, with "expressive form." Langer, "The Art Symbol," 127. But for the sake of unity, I reference here all acts that join a feeling or thought with a perceivable (or imaginary) shape—which either represents, connotes, or references something—as *symbolic formation.* In *Philosophy in a New Key*, Langer chooses an even broader description and characterizes the function of the symbol as our ability "to holding on to the object" (Langer, *Philosophy in a New Key*, 135).
17. Langer's translation appeared in 1946, helping Cassirer's thoughts to gain traction in the US. This was, however, hindered by Cassirer's untimely death in 1945. See Ernst Cassirer, *Language and Myth*, trans. Susanne K. Langer (New York: Dover, 1946).
18. See *Sprache und Mythos* (ECW 16), 302–3, and later on *Myth of the State* (ECW 25), 47.
19. Langer constantly reworked her definitional range of "symbol" and "feeling." See Langer, "On a New Definition of the 'Symbol,'" 1956, *Philosophical Sketches* (New York: Barnes and Noble, 1962), 59–61. Here she focuses on "symbol" as a "presentational symbol" that functions as a means of abstraction, in particular of state of affairs which rational thought cannot capture. According to Langer, "feelings" cover sensation, sensibility, emotional attitude, "even our general mental or physical condition"—in short, "anything that may be felt." Langer, "The Cultural Importance of Art," 1958, ibid., 80. Her terminology has been criticized, as to whether the term "feeling" was actually well chosen, since the range of this concept is far greater than its ordinary usage would indicate. Iris Yob, "The Form of Feeling," *Philosophy of Music Education Review* 1 (1993): 20. I want to stress, in particular, that Langer's concept of "feeling" easily includes the Kantian notion of "the life of the mind."
20. In other words, an artwork represents something "virtually" in providing an "illusion" of the very thing itself—not necessarily by means of mere imitation, but by symbolic representation. Langer's concept of "semblance" is thus very close to Friedrich Schiller's as he defends it. See Friedrich Schiller, *On the Aesthetic Education of Man*, 1795 (London: Dover, 2004), the 26th letter, 124–30. I discuss this in more detail regarding Schiller in Anne Pollok, "A Further Mediation and the Setting of Limits: The Concept of Aesthetic Semblance and the Aesthetic State," in *Friedrich Schiller: Über die ästhetische Erziehung des Menschen in einer Reihe von Briefen*, ed. Gideon Stiening, Klassiker Auslegen 69 (London, New York: de Gruyter, 2019), 219–35.
21. Langer, *Feeling and Form*, 52.
22. William Schultz, *Cassirer and Langer on Myth: An Introduction* (New York/London: Garland, 2000), 259.
23. Ibid.
24. Cassirer, *The Myth of the State* (ECW 25), 47–8.
25. Langer's claim is that the form of a feeling occurs not just arbitrarily through convention, but is somewhat adequate to the bodily structure of said feelings. Langer, *Feeling and Form*, 82.
26. Langer, "The Cultural Importance of Art," 87.
27. Ibid., 88; *Feeling and Form*, 401.
28. Langer, *Feeling and Form*, 401.

29. Ibid., 319.
30. Ibid., 409.
31. In the context of Langer's privileging of non-discursive language, discursive language could only evolve from a shared world of meaning, which is established by the primordial means of presentational symbolisms such as myth, ritual, and art. Cf. Alexander Durig, "What did Susanne Langer really mean?" *Sociological Theory* 12, no. 3 (1994): 256.
32. This characterization is coined in Eva Shaper, "Significant Form," *The British Journal of Aesthetics* 1, no. 2 (1961): 36.
33. German original, "*eine fortwirkende ideelle Grundkraft des geistigen Geschehens selbst*," Cassirer, *Freiheit und Form* (ECW 7), 389.
34. Cassirer, *The Philosophy of Symbolic Forms*, vol. 2 (ECW 12), 275.
35. Robert E. Innis, *Susanne Langer in Focus* (Bloomington: Indiana University Press, 2009), 250–52.
36. Susanne K. Langer, *Mind: An Essay on Human Feeling*, vol. 3 (Baltimore: Johns Hopkins University Press, 1982), 155.
37. Langer, *Mind*, vol. 3, 194.
38. Ibid., 204.
39. Ibid., 205.
40. Susanne K. Langer, "Scientific Civilization and Cultural Crisis," *Philosophical Sketches* (New York: Barnes and Noble, 1962), 91.
41. Langer, "Scientific Civilization," 92.
42. Ibid.
43. Ibid., 93.
44. Ibid.
45. Ibid., 95.
46. In this sense, a clear account on cultural mobility should offer a fruitful background for a concept of agency that allows for individual justification and responsibility (as in Hannah Arendt's *The Human Condition*, 1958). Thus, such a society would evade the problem of myth that Cassirer diagnosed in Nazi Germany, and how the cult destroys any sense of autonomous agency.
47. The argument Langer presents in her discussion of the sunken civilizations of Incas, Mayans, and Toltecs is, of course, more complex. These societies were highly civilized with an immense prowess in warcraft and technical advancement, but their cultures were not keeping up the pace. For the Incas, for instance, Langer points out their avoidance of strong conceptions of immortality. As "[n]othing is so directly opposed to cultural advance as fear to develop a line of thought or face one's own knowledge" (Langer, *Mind*, vol. 3, 187), the Inca's inability to face the realities (and possibilities, culturally speaking) of death became problematic. Ultimately, their achievements as a civilization grew too far apart from their "timidity" in their "religious conception" (ibid.). Such an "unbalanced mentality" (ibid., 191) made them easy targets for the Spanish invaders; in the end, they only had to finish a job that was already in the making. Langer seems to suggest here that one reason for the cultural weakness (and subsequent downfall) of a people might have been their indifference towards "religious thinking" in particular (181). She also characterizes Aztec culture as an "impressive civilization … without, apparently, any true humane culture underlying it" (178). Even though they reached "a stage of worshipping gods with names and … characters," Langer identifies as one reason for their downfall that they included too many foreign gods (192). Even if I agree with Langer's general idea that an imbalance between civilization and culture is dangerous, the condemning of

"liberal sexual practices" and "too much tolerance" (191) has a rather disturbing ring to it. This is an issue that is in dire need of further discussion.

48. Susanne K. Langer, *Philosophy in a New Key* (Cambridge, MA: Harvard University Press, 1957), 171.
49. Innis, *Susanne Langer in Focus*, offers a wonderful discussion of this in ch. 3 under the section headline "The Symbolic Lessons of Myth," 72–9, building on *Mind*, vol. 1, 148–49.
50. Langer, *Mind*, vol. 1, 148.
51. We find one of the most fascinating versions of immortality in Plato's *Symposium*, in which Diotima mentions offspring of the physical but also of the intellectual kind as a form of immortality, see *Symposium* 209a–210d.
52. As already discussed, these expressions are not purely individual. Like Cassirer, Langer assumes that we are not the creators of symbolic expression, but find ourselves in a universe of symbolic forms. "The ego, the individual mind, cannot create reality. [Hum]an is surrounded by a reality that he [she] did not make, that he [she] has to accept as an ultimate fact." This is as close as both of them get to Heideggerian *Geworfenheit*. Ernst Cassirer, "Language and Art I/II," *Symbol, Myth, and Culture. Essays and Lectures of Ernst Cassirer*, 1935–45 (New Haven, CT: Yale University Press, 1979), 195.
53. Langer, *Philosophy in a New Key*, 290.
54. Ibid., 293.
55. Ibid.
56. Ibid.
57. Ibid.
58. Ibid., 292.
59. Ibid.
60. I would like to thank an anonymous reviewer who directed my attention to a case where the Southern Illinois University punished black kneeling cheerleaders, citing the necessity to showcase "unity" in front of "national symbols." This is a distinctive moment where two different concepts of nation clash, and a call for "national unity" is all but harming the minorities. A just state should offer thoughtful consideration, and not force unity where there is none. Langer's cultural anthropology is sensitive to minorities as bearers of a plurality of traditions that are not of lesser cultural value than those of the majority. The term "nation" as a political entity should be understood much closer to Langer's concept of "civilization" than the idea of one dominating "culture." Cf. Langer, *Philosophy in a New Key*, 290.
61. Langer, *Philosophy in a New Key*, 290.
62. Langer, *Mind*, vol. 3, 148.
63. Ibid., 194.

BIBLIOGRAPHY

Arendt, Hannah. *The Human Condition*. Chicago: University of Chicago Press, 1958.

Bottici, Chiara. "Who is Afraid of the Myth of the State? Remarks on Cassirer's Forgotten Manuscript." *Social Imaginaries* 3, no. 2 (2017): 213–27. Available online: https://doi.org/10.5840/si20173223.

Cassirer, Ernst. *Axel Hägerström: eine Studie zur schwedischen Philosophie der Gegenwart*. Gothenburg: Elanders Boktryckeri Aktiebolag, 1939. Available online: https://gupea.ub.gu.se/handle/2077/52252 (accessed July 16, 2023).

Cassirer, Ernst. *Gesammelte Werke* (ECW 1–26). Edited by Birgit Recki et al. 26 vols. Hamburg: Meiner, 1995–2006.

Cassirer, Ernst. *Language and Myth*. Translated by Susanne K. Langer. New York: Dover, 1946.

Cassirer, Ernst. "Language and Art I/II." *Symbol, Myth, and Culture. Essays and Lectures of Ernst Cassirer*. 1935–45. Edited by Donald Phillip Verene. New Haven, CT: Yale University Press, 1979.

Cassirer, Ernst. *The Philosophy of Symbolic Forms: Three-Volume Set*. London: Routledge, 2021. Available online: https://doi.org/10.4324/9780429284922.

Dengerink Chaplin, Adrienne. *The Philosophy of Susanne Langer*. London: Bloomsbury, 2020.

Dryden, Donald. "Memory, Imagination, and the Cognitive Value of the Arts." *Consciousness and Cognition* 13 (2004): 254–67. Available online: https://doi.org/10.1016/j.concog.2004.01.003.

Durig, Alexander, "What Did Susanne Langer Really Mean?" *Sociological Theory* 12, no. 3 (1994): 254–65. Available online: https://www.jstor.org/stable/202124 (accessed September 9, 2022).

Innis, Robert E. "Art, Symbol, and Consciousness: A Polanyi Gloss on Susan [*sic*] Langer and Nelson Goodman." *The International Philosophical Quarterly* 17, no. 4 (1977): 455–76. Available online: https://doi.org/10.5840/ipq197717442.

Innis, Robert E. *Susanne Langer in Focus*. Bloomington: Indiana University Press, 2009.

Krois, John M. *Ernst Cassirer: Symbolic Forms and History*. New Haven, CT/London: Yale University Press, 1987.

Langer, Susanne K. "The Cultural Importance of the Arts." *The Journal of Aesthetic Education* 1, no. 1 (Spring 1966): 5–12. Available online: http://www.jstor.org/stable/3331349 (accessed July 16, 2023).

Langer, Susanne K. *Feeling and Form: A Theory of Art Developed from Philosophy in a New Key*. New York: Charles Scribner's Sons, 1953.

Langer, Susanne K. *Mind: An Essay on Human Feeling*, vol. 1. Baltimore: Johns Hopkins University Press, 1967.

Langer, Susanne K. *Mind: An Essay on Human Feeling*, vol. 3. Baltimore: Johns Hopkins University Press, 1982.

Langer, Susanne K. *Philosophical Sketches*. New York: Barnes and Noble, 1962.

Langer, Susanne K. *Philosophy in a New Key: A Study in the Symbolism of Reason, Rite, and Art*. 3rd edn. Cambridge, MA: Harvard University Press, 1957.

Langer, Susanne K. *Problems of Art*. New York: Charles Scribner's Sons, 1956.

Langer, Susanne K. "Scientific Civilization and Cultural Crisis." *Annals of the Japan Association for Philosophy of Science* 2, no. 2 (1962): 120–6. Available online: https://doi.org/10.4288/jafpos1956.2.120 (accessed July 16, 2023).

Matherne, Samantha. *Ernst Cassirer*. The Routledge Philosophers. London: Routledge, 2021.

Morawski, Stefan. "Art as Semblance." *The Journal of Philosophy* 81, no. 11 (1984): 654–63. Available online: https://doi.org/10.2307/2026172.

Pollok, Anne. "A Further Mediation and the Setting of Limits: The Concept of Aesthetic Semblance and the Aesthetic State." In *Friedrich Schiller: Über die ästhetische Erziehung des Menschen in einer Reihe von Briefen*, edited by Gideon Stiening, 219–35. Klassiker Auslegen 69. London/New York: de Gruyter, 2019.

Richter, Cornelia. "The Body of Susanne Langer's *Mind*." In *Embodiment in Cognition and Culture*, edited by John Michael Krois et al., 107–25. Amsterdam: John Benjamins, 2007.

Schiller, Friedrich. *On the Aesthetic Education of Man*. London: Dover, 2004.

Schultz, William. *Cassirer and Langer on Myth: An Introduction*. New York/London: Garland, 2000.

Shaper, Eva. "Significant Form." *The British Journal of Aesthetics* 1, no. 2 (1961): 33–43.

Skidelsky, Edward. *Ernst Cassirer. The Last Philosopher of Culture*. Princeton, NJ: Princeton University Press, 2008. Available online: https://doi.org/10.1093/bjaesthetics/1.2.33.

Yob, Iris. "The Form of Feeling." *Philosophy of Music Education Review* 1 (1993): 18–32. Available online: https://www.jstor.org/stable/40327050 (accessed September 9, 2022).

CHAPTER EIGHT

The Systematic Position of Art in Susanne K. Langer's and Ernst Cassirer's Thinking

CHRISTIAN GRÜNY

Ernst Cassirer and Susanne K. Langer, while being very close in their fundamental philosophical orientations, exhibit certain characteristic differences. These are summed up neatly by the following quotes.

> Cassirer: We would, then, have a philosophical systematization of spirit in which each particular form would obtain its sense purely from the position in which it stands, a systematization in which its content [Gehalt] and its significance would be designated by the wealth and particular nature of the relations and entanglements in which they relate to other spiritual energies and ultimately to allness [Allheit].[1]

> Langer: [F]or to be able to define "musical meaning" adequately, precisely, but *for an artistic, not a positivistic context and purpose*, is the touchstone of a really powerful philosophy of symbolism."[2]

On one side we have a systematic philosopher whose philosophy of symbols clearly aimed at an "integrative, reflexive, and comprehensive theory of our natural and mental relations to ourselves, the social, and the world,"[3] on the other a former logician who centers her theory of symbols around art and music in particular and measures its success by referring to music, albeit without curtailing its scope.

The role of art lies at the center of my comparison of the two philosophies. In contrast to Langer, art was important for Cassirer but has a rather nomadic existence in his extensive philosophical oeuvre.[4] According to Langer herself, it was both meeting Cassirer and an intensified encounter with art that led her away from focusing on symbolic logic to a comprehensive philosophy of symbols.[5] If we disregard the differences their respective systematic orientations imply, we could say that Langer produced the theory of art that Cassirer could not write. But there is more: the allusion to music in the title of the book in which she first developed her generalized theory of symbols and symbolism—the "new key" of philosophical thinking—is more than a casual metaphor. Rather, it represents the paradigm for her own version of systematic philosophy, which is far from Cassirer's Hegelianism.[6]

Langer calls Cassirer the philosopher "who hewed the keystone of the structure, in his broad and disinterested study of symbolic forms," while she "put that stone in place, to join and sustain what so far we have built."[7] This describes the continuity between the two thinkers. But there is more than a different degree of elaboration that distinguishes them. There is a difference in systematic reach on the one hand and differentiation when it comes to specifics on the other. Here their strengths and deficits are complementary. Discussing the conceptual fuzziness that Langer sometimes exhibits—i.e., the constant reference to Wittgenstein's or even Frege's "logical form," the peculiar concept of abstraction—and also some of her systematic deficits would exceed the scope of this chapter. If, however, we are confronted with the task of "taking up Cassirer's theoretical and methodical intuition and positioning it against the continuity of a hypertrophic insistence on systematic unity,"[8] as Dirk Rustemeyer contends, rather than developing new systems, as I am convinced, Langer can be an important ally.

In order to pursue this thought, I will first turn to the systematic position of art in Cassirer's philosophy and then to the role of art in Langer's philosophy of feeling. The background of Cassirer's thinking will serve to highlight certain aspects in Langer whose importance is not always recognized. In her chapter, Anne Pollok undertakes a similar endeavor with regard to the political side of their philosophies; this chapter is more systematically oriented.

ART AS SYMBOLIC FORM

In the introduction to the third volume of his *Philosophy of Symbolic Forms*, Cassirer undertakes a run through the history of philosophy in order to historically situate and justify his concept of symbolic forms. The very first sentence contains one of the lists that he comes back to again and again: "When we designate language, myth, and art as 'symbolic forms' …"[9] There is a reason why these lists abound in his work: while its final aim is erecting a system, its starting point is a kind of inventory of everything that might be addressed as a symbolic form. What this inventory contains is an empirical question, as it were, but the systematic connections between the different forms will have to be constructed rather than simply discovered. The success of any such system will depend both on the plausibility of the overall structure and on its ability to elucidate the specificity of each of the symbolic forms. Art is a special case: while there was never any doubt for Cassirer that it can claim the status of a symbolic form in its own right, its systematic status remains somewhat unclear.

The fact that Cassirer juxtaposes language, myth, and art in this way seems to suggest that they occupy a similar systematic position in his writings, which is clearly not the case. The two previous volumes of the *Philosophy of Symbolic Forms* had indeed been dedicated to the first two of these forms, language and mythical thinking. Opening the third volume and finding this list, one might expect it to be dedicated to art. Its title, however, is *Phenomenology of Cognition*; it is the attempt to provide a systematic genealogy of forms of cognition in their interconnectedness, situated in the tradition of Hegel's *Phenomenology of Spirit*. There is hardly a trace of art—apparently it has not found its place, after all.

The tripartite schema Cassirer uses to classify symbolic forms in their development employs different terms in various texts but essentially remains the same: expression—(re) presentation[10]—meaning, representing a successive uncoupling of sense from its medium starting from their initial identity. This entails a gain of reflective possibilities from the mythical identification of symbol and sense to the formalized and purely conventional signs of scientific formulas that have severed any intrinsic relation to concrete reality or experience.

In this schema, art is situated somewhere between myth and science. In contrast to them—but like language—it cannot be clearly mapped onto one of the three stages of symbolic sense. Art draws on the pictorial, which we already find in myth, objectifies it, and makes it manageable without losing the richness of its inner articulation. As the epitome of art, the image is neither an active agent as in mythical life nor a mere means to an end: "Artistic intuition does not look through the image at something else that is expressed and represented in it but enters the pure form of the image and remains in it."[11] And "only now is spirit able to enter into a truly free relationship with it."[12]

Accordingly, Cassirer finds in art an interlacing of expression and (re)presentation. This has some obvious implications regarding a philosophy of art because the alternative between an aesthetics of expression and an aesthetics of form becomes implausible. The affective and the formal dimensions of art come together in the formal (re)presentation of expression.[13] If "the 'subjective' and 'objective,' pure feeling and pure figure [*Gestalt*], merge into one another and acquire precisely in this merging a new, consistent existence and a new content,"[14] the alternative between an aesthetics of experience and one that focuses on works becomes just as problematic since both detachment and involvement are aspects of art. We will return to all these motifs in the section on Langer.

The transition one would expect from art to science is not as clear as the one from myth to art; in fact, it is not spelled out at all. The distinctness of the image grasped and isolated *as image* has a superficial similarity to the purity of scientific symbols that have severed all immediate ties to the world, but we can hardly assume a developmental link from one to the other. Rather, it is *language* that mediates between myth and science, and language is the overarching paradigm of the whole construction, while art is relegated to a peculiar and unclear intermediate status. In the notes on the metaphysics of symbolic forms this is stated unequivocally: "Unity of myth, language, science as moments of cognition—/they erect a world of 'objects'—/the aim at 'reality'—/art does not belong to *this* line—/it stands for itself."[15]

The developmental schema notwithstanding, symbolic forms cannot be ordered into a unilinear process where they neatly follow one another after all: "The totality [*Gesamtheit*] of the possible stages in the objectivization of spirit cannot be projected onto a single straight line without schematic picturing [*Abbildung*] obscuring essential features."[16] The different forms sometimes develop in parallel, intertwine in different ways and never completely supersede one another but continue to exist side by side.

At the end of *Language and Myth* (which Langer translated), Cassirer describes how language gradually disengages itself from myth and moves towards conceptual clarity and constructive freedom. There is a certain ambivalence to this movement: on the one hand there is a gain in availability and flexibility, on the other hand language pays "the price of forgoing the wealth and fullness of immediate experience. In the end, what is left of the concrete sense and feeling content it once possessed is little more than a bare skeleton."[17] This is where art (which in Cassirer's case mostly means poetry, so language once more) comes in, functioning as a counterpoint to rationalized language as an "intellectual realm in which the word not only preserves its original creative power, but is ever renewing it," recovering "the fullness of life," namely an "aesthetically liberated life."[18]

In this perspective, language and art are not merely parallel and independent forms but complementary endeavors that stem from the same source. Image and concept, which are unseparated and inseparable in myth, are differentiated and developed towards greater clarity and consistency. Both imply distance and freedom, but while language relies on standardization and operationalization, art retains the richness of presentation and makes it available for inspection and interpretation. In *An Essay on Man* Cassirer will write: "Language and Science are abbreviations of reality; art is an intensification of reality."[19] But this idea is not systematically developed.

Still, his *Essay on Man* is the only book that devotes some systematic attention to art while all but abandoning the grand system that he attempted to construct again and again. The topics now are myth and religion, language, art, history, and science. As usual, myth and science are situated at systematic opposites, but there is no attempt to find a specific order for the forms that lie in between (except for the idea that language/science and art are complementary opposites, which is only briefly mentioned in the chapter on art). Cassirer is still looking for the "unity of the *creative process*" but the individual forms are now merely regarded as "so many variations on a common theme."[20] The fact that the purportedly generic title of the book really only refers to roughly one half of humanity may serve as a general reminder that he was writing in a time where it seemed easier to postulate such a unity without having to accommodate potentially irreconcilable differences.

It is rather interesting to see that we find characterizations of art that could be direct quotes from Langer, as when he writes that art expresses "the dynamic process of life itself."[21] Even though this goes together well with his own theory and Langer is not explicitly mentioned, we might assume a mutual influence—the book was published two years after *Philosophy in a New Key* and three years after the two first met in person.[22]

The final chapter is marked by indecision: culture is described as "the process of man's progressive self-liberation,"[23] with language, art, religion, and science (surprisingly, myth and history are dropped) as "phases" of this process. Unfortunately, there is no hint as to how we should conceive this phase model, and it is not pursued any further. Finally, all we have are "different directions" obeying "different principles," "multiplicity and disparateness," and a vague appeal to the complementarity of all the different symbolic forms: "Each one opens a new horizon and shows us a new aspect of humanity."[24] The earlier assertion that "each becomes what it is only by demonstrating its own peculiar force against the others and in a struggle with them"[25] has given way to an image of peaceful coexistence where all we can say is that they somehow complement one another.

All this points to the conclusion that Cassirer's writings on art as a symbolic form remained sketchy precisely *because* he could find no systematic place for it. It comes as no surprise that Guido Kreis, whose book is committed to the systematic side of Cassirer's thought, all but ignores art.[26] We might ask whether the obligation to devise a system in itself might not obstruct the detailed investigation of art as a symbolic form and whether art might not resist being included into a developmental schema even if we can clearly differentiate it from language, myth, and science. Most of all, however, it is the unchallenged position of language at the center of the whole system that precludes this: in an "allness" that is constructed from and around language, art cannot find its place.

And finally: any theory that wants to really understand art as a human practice will have to go beyond a generic concept of art as such and also beyond the analysis of individual examples, which remain contingent even if their authors' names are Goethe or Shakespeare, and look at the arts in the plural. It will have to investigate their specific conditions of existence, their media, and their place in society. In fact, Cassirer shares this conviction: "The purpose of one art lies in that, which it is capable of by virtue of its specific signs, not in that, which other arts are capable of just as well or even better."[27] Despite this assertion he never really pursued this, but the quote does give a succinct description of Langer's approach, which I will turn to now.

(RE)PRESENTATION IN A NEW KEY

As is well known, *Philosophy in a New Key*, her philosophical bestseller, marked a turning point in Langer's thought: whereas her previous perspective on the history of philosophy had been informed by her training as a logician, she now turned to developing a more general theory of symbols and moved art into the center of her work. Even though the book contains only two chapters on the topic, one on music and one on art in general, the statement I quoted at the beginning makes clear that it already occupies a key position. The new mode of doing philosophy, she observes, is partly new for her as well, and her extensive readings lead her to a group of fairly heterogeneous philosophies of symbol, meaning, and/or semantics—from Frege through early Wittgenstein to Cassirer—but also to developmental psychology, biology, anthropology, and art theory. The study is centered around ritual and myth, language, and art, topics we are familiar with from Cassirer. Science only appears at the margins.

Langer's systematic point is not to produce a grand overview or system, or a systematic genealogy of symbolic forms; the main contribution of the book is the exploration of the complementary character of language and art we know from Cassirer's later writings. Langer goes one step further from distinguishing the two as symbolic forms to assigning them complementary logical *types* of symbols: discursive and presentational, which cannot be brought into a hierarchical or developmental order. Introducing the latter may be considered the true core of the book, and this is what she is primarily known for today. Discursive symbols are discrete, linear, systematically ordered, and generalize, they develop a vocabulary and can be translated, while presentational symbols are holistic, concrete, linked to the singular but articulated in a sophisticated way. In this way, the inner structure of the symbols becomes just as important as their functions.[28] Her approach is that of the semiotician rather than the creator of a grand philosophy of *Geist*, working bottom-up, as it were, instead of top-down.

She counters the prevalent focus on the discursive symbolisms of language and logic with an emphasis on presentational symbols and on art. But even before this, the functional nature of all meaning is illustrated by referring to a musical chord.[29] The shift from societies organized by ritual and consciousness dominated by myth, both of which contain discursive and presentational elements, to the "realism" prevalent today is the only historical development that we are presented with, and even this is not unambiguous: the alienation produced by rationalization has yet to be integrated into a new coherent worldview, which will necessarily contain elements of ritual and condensing of meaning reminiscent of mythical thinking.

Despite the different approaches her aim, like Cassirer's, is a non-reductive theory of the human mind that does not restrict rationality to that which is or can be linguistically articulated but includes all areas of experience and investigates their specific modes of articulation. We might say that she already recognizes the incipient linguistic turn as reductive and seeks to expand it into a fully developed symbolic turn. Hence, one of the crucial questions reads: "*Just how* can feelings be conceived as possible ingredients of rationality?"[30] Langer's answer is: by recognizing them as articulated modes of experience and by analyzing art and especially music as that cultural sphere that is dedicated to observing, (re)presenting, and developing this articulation.

This may serve as an explanation of the book's title, which describes the symbolic turn as a change of key. Interestingly, the title of the German translation completely ignores this reference to music and vaguely speaks of "philosophy on a new path."[31] Presumably the translator Ada Löwith did not trust the musical metaphor; possibly she thought it was too specific or would not be understood. I think this is a grave mistake because it obscures the fact that the experience of music forms the background for this philosophy as a whole. It makes sense to spell out the metaphor: a key is a systematic connection of tones that is the foundation for specific pieces. A change of key cannot be compared to a translation or a switching of languages. Rather, it is an internal shift, a restructuring that shapes every single event or lets it appear in a new light. One of the central concepts of the book remains that of *form*, and following gestalt theory an "unconscious appreciation of forms" is assumed to be "the primitive root of all abstraction, which in turn is the keynote of rationality"[32]—another musical metaphor. For Langer, the systematic neglect of art in Cassirer must be countered by its systematic distinction. In addition, music replaces literature as the primary point of reference.[33]

This orientation is crucial for *Philosophy in a New Key* as a whole, and it ultimately leads to a different type of philosophy that hasn't abandoned systematic aspirations but jettisons Cassirer's Hegelian background. The following book, *Feeling and Form*, is exclusively dedicated to art. There are strong indications that it must be placed at the center of Langer's oeuvre as a whole: in it, she calls the earlier book "a promise of a philosophy of art,"[34] i.e., its own prehistory, while the monumental *Mind* is described as a "philosophy of life guided by the vital image created by artists,"[35] i.e., as its development, which shifts the focus and expands the frame. To be able to do this, she transforms Cassirer's systematic approach so that and the great systematic questions somewhat recede into the background.

On the other hand she places her own philosophy more directly in Cassirer's succession: in the preface to *Philosophy in a New Key*, Whitehead, Cassirer, Schenker, Reid, Goldstein, "and many others"[36] are referred to as predecessors whose occasional mentions do not mirror their actual influence; *Feeling and Form* is dedicated to Cassirer's memory; *Mind* looks back at this book and squarely situates it in Cassirer's tradition: "There I have treated the arts in some detail under the rubric of non-discursive and non-systematic symbolic expression, an epistemological concept proposed and developed by Cassirer in his *Philosophy of Symbolic Forms*."[37] As we have seen, this is not really true for Cassirer: the philosophy of symbolic forms may provide the framework for introducing such a non-discursive and non-systematic type of symbols, but it does not actually do this. That was to be Langer's part.

All this is not to say that *Feeling and Form* does not raise systematic claims—its aim is "the construction of an intellectual framework for philosophical studies, general or detailed, relating to art."[38] The crucial question, however, is how this structure is to be arrived at. There are three main points: firstly, her goal is not a discussion of the prevalent general concepts but of specific problems and questions in the various arts from whose examination general concepts should be derived or transformed. Her criterion of any general concept is the question whether and to what extent it can inspire further research and discussions.[39] Secondly, Langer's starting point for a philosophy of art is the artist's studio. It aims at reflectively elucidating the perspective of the artists without losing track of the recipients. She calls precisely for that which Cassirer achieved for the sciences, especially physics, and could not do for the arts: "The philosopher must know the arts, so to speak, 'from the inside.'"[40] She is well aware that this is a formidable task that will hardly be accomplishable in its entirety. It does make a difference, however, whether artistic practice or its philosophical discussion are used as the main point of reference. Thirdly, and as a consequence of the first two points, the unity of art cannot simply be presupposed. The primary task is "to examine the differences, and trace the distinctions among the arts as far as they can be followed." Unity remains a regulative idea for this examination: "But there is a definite level at which no more distinctions can be made; everything one can say of any single art can be said of any other as well. *There lies the unity*."[41]

Still, one might doubt whether this presumed unity will really only be reached after close scrutiny of all artistic disciplines. After all, Langer leaves no doubt concerning her conviction about the crucial role of feeling for *all* art, or of art for the articulation and explication of feeling: "Art is the creation of forms symbolic of human feeling."[42] These forms disclose reality: our affectively charged, temporally articulated experience is made accessible and clarified by them instead of just being confirmed. They do this as a type of (re)presentation, so that "what art expresses is *not* actual feeling, but ideas of feeling."[43]

This conception clearly has its origin in the theory of musical signification, which has been generalized to encompass all art(s). *Philosophy in a New Key* paved the way for this move, although Langer was still slightly hesitant: "I strongly suspect, though I am not ready to assert it dogmatically, that the import of artistic expression is broadly the same in all arts as it is in music—the verbally ineffable, yet not inexpressible law of vital experience, the pattern of affective and sentient being."[44] Music provides the paradigm for this theory of art, and the differentiation between the different arts takes place within this paradigm.

What we also see is generalization from *feelings* to *feeling*, which Donald Dryden explores in detail in his chapter. In *Philosophy in a New Key*, the aim was to counter a conception of mind that reduced it to discursive reasoning and language by referring to an inner life of experience that already exceeds emotionality in the narrow sense. In the following books and papers, this concept is successively expanded until it is finally understood "in the widest sense, including emotions, sensations and even thoughts,"[45] encompassing the human mind as a whole from the perspective of processual vitality. The shift from thinking to feeling is somewhat analogue to the shift from language to art: switching points of reference like this is an attempt to view rationality in all its realizations and expressions in a different light because "art has no ready-made symbols or rules of their combination, it is not a symbolism, but forever problematical, every work being a new and, normally, entire expressive form."[46] If concrete, nuanced, meaningful, problematic articulation is placed at the center of attention instead of rule-based, general, referential linguistic form, tracing the process of differentiating and articulating the world appears more meaningful than cataloging its forms because such a catalog can never be complete or final—a lesson we might also draw from Cassirer's various attempts of producing such a complete register and his final renunciation in *An Essay on Man*·

Only a close examination of actual figurations will teach us what feeling can encompass, and for this we will have to turn to the various arts within the paradigm provided by music as the clearest example of a purely relational energetic organization, a topic that Lona Gaikis further explores in her chapter. In order to distinguish the arts, Langer introduces the concepts of virtuality and the primary illusions that are specific to each of them: visual art—or rather the visual arts—is described as virtual space, music as virtual time, dance as a gestural play of virtual forces, and literature as virtual life. This approach may seem rather formal, especially in the first two cases, but formal structure is always related back to a type of energetic organization that activates form or is its very mode of existence. Pure musical or pictorial form is thus never really pure but related to an energetic and affective process. As Langer notes, relying on numerous statements by artists, what counts is the "'life' of forms,"[47] not their putative purity. What this "life" consists of and how it unfolds must be observed individually, and hence the constant reference to feeling does not imply a kind of standardization or reduction of formal possibilities but rather its multiplication.

Her unfinished *magnum opus*, *Mind*, takes this life of forms as its starting point to perform an operation that remains unusual even today: in its systematic reconstruction of human mentality from elementary biological functions, it proposes "going back to the beginnings of thought about mental phenomena and starting with different ideas, different expectations, without concern for experiments or statistics or formalized language"[48] to escape the "idols of the laboratory." In order to do this, we need new images because "only an image can hold us to a conception of a total phenomenon, against which we can measure the adequacy of the scientific terms wherewith we describe it."[49] Rather than being presented as one of the outgrowths of the evolution of mind, art—now treated "as a unitary phenomenon"[50]—is used as a reservoir of such images, images that allow us to think the continuity between biological life and human mentality without negating the rupture that the emergence of the human mind entailed. Langer's crucial concept of the "act" as the basic unit in the formation of life would not be possible without such images.

Elaborating a naturalistic but non-reductive theory of the development of the human mind is closer to current debates than to Cassirer's neo-Kantian background, and he hardly would have agreed with Langer's conclusions.[51] For her, this move does not mean abandoning the idea of symbolic forms but providing them with a foundation in nature. If we look back at the development of her thinking as a whole and compare it to Cassirer's, we see a gradual movement of extension and generalization—the grand gesture that came naturally to Cassirer was clearly not her thing. She opens perspectives and deepens our understanding of the intricate workings of human culture and mind without filling in all the gaps, thus leaving us room to think. None of this could have been done without having one specific symbolic form as background and resource: art.

NOTES

1. Ernst Cassirer, *The Philosophy of Symbolic Forms*, vol. 1: *Language*, 1923 (London/New York: Routledge, 2021), 12.
2. Susanne K. Langer, *Philosophy in a New Key: A Study in the Symbolism of Reason, Rite, and Art* (Cambridge, MA: Harvard University Press, 1942), 219.
3. Guido Kreis, *Cassirer und die Formen des Geistes* (Berlin: Suhrkamp, 2010), 11. [All translations by the author.]
4. Cf. Marion Lauschke, *Ästhetik im Zeichen des Menschen. Die ästhetische Vorgeschichte der Symbolphilosophie Ernst Cassirers und die symbolische Form der Kunst* (Hamburg: Meiner, 2007). While Lauschke's detailed investigation reveals the richness of Cassirer's writings on art, it also shows that such a systematic reconstruction requires a great effort of collecting motifs and traces from all of Cassirer's works. Katharine Gilbert's early essay ("Cassirer's Placement of Art," in *The Philosophy of Ernst Cassirer*, ed. Paul Arthur

Schilpp [Evanston, IL: The Library of Living Philosophers, 1949], 605–30) fails to capture the complexity of the issue.

5. Cf. Susanne K. Langer, "On a New Definition of 'Symbol,'" in *Philosophical Sketches* (Baltimore: Johns Hopkins University Press, 1962), 54–65, 58.
6. Dengerink Chaplin, who devotes a whole chapter to Cassirer, exclusively focuses on the similarities between the two thinkers, yet disregards their differences: cf. Adrienne Dengerink Chaplin, *The Philosophy of Susanne K. Langer: Embodied Meaning in Logic, Art and Feeling* (London: Bloomsbury, 2020), ch. 5.
7. Susanne K. Langer, *Feeling and Form: A Theory of Art Developed from Philosophy in a New Key* (New York: Charles Scribner's Sons, 1953), 410.
8. Dirk Rustemeyer, *Diagramme. Dissonante Resonanzen: Kunstsemiotik als Kulturtheorie* (Weilerswist: Velbrück Wissenschaft, 2009), 51.
9. Ernst Cassirer, *The Philosophy of Symbolic Forms,* vol. 3: *Phenomenology of Cognition,* 1929 (London/New York: Routledge, 2021), 1.
10. This somewhat mannered translation of the German term "Darstellung," which is particularly important in Cassirer's philosophy, is meant to convey that it can be understood both ways, as *presentation* and *representation*, the point being that the two cannot be clearly separated in this volume. Rolf Lachmann uses the same term with a slightly different meaning in Chapter 6 of this volume.
11. Cassirer, *The Philosophy of Symbolic Forms,* vol. 3, 94.
12. Ernst Cassirer, *The Philosophy of Symbolic Forms,* vol. 2: *Mythical Thinking*, 1925 (London/New York: Routledge, 2021), 30.
13. Cf. Ernst Cassirer, "Das Symbolproblem im System der Philosophie" (1927), in *Aufsätze und kleine Schriften 1927–1931* (ECW 17) (Hamburg: Meiner, 2004), S. 253–82, 267–8. Here, Konrad Fiedler and Benedetto Croce are positioned as complementary thinkers (cf. Lauschke, *Ästhetik im Zeichen des Menschen*, 218–22).
14. Cassirer, *Philosophy of Symbolic Forms,* vol. 1, 24.
15. Ernst Cassirer, *Zur Metaphysik der symbolischen Formen* (ECN 1) (Hamburg: Meiner, 1995), 246.
16. Cassirer, *Philosophy of Symbolic Forms,* vol. 3, 62.
17. Ernst Cassirer, *Language and Myth* (New York: Harper and Brothers, 1946), 98.
18. Cassirer, *Language and Myth*, 98.
19. Ernst Cassirer, *An Essay on Man: An Introduction to a Philosophy of Human Culture* (New Haven, CT: Yale University Press, 1944), 143.
20. Cassirer, *An Essay on Man*, 70.
21. Ibid., 190.
22. As Dengerink Chaplin remarks, even the structure of *An Essay on Man* mirrors that of *Philosophy in a New Key* (cf. *The Philosophy of Susanne K. Langer*, 72–3).
23. Cassirer, *An Essay on Man*, 286.
24. Ibid.
25. Cassirer, *Philosophy of Symbolic Forms,* vol. 1, 11.
26. Kreis himself describes this difficulty as follows: "The main problem is that it remains unclear how the other symbolic forms are to be integrated into this system—or whether the systematic treatment of art, technology, law and morality does not call for a completely different systematic construction." (Kreis, *Cassirer und die Formen des Geistes*, 404).
27. Cassirer, "Das Symbolproblem im System der Philosophie," 270.

28. Cf. Langer, *Philosophy in a New Key*, ch. 4; Rolf Lachmann, *Susanne K. Langer. Die lebendige Form menschlichen Fühlens und Verstehens* (Munich: Fink 2000), 62–78. Nelson Goodman later developed this idea into the concept of (metaphoric) exemplification, albeit without explicit reference to Langer: cf. Nelson Goodman, *Languages of Art: An Approach to a Theory of Symbols* (Indianapolis, IN/Cambridge, MA: Hackett, 1976); Katrin Eggers, "'The Matrix of Mentality.' Susanne K. Langers Symboltheorie der Musik in Abgrenzung zu Nelson Goodman," *Musik & Ästhetik* 53 (2010): 20–36.
29. Cf. Langer, *Philosophy in a New Key*, 55–6.
30. Ibid., 100.
31. Susanne K. Langer, *Philosophie auf neuem Wege: Das Symbol im Denken, im Ritus und in der Kunst* (Frankfurt a.M.: Fischer 1965).
32. Langer, *Philosophy in a New Key*, 89.
33. This leads to a situation where Langer's aesthetics is referenced—and criticized—almost exclusively within the philosophy of music. In their criticism most authors all but ignore the larger context of her philosophy of symbols (cf., among many others, Malcolm Budd, *Music and the Emotions: The Philosophical Theories* (London/New York: Routledge, 1992), ch. 6; Stephen Davies, *Musical Meaning and Expression* (Ithaca, NY/London: Cornell University Press, 1994), 123–33).
34. Langer, *Feeling and Form*, xii.
35. Susanne K. Langer, *Mind: An Essay on Human Feeling*, vol. 1 (Baltimore: Johns Hopkins University Press, 1967), xix.
36. Langer, *Philosophy in a New Key*, ii.
37. Langer, *Mind*, vol. 1, xv.
38. Langer, *Feeling and Form*, ix.
39. Cf. ibid., 7–11.
40. Ibid., ix.
41. Ibid., 103.
42. Ibid., 40. This sentence, which sums up her philosophy of art as a whole, also clearly points to all the misunderstandings it gives rise to—after all, this is neither a theory that takes emotions as its main point of reference nor a semiotic aesthetics that conceives aesthetic experience as a kind of decoding of signs.
43. Ibid., 59.
44. Langer, *Philosophy in a New Key*, 257.
45. Langer, *Mind*, vol. 1, 5.
46. Ibid., 81.
47. Ibid., 79.
48. Langer, *Mind*, vol. 1, 52.
49. Ibid., xviii.
50. Ibid., 74.
51. For Cassirer, a naturalistic approach would have entailed a causal explanation of symbolic forms, which would have been tantamount to a *reduction* of mind to natural processes and would ultimately have led to a circular argument where the very categories that were to be explained would be used in the explanation. Langer disagreed on both counts (cf. Lachmann, *Susanne K. Langer*, 109–11).

BIBLIOGRAPHY

Budd, Malcolm. *Music and the Emotions: The Philosophical Theories*. London/New York: Routledge, 1992.

Cassirer, Ernst. *An Essay on Man. An Introduction to a Philosophy of Human Culture*. New Haven, CT: Yale University Press, 1962.

Cassirer, Ernst. *Language and Myth*. New York: Harper and Brothers, [1925] 1946.

Cassirer, Ernst. "Das Symbolproblem im System der Philosophie." In *Aufsätze und kleine Schriften 1927–1931* (ECW 17), 253–82. Hamburg: Meiner, [1927] 2004.

Cassirer, Ernst. *The Philosophy of Symbolic Forms,* vol. 1: *Language*. London/New York: Routledge, [1923] 2021.

Cassirer, Ernst. *The Philosophy of Symbolic Forms,* vol. 2: *Mythical Thinking*. London/New York: Routledge, [1925] 2021.

Cassirer, Ernst. *The Philosophy of Symbolic Forms,* vol. 3: *Phenomenology of Cognition*. London/New York: Routledge, [1929] 2021.

Cassirer, Ernst. *Zur Metaphysik der symbolischen Formen* (ECN 1). Hamburg: Meiner, 1995.

Davies, Stephen. *Musical Meaning and Expression*. Ithaca, NY/London: Cornell University Press, 1994.

Dengerink Chaplin, Adrienne. *The Philosophy of Susanne K. Langer: Embodied Meaning in Logic, Art and Feeling*. London: Bloomsbury, 2020.

Eggers, Katrin. "'The Matrix of Mentality Susanne K. Langers Symboltheorie der Musik in Abgrenzung zu Nelson Goodman." *Musik & Ästhetik* 53 (2010): 20–36.

Gilbert, Katharine. "Cassirer's Placement of Art." In *The Philosophy of Ernst Cassirer*, edited by Paul Arthur Schilpp, 620–4. Evanston, IL: The Library of Living Philosophers, 1949.

Goodman, Nelson. *Languages of Art: An Approach to a Theory of Symbols*. Indianapolis, IN/Cambridge, MA: Hackett, 1976.

Kreis, Guido. *Cassirer und die Formen des Geistes*. Berlin: Suhrkamp, 2010.

Lachmann, Rolf. *Susanne K. Langer. Die lebendige Form menschlichen Fühlens und Verstehens*. Munich: Fink, 2000.

Langer, Susanne K. *Feeling and Form: A Theory of Art Developed from Philosophy in a New Key*. New York: Charles Scribner's Sons, 1953.

Langer, Susanne K. *Mind: An Essay on Human Feeling*, vol. 1. Baltimore: Johns Hopkins University Press, 1967.

Langer, Susanne K. "On a New Definition of 'Symbol.'" In *Philosophical Sketches*, 54–65. Baltimore: Johns Hopkins University Press, 1962.

Langer, Susanne K. *Philosophie auf neuem Wege: Das Symbol im Denken, im Ritus und in der Kunst*. Frankfurt a.M.: Fischer, 1965.

Langer, Susanne K. *Philosophy in a New Key: A Study in the Symbolism of Reason, Rite, and Art*. Cambridge, MA: Harvard University Press, 1942.

Lauschke, Marion. *Ästhetik im Zeichen des Menschen. Die ästhetische Vorgeschichte der Symbolphilosophie Ernst Cassirers und die symbolische Form der Kunst*. Hamburg: Meiner, 2007.

Rustemeyer, Dirk. *Diagramme. Dissonante Resonanzen: Kunstsemiotik als Kulturtheorie*. Weilerswist: Velbrück Wissenschaft, 2009.

BIBLIOGRAPHY

CHAPTER NINE

The Meaning of "feeling" in Susanne K. Langer's Project of *Mind*

DONALD DRYDEN

INTRODUCTION: WHAT'S IN A NAME?

Susanne Langer's use of the term "feeling," as it appears in the title of *Mind: An Essay on Human Feeling* and figures throughout all three volumes of that work, is the potential source of a fundamental misunderstanding of her entire project. Langer herself, however, bears some responsibility for this problem, for she used the general term "feeling" or "feelings" in connection with the meaning of art in all of her writings, often interspersed with references to "emotions," "moods," and "passions," as well as to "affective," "impulsive," "instinctive," and "sentient" life. This, of course, is entirely in keeping with one common meaning—given in the *Oxford English Dictionary*—of "feeling" as "a generic term comprising sensation, desire, and emotion, but *excluding perception and thought*" (emphasis added).[1]

By the time Langer wrote *Feeling and Form* (1953), however, she was using a much wider—though still largely implicit—definition of "feeling" that anticipated the project of *Mind* in applying the term to "*our whole subjective reality*, woven of thought and emotion, imagination and sense perception" (emphasis added).[2] Langer's definition was not without precedent, for at least two nineteenth-century philosophers also used "feeling" as a generic term for conscious experiences, including perception and thought. In his *Lectures on the Philosophy of the Human Mind* (1820), Thomas Brown noted that "consciousness" "is only a general term for all our feelings, of whatever species these may be—sensations, thoughts, desire—in short, all those states or affections of mind in which the phenomena of mind consist."[3] John Stuart Mill, in *A System of Logic* (1843), had similarly declared that,

> [A] Feeling and a State of Consciousness are, in the language of philosophy, equivalent expressions: everything is a Feeling, of which the mind is conscious: everything which it feels, or, in other words, which forms a part of its own sentient existence … Feeling, in the proper sense of the term, is a genus, of which Sensation, Emotion, and Thought, are subordinate species.[4]

Mill also noted a common meaning of "feeling" "in popular language" that restricted it to "sensitive" or "emotional" experience, but he stated that "this is an admitted departure from correctness of language."[5] By the middle of the twentieth century, however, this "popular" meaning had all but eclipsed the wider definition, which was largely forgotten; and Langer's failure to address this problem further contributed to the misunderstanding of her writings on art and mind.

In order to understand the meaning that Langer came to attach to the term "feeling," therefore, we have to go back to those earlier thinkers who acknowledged the problem of choosing a generic term for what Thomas Brown had called "all those states or affections of mind in which the phenomena of mind consist."[6] Among those who wrote about this question, the one who holds the key to understanding Langer's use of "feeling" is William James. In his early writings on psychology near the end of the nineteenth century, James discussed the problem of finding "some general term by which to designate all states of consciousness merely as such, and apart from their particular quality or cognitive function."[7] In contrast to both Brown and Mill, James rejected "feeling" as a candidate for that purpose, and he initially chose "thought" and "thinking" as his preferred generic terms. In fact, Langer makes a brief reference to James when she states, in a footnote early in the first volume of *Mind*, that "William James ... used 'thinking' in the sense in which I use 'feeling'";[8] but she never explained the wider context in which James made his decision.

How James thought about this problem and how he solved it will help us understand that Langer, too, would have faced a similar problem from the time she began the work that became *Feeling and Form* (1953). Until she had faced the difficulties of developing a detailed and comprehensive theory of the arts, she was able to get by with using the term "feeling" in the context of suggestive but indefinite generalities when discussing meaning in the arts. But a deepening understanding of the meaning of art in general, and the several orders of art in particular, would have forced her to recognize the need for a more careful vocabulary, including a generic term for what James had called "mental states at large, irrespective of their kind."[9] From her writings it is evident that she chose "feeling" for that role, but she never explicitly discussed the problem nor the reasons for her choice, as James had done at some length in *The Principles of Psychology* (1890).

In what follows, I will take a closer look at James's discussion of the difficulties he encountered in trying to decide on a generic term for conscious experiences, and how his solution to the problem continued to develop during the ten years following *The Principles of Psychology* (1890). With James's deliberations as a point of reference, I will follow the development of Langer's understanding of art and feeling until, with *Feeling and Form* (1953), she arrived at an understanding of feeling that prepared the way for the project of *Mind*.

WILLIAM JAMES ON NAMING STATES OF CONSCIOUSNESS

The clue Langer offers us for understanding the concept of feeling that underlies the entire *Essay* appears in a footnote in the first volume of *Mind*. Referring to *The Principles of Psychology*, Langer writes:

> William James ... used "thinking" in the sense in which I use "feeling." ... Remarking on the difficulty of finding a generic term for "mental states at large, irrespective of their kind," he said: "My own partiality is for either FEELING or THOUGHT," but he finally rejected "feeling" because it had kept particularly bad company in his day.[10]

This brief quotation gives us only the conclusion of a much longer discussion in which James presented the dilemma he found himself facing, the various solutions he considered and rejected, and his reasons for reaching the decision Langer cites in her footnote. Without this larger background, however, few readers will be able to understand just what James—and therefore Langer herself—had in mind, for all the considerations that accompanied his deliberations are omitted from Langer's footnote, which simply states that James "used 'thinking' in the sense in which I use 'feeling.'"[11] But why did James feel the need for a term to which "FEELING or THOUGHT" provided the best solution? What other candidates did he consider, and why did he reject them? Most importantly, what does this wider background tell us about what James took to be the subject matter of psychology? By looking at the context of Langer's brief quotation, we should be able to learn more about how Langer herself thought about these questions, how she came to understand the term "feeling," and what this might tell us about the larger work she subtitled *An Essay on Human Feeling*.

In *The Principles of Psychology* (1890) William James defines psychology as "the Science of Mental Life," the phenomena of which include "such things as we call feelings, desires, cogitations, reasonings, decisions, and the like."[12] He realizes, however, that any writer who deals with this wide variety of mental phenomena faces a problem: the lack of a suitable generic term that refers to the various mental states indiscriminately, much as biology has the term "plant" to refer to "plants in general" but not to any one kind of plant in particular.

As James put it: "We ought to have some general term by which to designate all states of consciousness merely as such, and apart from their particular quality or cognitive function."[13] He considers a number of candidates for this role but finds problems with each of them:

> "Mental state," "state of consciousness," "conscious modification," are cumbrous and have no kindred verbs. The same is true of "subjective condition." "Feeling" has the verb "to feel," both active and neuter, and such derivatives as "feelingly," "felt," "feltness," etc., which make it extremely convenient. But on the other hand it has specific meanings as well as its generic one, sometimes standing for pleasure and pain, and being sometimes a synonym of "*sensation*" as opposed to *thought*; whereas we wish a term to cover both sensation and thought indifferently. [James's emphasis][14]

He also notes, without further explanation, that the term "feeling" had acquired a very "opprobrious connotation."[15]

Finally, James considers the term "thought," which he says "would be by far the best word to use," provided, however, "it could be made to cover sensations." Thought "immediately suggests the omnipresence of cognition (or reference to an object other than the mental state itself), which we shall soon see to be of the mental life's essence." But there is still the problem of extending "thought" to cover *sensations*: "Can the expression 'thought of a toothache' ever suggest to the reader the actual present pain itself? It is hardly possible."[16]

Every term, it seems, has disadvantages, and "in this quandary," James concludes, "we can make no definitive choice, but must, according to the convenience of the context, use sometimes one, sometimes another of the synonyms that have been mentioned. *My own partiality is for either FEELING or THOUGHT*. I shall probably often use both words in a wider sense than usual," leaving it to the reader to let the context determine when "mental states at large, irrespective of their kind, are meant" [James's emphasis].[17]

Later in the book, James opens what is probably the most famous chapter in *The Principles*, "The Stream of Thought," by announcing that he will be presenting what we would now call a phenomenology of conscious experience: "We now begin our study of the mind from within," using "the empirical method of investigation,"[18] which starts with what we actually find when we examine our own lived experience using as few preconceptions as possible: "Consciousness ... is of a teeming multiplicity of objects and relations"; and "the only thing which psychology has a right to postulate at the outset is the fact of thinking itself ... *The first fact for us, then, as psychologists, is that thinking of some sort goes on*" (James's emphasis).[19] And here he reminds us: "I use the word '*thinking*,' in accordance with what was said on p. 186, for *every form of consciousness indiscriminately*" (James's emphasis).[20]

James argues that this "ongoingness" of conscious experience is "sensibly continuous," which he defines as being "without breach, crack, or division"; and he adds that conscious experience feels continuous because "the changes from one moment to another in the quality of the consciousness are never absolutely abrupt"; and even where there is a time gap, as during sleep, the consciousness that reappears after the interruption still "feels as if it belonged together with the consciousness before it, as another part of the same self."[21]

James concludes his discussion of the "sensible continuity" of conscious experience with the metaphor that gives the chapter its title:

> Consciousness, then, does not appear to itself chopped up in bits. Such words as "chain" or "train" do not describe it fitly as it presents itself in the first instance. It is nothing jointed; it flows. A "river" or a "stream" are the metaphors by which it is most naturally described. *In talking of it hereafter, let us call it the stream of thought, of consciousness, or of subjective life.* [James's emphasis][22]

In preparing an abridged version of *The Principles of Psychology*, which he titled *Psychology: Briefer Course* (1892), James made some changes to key passages in the text. To begin with, he redefined psychology as "*the description and explanation of states of consciousness as such*" (James's emphasis);[23] and in a number of other places throughout the book he replaced "thought" and "thinking" with "consciousness" and related terms. He no longer explicitly discussed the problem of choosing a generic term for "conscious states at large," mentioning only that, among the data of psychology, are "*thoughts and feelings*, or whatever other names transitory *states of consciousness* may be known by" (James's emphasis).[24] James also revised the title of the best-known chapter in the book, renaming it "The Stream of Consciousness" and emphasizing that what we find when we turn our attention of our own lived experience is "that *consciousness of some sort goes on*": "States of mind succeed each other ... Now we are seeing, now hearing, now reasoning, now willing; now recollecting, now expecting; now loving, now hating; and in a hundred other ways we know our minds to be alternately engaged" (James's emphasis).[25]

In his *Talks to Teachers on Psychology* (1899), seven years later, James revisited the same material once again; and this time he was even more succinct and to the point:

> The *immediate* fact which psychology, the science of mind, has to study is also the most general fact. It is the fact that in each of us, when awake (and often when asleep), *some kind of consciousness is always going on*. There is a stream, a succession of states, or waves, or fields (or of whatever you please to call them), of knowledge, of feeling, of desire, of deliberation, etc., that constantly pass and repass, and that constitute our inner life. The existence of this stream is the primal fact, the nature and origin form the essential problem, of our science. [James's emphasis][26]

These "fields of consciousness," James continued, "are always complex," for most of them contain "sensations of our bodies and of the objects around us, memories of past experiences and thoughts of distant things, feelings of satisfaction and dissatisfaction, desires and aversions, and other emotional conditions, together with determinations of the will, in every variety of permutation and combination." Furthermore, "in most of our concrete states of consciousness all these different classes of ingredients are found simultaneously present to some degree, though the relative proportion they bear to one another is very shifting."[27]

In the first volume of *Mind*, therefore, when Langer states that "William James ... used 'thinking' in the sense in which I use 'feeling,'"[28] she is affirming unequivocally that her definition of "feeling" is equivalent to 'thought' as James used it in *The Principles*—as a generic term for conscious experiences, which she had described in *Feeling and Form* as "our whole subjective reality, woven of thought and emotion, imagination and sense perception."[29] As the psychologist Carroll Pratt stated in his review of the first volume of *Mind*: "Mind is feeling, and for Langer feeling would seem to be coextensive with what an older psychology called consciousness."[30] Pratt's reference to "an older psychology," is especially ironic, given psychology's "rediscovery" of consciousness twenty-five years later, in the 1990s.

THE DEVELOPMENT OF LANGER'S UNDERSTANDING OF ART AND FEELING

A review of Langer's published writings shows that her understanding of the meaning of both art and feeling remained largely underdeveloped before *Feeling and Form* (1953).[31] Even in *Philosophy in a New Key* (1942), which offered some of the resources for a more fully developed philosophy of art, Langer confined much of her discussion of art to significance in music and made no mention of the additional meanings that the term "feeling" would eventually come to encompass in her later work. It was not until *Feeling and Form* that Langer finally reached her clearest, most stable, and comprehensive formulation of meaning in the arts, while at the same time consolidating her understanding of the broad scope of the term "feeling" that she would later use to define the purpose of *Mind*, as indicated by its subtitle, *An Essay on Human Feeling*.

The Practice of Philosophy (1930) was the first published work in which Langer developed the idea that the key to understanding human mentality and all of its cultural expressions lies in our capacity for apprehending *forms* or patterns in the material furnished by experience, and using these patterns to find *analogies*, whereby we come to understand one thing in terms of another. The recognition of analogy—or the correspondence of configuration between patterns—is, in turn, the basis of the human capacity for *symbolization*.[32] Here Langer drew upon her reading of Cassirer's *Philosophie der symbolischen Formen*, the first volume of which had appeared in German in 1923 (thirty years before it was translated into English), and which she had already cited in her dissertation.[33]

It was Cassirer's work that suggested the general thesis that abstractions, and the patterns they make available to human understanding, are dependent on the symbolic materials that are used to formulate them, and that the possibilities for the appreciation of patterns must include far more than what is available through the resources of language alone—even if language is broadly defined to include the notational systems of formal logic and mathematics. In *The Practice of Philosophy*, Langer applied this fundamental insight to the further development of her understanding of art. The capacity of the human mind to apprehend patterns in experience must extend to configurations that are too intricate to be adequately expressed in the medium of language. People who are responsive to the arts, for example, live "through the eye, the musical hearing, the bodily senses," and "see more meaning in artistic wholes, i.e., in things, situations, feelings, etc., than they can ever find in propositions."[34] If we define reason and logical insight in the broadest sense as the appreciation of patterns, then the apprehension of artistic significance is no more "irrational" or "illogical" than the process of understanding propositional knowledge.

Of particular interest here, however, is that Langer has almost nothing specific to say about what the "non-discursive"[35] patterns embodied in works of art might be giving us knowledge *of*. In a painting, for example, she states that "the balance of values, line and color and light, and I know not what other elements, is so highly adjusted that no verbal proposition could hope to embody its pattern."[36] As for the meaning of the pattern, however, Langer says only that the painting enables us to apprehend "equally indescribable meanings in our world of experience."[37] But *what kinds* of experience? In answer to this question, she offers only a tentative, speculative question: "Could it be that the final object of musical expression is the endlessly intricate and universal pattern of emotional life?"[38] This is as far as she was willing to go in *The Practice of Philosophy*.

In *Philosophy in a New Key* (1942), Langer discusses how the material furnished by our senses, especially the *Gestalten* or fundamental forms of visual and auditory perception, provides the basis for a *presentational* or *non-discursive symbolism* that is "peculiarly well suited to the expression of ideas that defy linguistic 'projection.'"[39] But ideas about *what sorts of things*? Again, Langer answers with another suggestive, speculative question: "May not the order of perceptual forms ... be a possible principle for symbolization, and hence the conception, expression, and apprehension, of impulsive, instinctive, and sentient life? May not a non-discursive symbolism of light and color, or of tone, be formulative of that life?"[40]

Although Langer devotes an entire chapter in *Philosophy in a New Key* to significance in music, she has very little to say, once again, about *what*, specifically, music—or any of the other arts—might *mean*. She does, however, point a number of times in the direction of a vague and largely undifferentiated region of possibilities for what the arts might be *about*. Here are just a few examples. She reiterates the failure of language "to convey the ever-moving patterns, the ambivalences and intricacies of inner experience, the interplay of feelings with thoughts and impressions, memories and echoes of memories, transient fantasy, or its mere runic traces, all turned into nameless, emotional stuff."[41] She calls music the "formulation and representation of emotions, moods, mental tensions and resolutions—a 'logical picture' of sentient, responsive life."[42] She states that "the content of art" is "the life of feeling, impulse, passion"[43] and that music provides "a revelation of emotions, moods, or subtle nameless affects."[44] Langer concludes by asking whether "the realm of emotional experience," which is "the field of musical meanings," is "ultimately the subject-matter of all art."[45]

In *Philosophy in a New Key*, then, Langer's frequent references to "emotions," "moods," "passions," and "subtle nameless affects," as well as to patterns of "affective," "impulsive," "instinctive," and "sentient" life, often interspersed with the more general term "feeling" or "feelings" only reinforce the common association of "feeling" with emotional or affective experience, to the exclusion of perception and thought.

In *Feeling and Form* (1953), Langer finally arrived at a definition of "feeling" as a generic term for conscious experiences, now enriched by an understanding of many of the dimensions of feeling that were missing from her earlier discussions of meaning in art. In the opening chapters, however, we are still in the realm of suggestive but indefinite generalities whenever it comes to the meaning of art itself. In setting the stage for the introduction of her own thesis, for example, Langer presents the claim of a German philosopher named Otto Baensch that works of art present us directly with *feelings* that appear to be "objectively given" with the work itself, "as one of its attributes, belonging to it like any other attribute we perceive it to have."[46] The *mood* of a landscape painting, for example, "surrounds, fills, and permeates it like the light that illumines it, or the odor it exhales. The mood belongs to our total impression of the landscape, and can only be distinguished as one of its components by a process of abstraction."[47] These feelings "are contained in the sensory qualities as well as in the formal aspects, and despite all their own variety and contrasts, they melt and mingle in a total impression which is hard to analyze."[48]

Here again, in quoting extensively from Baensch's article in one of the book's introductory chapters, Langer continues to reinforce the association of "feeling" with "emotion." For example, Langer quotes Baensch as saying that the purpose of art is to bring "the world's emotional content" into "the realm of objectively valid cognition," which he underscores by immediately adding that what art makes us aware of "is always of an emotive character."[49] This makes it more difficult to appreciate that, with *Feeling and Form*, Langer's definition of "feeling" has become equivalent to "thought" as James used it in *The Principles*—as a generic term for conscious experiences.

In the fifteen chapters that make up the largest part of the book, however, it becomes abundantly clear that Langer has broadened her understanding of feeling considerably, and that each of the major orders of art deals primarily with just one of the many possible aspects that characterize what James called the "much-at-onceness" of conscious experience.[50] She finally affirms this explicitly in the first volume of *Mind*, although she buried the affirmation in a footnote, as we have seen.[51]

This survey of Langer's writings that touch on feeling and art has revealed how little her understanding of feeling changed—at least explicitly—until the appearance of *Feeling and Form*. Even then, the increasing scope of Langer's definition remains largely implicit, although it is unmistakable, if still unacknowledged, in the chapters in which she discusses the meaning of each of the "great orders of art"[52]—painting, sculpture, architecture, music, dance, literature, drama, and film—and argues that each one of them explores "a special dimension of experience" that it alone is uniquely suited to express.[53] The consequences for Langer's definition of "feeling" are inescapable, for if all art is "the creation of forms symbolic of human feeling,"[54] and if each of the arts explores a different dimension of subjective experience, then feeling itself must encompass at least these eight dimensions, which—taken together—far exceed the bounds of emotional or affective experience. Indeed, Langer finally states in *Feeling and Form* that she is using the term "feeling" to apply to "*our whole subjective reality*, woven of thought and emotion, imagination and sense perception" (emphasis added).[55] Four years later, in *Problems of Art* (1957), she refers to "feeling" as "what is sometimes called 'inner life,' 'subjective reality,' 'consciousness'—there are many designations for it";[56] and in *Philosophical Sketches* (1962), five years before the publication of the first volume of *Mind*, she provides two examples of *feeling* as including ordinary *perceptual experience:* "*Vision* is the way the optic apparatus *feels* the impingement of light, and *hearing* is the way the auditory structures *feel* sound waves" (emphasis added).[57]

DIMENSIONS OF CONSCIOUS EXPERIENCE: THE MEANING OF MUSIC

At the close of the third and final introductory chapter of *Feeling and Form*, Langer offers a "tentative definition" on which the subsequent chapters are built: "*Art is the creation of forms symbolic of human feeling*," (emphasis added)[58] where feeling must now be defined comprehensively as consciousness or "inner life"—"our whole subjective reality, woven of thought and emotion, imagination and sense perception."[59] In the succeeding chapters, she focuses on eight specific dimensions of feeling, each of which is the main province of one of "the great orders of art."[60]

In Langer's view, each of these orders creates the semblance of a different aspect or dimension of conscious experience. Each of the arts "begets a special dimension of experience that is a special kind of image" of some aspect of subjective reality.[61] Langer calls this the *primary illusion* of an art—"a special sort of appearance, in terms of which all its works are made."[62] The primary illusion created by a literary work, for example, is "the appearance of 'experiences,' the semblance of events lived and felt,"[63] a *virtual past*, or *virtual history* "in the mode typified by memory."[64] The primary illusion of the plastic arts—painting, sculpture, and architecture—is "*virtual space* in its several modes."[65] And the primary illusion of music is *virtual time*, "an auditory apparition" of felt time,[66] or "time as we know it in direct experience."[67]

Music, like all the arts, "is the creation of forms symbolic of human feeling."[68] But what kinds of forms? And what kinds of feeling? Langer states that "the materials of music ... are sounds of a certain pitch, loudness, overtone mixture, and metronomic length";[69] but a musical composition itself, as a created form symbolic of human feeling, is more than "arranged tonal material."[70] As a work of art, something emerges from the process of arranging tonal materials that was not there before. We hear "tonal forms, moving, mingling, resolving, having direction and energy";[71] we hear "movement and rest, swift movement or slow, stop, attack, direction, parallel and contrary motion, melody rising or soaring or sinking, harmonies crowding or resolving or clashing; moving forms in continuous flux."[72] These are the elements of music, and they may be made out of "harmonic or melodic material, shifts of range or of tone color, rhythms or dynamic accents or simply changes of volume."[73]

The elements of music are, "like all artistic elements, something virtual, created only for perception"[74]—in the case of music, created for the ear alone. What we hear, Langer maintains, are what the German music critic Eduard Hanslick described as "*tönend bewegte Formen*," which Langer translates as "sounding forms in motion"[75] or "sonorous moving forms"[76] and identifies as "the essence of music."[77] "Music flows; a melody moves; a succession of tones is heard as a progression. The differences between successive tones are steps, or jumps, or slides. Harmonies arise, and shift, and move to resolutions. A complete section of a sonata is quite naturally called a 'movement.'"[78]

And yet, "in all the progressive movements we hear, there is actually nothing that moves."[79] Music "presents us with an obvious illusion, which is so strong that despite its obviousness it is sometimes unrecognized because it is taken for a real, physical phenomenon: that is the appearance of *movement*."[80] The forms and motions we hear when listening to music "are only seemingly there; they are elements in a purely auditory illusion."[81] "The forms are virtual, their motion is virtual, and the whole composition is a semblance."[82]

But what does it all mean? If the "sonorous moving forms" of music—though a purely apparent, virtual creation—are supposed to be symbolic of human feeling, what kinds of feeling—what aspect or dimension of our lived experience—do they express?

In *The Practice of Philosophy*, Langer had offered the suggestion that "the final object of musical expression" might be "the endlessly intricate and universal pattern of emotional life."[83] In *Philosophy in a New Key*, she defined "the field of musical meanings" as "the realm of emotional experience"[84] and called music the "formulation and representation of emotions, moods, mental tensions and resolutions—a 'logical picture' of sentient, responsive life" more generally.[85] In *Feeling and Form*, however, Langer has reached a more detailed understanding of the specific aspect of conscious experience that music is uniquely suited to express, or formulate for our conception: The moving tonal forms created by music present us with an auditory image or semblance of *time*, "more precisely, of what one might call 'felt time'"[86]—time as an essential characteristic of each and every one of our lived experiences. Furthermore, the experience we have of the passage of time will vary with different kinds of experience. Therefore, with different kinds of experiences come different qualities of lived time—a fact to which the seemingly limitless number and variety of compositions that make up the history of music bear ample witness. If conscious experience is like an ever-flowing stream, as James observed, then music shows us the many different currents, ripples, eddies, torrents, cascades, and meanderings that make up the stream.

DIMENSIONS OF CONSCIOUS EXPERIENCE: THE MEANING OF LITERARY WORKS[87]

In contrast to music, literary art (poetry and prose fiction) uses the materials of discourse to create an image of human experience in the mode typified by memory, which represents "life as a realm of events—completed, lived, as words formulate them—events that compose a Past."[88]

Langer points out that actual experience, as it unfolds in the present, is a "chaotic advance,"[89] dominated by the exigencies of practical action, in which thoughts, fantasies, beliefs, and expectations are usually "fragmentary, transient and often indefinite";[90] and that experience takes on form and character only retrospectively, in the process of recounting it to ourselves and to others—that is, in remembering and retelling. "Memory is a special kind of experience, because it is composed of selected impressions, whereas actual experience is a welter of sights, sounds, feelings, physical strains, expectations, and minute, undeveloped reactions. Memory sifts all this material and represents it in the form of distinguishable events,"[91] shaping experience "into a distinct mode, under which it can be apprehended and valued";[92] and language plays an essential role in this process.

To begin with, in recounting our experiences to ourselves or to others, language alone must be made to stand in for the more complex experiences of actual remembering; and we must therefore choose and order our words, and structure the resulting narrative, in such a way that it captures and conveys much of the appearance of actual lived experience.

Furthermore, the way we formulate our experiences—the language we use in remembering and retelling them—expresses their effect on our sensibility, for every experience leaves the mark of its appearance and emotional value on the choices of language we make in recounting it. The things that we notice and emphasize; the things that we leave out or push into the background; the associated thoughts, feelings, and impressions that come to mind; the choice and order of words; the length, rhythm, and complexity of the sentences—all these and many other factors bespeak a particular "apprehensive condition of the soul,"[93] a unique mode of thinking and feeling that enters into the very events that figure in the telling.

Because language is an essential ingredient in the creation of every actual subjectivity—the individual consciousness, that is, of every actual human person—language alone can be used to create the semblance of the unique mode of apprehension that is characteristic of a particular human sensibility. In this way every literary work creates what Langer calls a "virtual subjectivity."[94] And the literary artist is someone with a heightened sensitivity to the power of language to create the semblance of "a reality lived and remembered,"[95] which is *virtual memory*, "or history in the mode of an experienced Past,"[96] expressing the "logic" of a particular human sensibility. In this way, the art of literary composition, like all the arts, "makes [the logic of consciousness] apparent, objectively given so we may reflect on it and understand it."[97]

CONCLUSION: FEELING AS CONSCIOUS EXPERIENCE

Early in the *Essay on Human Feeling*, Langer states unequivocally that "William James," in *The Principles of Psychology*, "used 'thinking' in the sense in which I use 'feeling.'"[98] To understand what Langer means by "feeling," therefore, we need only look at how James used "thought" and "thinking" in his early writings on psychology. In *The Principles of Psychology* (1892), he used "thought" and "thinking" as generic terms for "mental states at large irrespective of their kind";[99] but in later writings he often replaced these with "consciousness" and related terms, redefined psychology as "the description and explanation of *states of consciousness*,"[100] and renamed his famous chapter "The Stream of Consciousness," thereby making his meaning unmistakable.

Despite a widespread understanding, reflected in ordinary usage, of *feeling* as "comprising sensation, desire, and emotion, but excluding perception and thought,"[101] Langer clearly intended the term to apply to "our whole subjective reality, woven of thought and emotion, imagination and sense perception,"[102] although there are many places in her writing where she undercut her explicit intent and unwittingly encouraged its more limited meaning.

In a well-known essay called "What Is It Like to Be a Bat?" the philosopher Thomas Nagel offered a surprisingly simple but illuminating definition of *consciousness*: "The fact that an organism has conscious experience *at all* means, basically, that there is something it is like to *be* that organism … something it is like *for* the organism" (Nagel's emphasis).[103] When Langer defines works of art as "perceptible forms expressive of human feeling,"[104] what she is saying in effect is that every work of art shows us *what it is like* to be a human organism having a particular kind of conscious experience. Or, as Langer put it: "Feeling [i.e., conscious experience] is *like* the dynamic and rhythmic structures created by artists" in every medium of expression.[105] A work of art "presents a form … which we recognize intuitively as something very much *like* feeling [i.e. conscious experience]," and in this way it teaches us, "without effort or explicit awareness, what feeling [i.e., conscious experience] is *like*" (emphasis added).[106] Notice that, in every case, the "what-it-is-like" of conscious experience can *only* be expressed and apprehended through some kind of *nonpropositional* formulation, which for Langer is most commonly found in the arts.[107]

To listen to a particular piece of music is to know—"without effort or explicit awareness"—*what it is like* to be experiencing the passage of time for just one possible kind of experience; and different pieces of music present the passage of time for very different kinds of experiences. Similarly, to read a particular novel, poem, or short story is to know *what it is like* for a particular person with a particular sensibility to be remembering and retelling events from the lived experience of their past. More generally, each of the works belonging to every one of the major orders of art shows us *what it is like* to be a human being with respect to one special dimension of one particular kind of conscious experience. Taken together, the history of the arts provides a window into the limitless variety of possible states of human consciousness that these created forms have been able to formulate and convey.

With *Feeling and Form*, then, Langer arrived at a definition of "feeling" that is largely synonymous with the subjective aspects of experience, or *consciousness*, broadly construed, although she conveys this largely by *showing* how the major orders of art give us insight into the different dimensions of conscious experience, or feeling, which, in all its limitless variety and particularity, "defies discursive formulation, and therefore verbal expression."[108] Through the arts, however, we *see*, as we can understand in no other way, that "the ways we are moved are as various as the lights in a forest," and that "all these inseparable elements of subjective reality compose what we call the 'inward life' of human beings."[109] With the achievement of this understanding of feeling, Langer was ready to begin the project of *Mind*, which would occupy her for the second half of her intellectual career.

NOTES

1. "Feeling, n." December 2021. In *Oxford English Dictionary* (Online). https://www.oed.com/.
2. Susanne K. Langer, *Feeling and Form: A Theory of Art Developed from Philosophy in a New Key* (New York: Charles Scribner's Sons, 1953), 127.
3. Thomas Brown, *Lectures on the Philosophy of the Human Mind*, vol. 1 (Edinburgh: W. & C. Tait, 1820), 256.
4. John Stuart Mill, *A System of Logic, Ratiocinative and Inductive* (New York: Harper & Brothers, 1846), Book I, chapter 3, section 3, 34. Langer could have quoted Mill's entire passage almost verbatim in support of her own definition of "feeling."
5. Ibid.
6. Brown, *Lectures on the Philosophy of the Human Mind*, vol. 1, 256.
7. William James, *The Principles of Psychology*, vol. 1 (Cambridge, MA: Harvard University Press, 1981), 15.
8. Susanne K. Langer, *Mind: An Essay on Human Feeling*, vol. 1 (Baltimore: Johns Hopkins University Press, 1967), 21.
9. James, *The Principles of Psychology*, vol. 1, 186.
10. Langer, *Mind*, vol. 1, 21.
11. Ibid.
12. James, *The Principles of Psychology*, vol. 1, 15.
13. Ibid., 185.
14. Ibid.
15. Ibid., 186.

16. Ibid.
17. Ibid.
18. Ibid., 219.
19. Ibid.
20. Ibid.
21. Ibid., 231.
22. Ibid., 233.
23. William James, *Psychology: Briefer Course* (Cambridge, MA: Harvard University Press, 1984), 9.
24. Ibid., 10.
25. Ibid., 140–1.
26. William James, *Talks to Teachers on Psychology* (Cambridge, MA: Harvard University Press, 1983), 19.
27. Ibid., 20.
28. Langer, *Mind*, vol. 1, 21.
29. Langer, *Feeling and Form*, 127.
30. Carroll Pratt, "Review of *Mind: An Essay on Human Feeling*, vol. 1, by Susanne K. Langer," *Contemporary Psychology* 13, no. 9 (September 1968): 447.
31. Langer's unpublished doctoral dissertation, "A Logical Analysis of Meaning" (1926), included an appendix titled "'Meaning' in Art," in which she offered a brief sketch of a theory based on "that correlation between elements of different [sensory] systems, known as 'synaesthesia'" (Susanne K. Langer, "A Logical Analysis of Meaning," [PhD diss., Radcliffe College, Cambridge, MA, 1926], 166); but this conjecture never reappeared in any of her subsequent writings.
32. The thesis that analogy of form, similarity of pattern, or correspondence of configuration between relational structures is the basis of all meaning situations first appeared in Langer's doctoral dissertation (see note 28) and was the "generative idea" for what eventually became the general theory of symbolization she introduced in *The Practice of Philosophy* (1930) and further elaborated in *Philosophy in a New Key* (1942), from which it later developed into the basis for her mature philosophy of art.
33. Langer, "A Logical Analysis of Meaning," 9.
34. Susanne K. Langer, *The Practice of Philosophy* (New York: Henry Holt, 1930), 162.
35. Langer, *The Practice of Philosophy*, 162.
36. Ibid., 160.
37. Ibid.
38. Ibid., 161.
39. Susanne K. Langer, *Philosophy in a New Key: A Study in the Symbolism of Reason, Rite, and Art* (Cambridge, MA: Harvard University Press, 1957), 93.
40. Langer, *The Practice of Philosophy*, 98.
41. Ibid., 100–1.
42. Ibid., 222.
43. Ibid., 223.
44. Ibid., 235.
45. Ibid., 257–8.
46. Otto Baensch, "Art and Feeling," in *Reflections on Art: A Source Book of Writings by Artists, Critics, and Philosophers*, ed. Susanne K. Langer (Baltimore: Johns Hopkins University Press, 1958), 11.
47. Baensch, "Art and Feeling," 11.

48. Ibid., 13.
49. Baensch, "Art and Feeling," quoted in Langer, *Feeling and Form*, 19.
50. William James, *Some Problems of Philosophy* (Cambridge, MA: Harvard University Press, 1979), 32.
51. Langer, *Mind*, vol. 1, 21.
52. Langer, *Feeling and Form*, 133.
53. Susanne K. Langer, *Problems of Art: Ten Philosophical Lectures* (New York: Charles Scribner's Sons, 1957), 81.
54. Langer, *Feeling and Form*, 40.
55. Ibid., 127.
56. Langer, *Problems of Art*, 112.
57. Susanne K. Langer, *Philosophical Sketches* (Baltimore: Johns Hopkins University Press), 8.
58. Langer, *Feeling and Form*, 40. In this definition Langer refers to a work of art a *symbol* of feeling. She later recognized, however, that it cannot be a symbol in the usual sense because it has no conventional *reference*, and so, she decided, cannot be properly said to have a *meaning*. Referring to Ernest Nagel's definition of a symbol as "any occurrence (or type of occurrence), usually linguistic in status, which is taken to signify *something else* by way of tacit or explicit conventions or rules of language" (Ernest Nagel, "Symbolism and Science," in *Logic without Metaphysics, and Other Essays in the Philosophy of Science*, by Ernest Nagel [Glencoe, IL: The Free Press, 1956], quoted in Langer, *Problem of Art*, 130; emphasis added), Langer notes that a work of art does not point beyond itself to something known by other means, for what is expressed in a work of art "cannot be grasped apart from the sensuous or poetic form that expresses it" (Langer, *Problems of Art*, 134). For these reasons, Langer decided that *expressive form* was a better term than "symbol," and that it was preferable to speak of the *import*, rather than the meaning, of the work (ibid., 126–7). Hence she later offered a revised definition of art as "*the creation of perceptible forms expressive of human feeling*" (Langer, *Problems of Art*, 80; original emphasis). Any further consideration of why a work of art is not properly called a *symbol*, however, is beyond the scope of this chapter.
59. Langer, *Feeling and Form*, 127.
60. Ibid., 133.
61. Langer, *Problems of Art*, 81.
62. Ibid., 144.
63. Langer, *Feeling and Form*, 212.
64. Ibid., 264.
65. Ibid., 102 (original emphasis).
66. Langer, *Problems of Art*, 37.
67. Langer, *Feeling and Form*, 112.
68. Ibid., 40.
69. Langer, *Problems of Art*, 39.
70. Ibid.
71. Ibid., 144.
72. Susanne K. Langer, "The Primary Illusions and the Great Orders of Art," *The Hudson Review* 3, no. 2 (Summer 1950): 220.
73. Langer, *Feeling and Form*, 107.
74. Ibid.
75. Ibid.
76. Langer, *Problems of Art*, 39.

77. Langer, *Feeling and Form*, 107.
78. Langer, *Problems of Art*, 36.
79. Langer, "The Primary Illusions and the Great Orders of Art," 220.
80. Langer, *Problems of Art*, 36 (original emphasis).
81. Langer, *Feeling and Form*, 108.
82. Langer, "The Primary Illusions and the Great Orders of Art," 221.
83. Langer, *The Practice of Philosophy*, 161.
84. Langer, *Philosophy in a New Key*, 257–8.
85. Ibid., 222.
86. Langer, *Problems of Art*, 37.
87. For an extended discussion of Langer's theory of literary art and its relation to the psychology of human memory, see Donald Dryden, "Memory, Imagination, and the Cognitive Value of the Arts," *Consciousness and Cognition* 13, no. 2 (June 2004): 254–67.
88. Langer, *Feeling and Form*, 306.
89. Ibid., 266.
90. Ibid., 212.
91. Ibid., 263.
92. Ibid., 262.
93. Langer, *Mind*, vol. 1, 118.
94. Langer, *Feeling and Form*, 257.
95. Ibid., 273.
96. Ibid., 279.
97. Langer, *Problems of Art*, 73.
98. Langer, *Mind*, vol. 1, 21.
99. James, *The Principles of Psychology*, vol. 1, 186.
100. James, *Psychology: Briefer Course*, 9.
101. "Feeling, n." December 2021, in *Oxford English Dictionary* (Online). https://www.oed.com/.
102. Langer, *Feeling and Form*, 127.
103. Thomas Nagel, "What Is It Like to Be a Bat?" in *Mortal Questions*, by Thomas Nagel (New York: Cambridge University Press, 1979), 166.
104. Langer, *Problems of Art*, 80.
105. Langer, *Mind*, vol. 1, 64 (original emphasis).
106. Ibid., 67.
107. Rolf Lachmann ("Susanne K. Langer's Foray into Art as a 'Phenomenology of Feeling'," Chapter 6, this volume (93–103) provides an illuminating discussion of why Langer believed that art was uniquely suited to convey the "what-it-is-like" of feeling, or conscious experience in all its multiplicity and variety, and why philosophical phenomenology (as represented, for example, by the work of Edmund Husserl and his disciples) was inadequate to the task.
108. Langer, *Problems of Art*, 22.
109. Ibid.

BIBLIOGRAPHY

Baensch, Otto. "Art and Feeling." In *Reflections on Art: A Source Book of Writings by Artists, Critics, and Philosophers*, edited by Susanne K. Langer, 10–36. Baltimore: Johns Hopkins University Press, 1958.

Brown, Thomas. *Lectures on the Philosophy of the Human Mind,* vol. 1. Edinburgh: W. & C. Tait, 1820.

Dryden, Donald. "Memory, Imagination, and the Cognitive Value of the Arts." *Consciousness and Cognition* 13, no. 2 (June 2004): 254–67.

"Feeling, n." December 2021. In *Oxford English Dictionary* (Online). https://www.oed.com.

James, William. *The Principles of Psychology,* vol. 1. 1890. Cambridge, MA: Harvard University Press, [1892] 1981.

James, William. *Psychology: Briefer Course.* Cambridge, MA: Harvard University Press, [1892] 1984.

James, William. *Some Problems of Philosophy.* Cambridge, MA: Harvard University Press, [1911] 1979.

James, William. *Talks to Teachers on Psychology.* Cambridge, MA: Harvard University Press, [1899] 1983.

Langer, Susanne K. *Feeling and Form: A Theory of Art Developed from Philosophy in a New Key.* New York: Charles Scribner's Sons, 1953.

Langer, Susanne K. "A Logical Analysis of Meaning." PhD diss., Radcliffe College, Cambridge, MA, 1926. Available online: https://langercircle.sites.uu.nl/wp-content/uploads/sites/662/2021/02/Langer-Susanne-K.-1926-A-Logical-Analysis-of-Meaning-doctoral-dissertation-scanned-PDF-from-photocopy.pdf (accessed July 16, 2023).

Langer, Susanne K. *Mind: An Essay on Human Feeling*, vol. 1. Baltimore: Johns Hopkins University Press, 1967.

Langer, Susanne K. *Philosophical Sketches.* Baltimore: Johns Hopkins University Press, 1962.

Langer, Susanne K. *Philosophy in a New Key: A Study in the Symbolism of Reason, Rite, and Art.* 1942. 3rd edn. Cambridge, MA: Harvard University Press, 1957.

Langer, Susanne K. *The Practice of Philosophy.* New York: Henry Holt, 1930.

Langer, Susanne K. "The Primary Illusions and the Great Orders of Art." *The Hudson Review* 3, no. 2 (Summer 1950): 219–33.

Langer, Susanne K. *Problems of Art: Ten Philosophical Lectures.* New York: Charles Scribner's Sons, 1957.

Mill, John Stuart. *A System of Logic, Ratiocinative and Inductive.* New York: Harper & Brothers, 1846.

Nagel, Thomas. "What Is It Like to Be a Bat?" In *Mortal Questions*, 165–89. New York: Cambridge University Press, 1979.

Pratt, Carroll. "Review of *Mind: An Essay on Human Feeling, volume 1*, by Susanne K. Langer." *Contemporary Psychology* 13, no. 9 (September 1968): 446–7.

CHAPTER TEN

Psychological Dimensions, Cultural Consequences, and their Breakings in Susanne K. Langer's Symbolic Mind

ROBERT E. INNIS

This chapter has two purposes: (a) to sketch the heuristic value of the different range of psychological resources Susanne Langer utilized to trace the pivotal transitions to specifically human symbolic minding, and (b) to highlight the tensions between symbolization and its forms as an achievement of a kind of semiotic freedom, that is, an accomplished critical awareness of domains of meaning and forms of life, both life affirming and life denying, grounded in the conscious use of signs and symbols. The transitions to human minding and forms of action occur in the body-based polysensory domain of affectively charged perceptual and image-schematic orderings and configurations of symbolic pregnancies immanent in the flux of experience. Langer shows how, following the advent of language, humans created further symbolic structures with unique semiotic logics, powers, and functions: ritual, myth, visual art, music, dance, and so forth with their presentations of the morphology of feeling. In her great trilogy, *Mind*, Langer traces the contours of a model of mind grounded in a scientifically precise and philosophically and semiotically rich generalized notion of feeling that culminates in the analysis of the clash and breaking of cultural forms and world views.[1] She shows how the varieties of felt significances embodied in these symbolic structures entail deep affective commitments, effects, and consequences. Human symbol systems are mediating devices for processes of *individuation* as well as for interactive *involvement* with others, a central distinction that appears in the chapters by Dryden, Lachmann, Pollok, and Nocek. The psychological outcomes of these processes, effected by symbolic activities, are not easily reconcilable. They are marked by forms of felt imbalance, experiences of "breakings" of their experiential contexts by encounters with other contexts, leading to consequential intra-psychic and intra-social conflicts. Such a breaking is exemplified in the drive to higher forms of scientific and formal rationality and abstraction that characterize modernity. Could Langer's concepts of feeling tones, experiential rhythms, balance, and sense of livingness and a concomitant normative concept of *aesthetic rationality* and *affective semiosis* offer a way of understanding the sociocultural effects of these "breakings" as a permanent feature of the symbolic mind?

THE ORDERING OF FELT LIFE

In Susanne Langer's indispensable *Feeling and Form* we find the following psychologically rich passage.

> Sentient beings react to their world by constantly changing their total condition. When a creature's attention shifts from one center of interest to another, not only the organs immediately involved ... but hundreds of fibers in the body are affected. Every smallest shift of awareness calls out a readjustment, and under ordinary circumstances such readjustments pass easily into another ... It is perception molded by imagination that gives us the outward world we know. And it is continuity of thought that systematizes our emotional reactions into attitudes with distinct feeling tones, and sets a certain scope for an individual's passions. In other words: by virtue of our thought and imagination we have not only feelings, but a life of feeling.[2]

Such a life of feeling, she continues, is "a stream of tensions and resolutions"[3] transforming and congealing reactions into attitudes or habits, which can be, or fail to be, rational. It is "in" this stream that we live. This foregrounding of felt tensions by Langer encompasses "all emotion, all feeling tone, mood, and even personal 'sense of life' or 'sense of identity.'"[4]

Langer's general point, made in the context of developing a theory of art, is that "all the forms of feeling are important, and the joyous pulse of life needs to be made apparent as the most involved passions, if we are to value it."[5] While art, as such, creatively explores and reveals the possible forms that feeling can take by being embodied in mediums of all sorts, its ultimate import is normative. It points to a universal or general task to build the forms of the world in such ways that the joyous pulse of life is fostered and the most involved passions preventing such a joyous pulse minimized, even if, in light of the ineluctable tragedy of human existence, which Langer admitted with open eyes, they cannot be eliminated. It is not just art but all material and symbolic products that formulate "felt life" and mold "the objective world for the people" and thereby ideally create a "defense against outer and inner chaos."[6] The symbolic animal strives toward possible inner and outer orders while at the same time finding itself perplexed and incapable of ordering or stabilizing these orders themselves. Langer offers analytical tools for understanding this dilemma, specifying its origins and avenues of avoidance or resolution.

Langer's work shows us that the cultural theorist or philosopher of culture, which she paradigmatically was, must first of all have the eye of an artist. What does such an eye uncover? Langer answers: "The artist's eye sees in nature, and even in human nature betraying itself in action, an inexhaustible wealth of tensions, rhythms, continuities and contrasts ... those are the 'internal forms' which the 'external forms' ... express for us."[7] For Langer, all cultural forms are to be seen as the objectification of feeling and the subjectification of nature, "one vast phenomenon of 'felt life' stretching from the elementary tonus of vital existence to the furthest reaches of mind."[8] The field of felt life is marked by what she calls 'gradients' that inform the flux of experience and the objects that animate it. Such a world, with its many fields and their thematic cores, Langer writes, involves an "endless rhythm of individuation and involvement."[9] Individuation is a sign of increasing interiority and consciousness of the creation of a unique self, the development of the awareness of being a subject of acts that determine who we are and who we want to be. Involvement is the inextricable social web in which we are caught and the constraints on, as well as the enabling conditions of, the actions by which we respond to the many situations in which we find ourselves.

Langer points out in volume 2 of *Mind* that these two poles marking the building of the human world through symbolic transformation and its objectifying powers remain in perpetual tension and as a consequence are subject to wild swings and imbalances.

> Society, like the spatiotemporal world itself, is a creation of man's specialized modes of feeling—perception, imagination, conceptual thought and the understanding of language. The rise of his typical way of life as a member of a continuous recognized society, built up on the ancient and gradual separation of the evolving Hominidae from all other, differentially evolving primate lines, in its advance constantly epitomizes the great shift from beast to man.[10]

Such a momentous shift constantly upsets what Langer calls the "ethnic balance," the equilibrium between agency of the individual and the individual's chosen or historically placed responsibility to its group, whatever that would be. This is the fateful permanent polarity between individuation and involvement.

How is such a balance to be pursued or to some degree attained? In volume 3 of *Mind* Langer argued that the "primal and perennial work of social organization is not to fix the bounds of behavior as permanent lines, which would make all evolutionary process impossible, but to retrieve the vital balance every time some act, public or private, has upset it."[11] Social life's integrity, our ability to live together in some form of unity or comity, is dependent upon finding this vital balance. Of course, since social organization is a dynamic matrix of agents defined by their acts, both individual and institutional, these multiple centers with their principles of interest are subject to constant tendencies to imbalance, splitting, and lack of coherence. Thus, for Langer, the speciation of humanity takes place through a sequence of seemingly permanent crises: speech, fantasy, ritual, and the "breaking" of tribal consciousness and commitments, crises that, in a kind of analogue to the Hegelian ruse of reason, mediate the "cultural move to civilization."[12] Langer's work shows that the study of these crises and the search for their grounds is a principal task of the cultural sciences, especially those that follow the heuristic lead of her complex model of the symbolic mind with special attention to its psychological roots. These roots are twofold and are intertwined: bio-psychological and semiotic.

ROOTS OF THE SYMBOLIC WORLD

The movement to what Langer called the "symbol world," the world of human culture and its symbolic forms, is the achievement of a specific type of organism whose semiotic powers were made possible by the evolution of bio-organic structures of the central nervous system and the transformational multifaceted development of the hand.[13] These structures underlie and define new forms of sensibility and perceptual and action possibilities. They are taken up into the distinctively human dynamic matrices of activities that make up the *animal symbolicum* who never operates, even in the formal sciences, in a purely transparent or seemingly bodiless world of meanings.

Already in *Philosophy in a New Key* Langer had traced the roots of minding, explored in detail in *Mind*, down to an ultimate level where "all sensitivity bears the stamp of mentality."[14] In the case of humans, this sensitivity has taken on the property of symbolic transformation, which, rather than being something that breaks the continuum of nature, is, in Langer's conception, a "natural activity, a high form of nervous response, a characteristic of man among animals."[15] In the human case, as a result of a long biological trajectory that Langer recounted in *Mind*, "all conscious experience is symbolically conceived experience; otherwise, it passes 'unrealized'."[16] Relying in volume 1 of *Mind* on rich psychological sources dealing with images, Langer foregrounded a fundamental feature of this process of realization:

> As most of our awareness of the world is a continual play of impressions, our primitive intellectual equipment is largely a fund of images, not necessarily visual, but often gestic, kinesthetic, verbal or what I can only call "situational" ... [We] apprehend everything which comes to us as impact from the world by imposing some image on it that stresses its salient features and shapes it for recognition and memory.[17]

These salient features make up diversified realms of forms. In human processes of minding "meaning ... accrues essentially to forms."[18] As a result, Langer considered even the primary world of the senses to be primarily a domain of symbols, that is, experiential configurations with "symbolic pregnancy," a form of significance exhibited in the luring or repelling affective tones or felt qualities of objects and experiential situations.

This felt sense of significance is the point of origin of Langer's key notion of presentational symbolization, as opposed to discursive symbolization effected by and rooted in language, not just as a "new departure in semantic"[19] and the source of wealth of developmental possibilities in art, but as a feature of a permanent stratum of consciousness that has remarkable and at times catastrophic cultural and social consequences. Langer saw, along with many others, that at the most fundamental level, including the historical, the "recognition of vague, vital meaning in physical forms—perhaps the first dawn of symbolism gave us our idols, emblems, and totems ... The momentous discovery of nature-symbolism, of the pattern of life reflected in natural phenomena, produced the first universal insights."[20] But they also produced the concomitant forms of feeling or patterns of sentience that mark a participatory form of mythic and ritualistic consciousness that identify by a process of reification the physical forms themselves with what they symbolize. Such a participatory form can be transferred to other domains, especially the political and social orders that are transformed or divided into tribes, each with their paradigmatic idols, emblems, totems, and affective relations to the deep significances ascribed to the cycles of nature that govern the fundamental rhythms of life. This transference can occur within what are otherwise highly symbolically developed civilizations and are not restricted to what are incorrectly thought of as "primitive" or "surpassed" cultures.

Langer's *Philosophical Sketches* (1962) is a set of studies that anticipate and present with remarkable clarity and accessibility—and with an avoidance of scholarly disputation—the main themes and analytical lines of the *Mind* trilogy. I want to indicate how in these studies from more than a half-century ago Langer already showed how to identify and analyze present as well as permanent cultural consequences of this transference by means of her model of the symbolic mind as a complex developmental branching of feeling. The combination of bio-psychological and semiotic frameworks constitutes a challenging and pressingly pertinent account of the linkages between scientific-technological civilization and cultural crisis. The consequences of these linkages surround us on all sides.

BETWEEN CULTURE AND CIVILIZATION

In the essay "Scientific Civilization and Cultural Crisis" in *Philosophical Sketches* Langer writes: "Every human life has an undercurrent of feeling that is peculiar to it. Each individual expresses this continuous pattern of feeling in what we can call its 'personality,' reflected in behavior, speech, voice, and even physical bearing (stance and walk) as his *individual style.*" It is the same, she contends, with every human society with "its undercurrent of feeling which is not individual, but general."[21] This societal style is assimilated in their own way by each individual in the processes of individuation.

In 1961, when this paper was presented in Japan, Langer contended that the undercurrent of feeling in the world was "confused, uncertain, strained."[22] There was, she claimed, a mix of pride and fear, a faith in science combined with an irrationalism that made such faith shaky, "a growing sense of world society, human rights, and the equal dignity of all mankind" accompanied at the same time by a "prevailing hostility and jealousy" that defined a "protracted 'cold war'."[23] She called this state one of "emotional instability" brought on by a sense of loss of things of "undoubted and irreplaceable value": social orders of rank and status, religious faith with its institutions, recognized art traditions, "ways of life in which people have long felt secure and useful. Such losses are not to be taken lightly."[24]

What is the nature of the distinction between civilization and culture, as Langer is using these terms?

One of Langer's ways of characterizing culture is that it is "the symbolic expression of developed habitual ways of feeling."[25] By feeling Langer means, in a way taken up in *Mind*, "everything that can be felt,"[26] a clearly bodily or somatic phenomenon. Langer includes such things as "rhythms of attention and the strain of thought, bodily relaxation or tension that cannot be reduced to any particular sensation, attitudes of mind, the general activity of our imagination, confidence in the goodness of life, or fundamental annoyance, boredom, cynicism, or ... the countless modes of humor."[27] These are clearly aspects of individuation. Cultures and the persons in them are marked by characteristic patterns of feeling, governing both acts of agents and the systems of artifacts beyond practical needs. Cultures belong to the expressive domain, rather than the efficaciously ordering domain of civilization. Cultures are in one sense natural communities underpinned by a unifying continuity of vital feeling.

While the conservative nature of cultural communities would incline toward ossification or stability, there is need of pioneering individuals to cut the channels for "new elements of feeling and carve out a frame for new attitudes and moral sentiments."[28] Cultural frames are therefore perilous in two ways: blocking advance by a kind of habitual inertia, yet subject to a kind of slow organic growth periodically punctured by interruptions.

Civilization for Langer is different from culture. In Langer's conception it is not identical with the "symbolic aspect of behavior" as a field of meanings and existential and social interpretations, but "the pattern of the practical implementation of life ... the practical organization of life, public and private," exemplified in contracts, liabilities, legal regulations, technological systems, exploitation of materials, money, and so forth. All of these, clearly, are due to our symbolic capacities and fatefully to the rise of cities and their "ferment of novelty" as opposed to country life, or non-urban life, and its regular and repetitive patterns.[29] Langer asserts, however, that civilization "strains and drains the life which engenders and supports it."[30] Its ability to be transplanted is at one and the same time the source of its power and the cause of its disturbing consequences, where actuality and symbolic tradition seem to go their separate ways with a consequent loss of community feeling and even of sacredness. Civilization, with its diverse components spread over the conduct of life, "descends like an iron grill to crush the heritage of feeling and faith and the beauty of life."[31]

Langer sees science, including what she calls the "so-called" social sciences, as the driving force in the creation of the present world. The consequences of this phenomenon, however, are not benign. The practical means and techniques of life have, in her evaluation, outrun our stage of thinking by reason of their global spread and reach. The consequences of the great transition to a world society are in her opinion not primarily economic but the dissolution or fragmentation of "self-sufficient cultural groups"[32] and of the symbolic coherences that mark shared cultures. This is a core feature of spreading globalization, only partially mitigated by attempts at tolerant "multi-culturalism." Langer, already in 1961, saw that the world was in a "socially anomalous state" between a kind of tribalism and a world of "global industrial organization" with its multiple goals, forms, and material and social constraints.

What type of culture with its forms of feeling is emerging out of this situation which is marked by diverse fields of possibilities? Langer warns us that the emergence of a real culture cannot be forced.[33] The sense of shared patterns of feeling and constitutive symbolic commitments that are articulated as new life symbols, the "objective record of developed feeling,"[34] arise slowly and not by deliberate action. Indeed, one of the functions of art is to capture and express, in various ways, the forms of feeling that participants in the culture strive to record as well as create, what she calls "the deeply and tacitly felt life of overt action, institutions, ways of living, things produced."[35] Science for Langer has a kind of universality, "not native or exotic ... it belongs to humanity and is the same wherever it is found"[36] and predicts, seemingly paradoxically, that it is destined to ignite a "truly artistic imagination" that will "give shape to a new feeling." Such a new feeling, Langer hypothesizes, would ideally emerge without being under central control and mediate the problems of "our runaway technological civilization" that is marked by "outward violence and inward uncertainty ... the price of our first truly international possession—scientific thought."[37]

THE SYMBOLIC SCOPE OF INDIVIDUATION AND INVOLVEMENT

Langer's analyses and descriptions bear in essential ways upon the psychological foundations that mark, as she puts it in the essay "Man and Animal: The City and the Hive," the deep gulf between "the highest animal and the most primitive normal human being."[38] A further look into Langer's account of this deep gulf and its links to its evitable and inevitable cultural consequences of outward violence and inward uncertainty leads us again to other aspects of the processes of individuation and involvement that are distinctively human. They are rooted in the symbolic capacity and its specific power of ideation that marks human beings. Linking language, imagination, and speculation, such ideational powers increase the range and scope of awareness, gives us a past with a narrative line constructed in memory, and a future constructed as a work of imagination. This is the human open ambient, a "symbolically conceived place" in which we situate ourselves and where the foreknowledge of death is "part of our knowledge of life."[39]

Such a "place" or "placement" in the symbolically transformed biological order makes human individuation something radically distinctive. Life itself, Langer notes, is a process of individuations, a *series* of individuations.[40] Each of us as a "higher" animal is "one final individuation from the great human stock" in a process of individuation that ends in death, the foreknowledge of which is "the price of the great gift of symbolism."[41]

Knowing that death is inevitable leads to the uniquely human aim of self-realization and not mere survival, the concentration of "as much life as possible" in our brief life span.[42] At the same time self-realization alone is not a sustainable ground for social structure. The greatest possible individuation is, indeed must be, tempered by the insight that one's life is rooted in the life of the human stock and that one's individuality is infinitesimal. Of course, such an insight can be shattering and the claim that fullness of personal life is not everything strongly resisted. When Langer writes that "involvement with the whole human race, past and present"[43] is needed, what is the nature of this "need"?

Langer is not alluding to a moral need or demand, but pointing to a natural need rooted in the very nature of individuation of human beings as symbolic animals. This individuation involves our recognition of and dependence upon or rootedness in "the great life of the stock," of which we are an "*expression*." Langer sees each individual as an embodied *representation* of the whole human past, with the consequence that a "single ruined life is the bankruptcy of a long line."[44] Such is the ultimate context of Langer's concept of involvement.

The sense of involvement is the social sense, which Langer characterizes as "the direct feeling of needing our own kind, caring what happens. Social sense is an instinctive sense of being somehow one with all other people ... Human society rests on this feeling ... the feeling of involvement."[45] This feeling can be stressed and strained by processes of what I would call hyper-individuation and its various forms of self-assertion. Such processes involve the diminished power of social symbols that situate us in what Langer calls a "greater life."[46] The role of such symbols in human life is to "constantly express our faith in the continuity of human existence," illustrated in buildings, the establishing of laws and institutions, and precedents of other sorts. While they could be considered conveniences of the day, they are at the same time "symbols of more than their day."[47] When these symbols lose their power to unify associated individuals by eliciting and steering their symbolic involvements with and bonds to one another—the handshake, protocols and rituals, the authority of sanctions and honor—there is a failure to recognize, Langer holds, that "in such bonds lies our ability to be individuals" in an *organization*, a "symbolic structure, a mental reality"[48] and not just masses in a hive subject to a genetically imposed or grounded unity.

The individuation-involvement duality does not mark a simple conceptual difference or distinction. They are not states, but relational processes, with degrees and varying directions. Individuation for human beings occurs on physical, vital, and psychic/ideational levels just as its converse involvement does. For humans, the "all-important humanizing habit of speech" powers and underlies "the deepest and most momentous specialization," leading to the exploitation and development of symbolic expression and understanding.[49] Mental individuation, rooted in symbolization, generates the specific dimension of the human ambient and is the measure of our freedom to determine our environment and also the source of the perilous nature of our being determined by it with severe social and individual consequences.

Langer follows Cassirer and others in making *humanity* the rich symbol linking individuation and involvement as processes undergone within social units, not just by separated individuals. Langer points out that humanity is more than a species; it is a society with a continuity held by commitments resident in the associated memories of its members. Involvement as essentially social entails commitments to society's symbolic memory, including commitments to the correctness of our memories of the types and consequences of the commitments circulating in it.

Langer notes that the feeling of loneliness is an index of the non-self-sufficiency of humans. We are certainly not emotionally self-sufficient and to try to be is clearly hazardous. At the same time, not all emotional commitments are to individuals. We are provided with "symbols of our participation in the greater life of mankind, symbols of humanity and of our involvement in it."[50] In such cases natural ties get replaced, to some extent, with symbolic ones: "obligations, recognition of hereditary commitments, pieties, sanctions, honors, and above all the diverse rites of holy communion,"[51] that is, rituals of being together in affirming a bond, whether sacred or otherwise.

Langer poses the charged question of how far we can or should carry the process of individuation. Her answer: "just as far as symbols of our social involvement hold the balance against it."[52] She then asks a further question of pressing pertinence, not just for the cultural situation in which she was writing but for us now. The question is clearly part of the whispering future in Langer's work: "What has happened to the relations of individuals to society that makes us aware of them as never before, and makes us feel vaguely if not acutely that something is wrong between them?"[53] It is true, as Langer remarks, that human beings "not only feel but conceive their identity."[54] As "symbol-mongers" they belong, willy-nilly, to society, tribe, group, church, and so forth, that function as a "greater body": "the symbolic office of the greater body to which he gives himself is manifested only in his emotion toward it" which is not purely or mainly practical or instrumental.[55]

Langer's analysis engages present cultural and social reality by tracing the roots of how "the operation of the individuating principle in the greater whole, the society, begins to outrun the tempo of man's symbol-making capacity … the emotional effect on people as individuals is that the holiness goes out of all institutions."[56] One exercises one's individual choices and preferences with the result being the breaking up of society into its ultimate units with no shared overarching symbolic canopy rooted in shared feeling with historical lineages embodied in institutions based on assumed natural social articulations. Thus, institutions arise, although not always beneficially, through "conscious planning and ruling."[57]

This is Langer's way of characterizing the turbulent shift that has occurred with the rise of science and science-based technologies: our inherited symbols of humanity are failing with a loss of the sense of involvement and a "meaningless rat race" in which human beings are "reduced to nothingness, alone in life and in death."[58] She admits that civilized life entails "onerous things" in that our strongest bonds to one another involve "the acceptance of commitments we did not make," commitments she sees being repudiated, with the consequent loss of emotional security attendant upon the "shattering of natural symbols" of traditional society.[59] Such a shattering, with its many root causes, Langer sees as a maelstrom moment in which human beings have become unanchored through the breaking of long cultural traditions.

Langer admits that she has no answer to demands of this moment, which is one we find ourselves in, or at least any easy answer, but does claim a central role for philosophical analysis of the growth of culture as the "modification and stretching that is mentality itself."[60] Langer holds that feeling and imagination are "greater, older mental functions"[61] that are subject to disturbance ranging from brief disorientation to serious long-term imbalance. Because of her foregrounding of the role of images in human world-building Langer claims that the action of imagination wields the greatest force on feelings, going beyond immediate experience and holding it together in a way different from other animals. It is imagination that supplements the fragmentary world of sensation to construct a framework of images, notions, figments of all sorts around the perceptual cores of experience that make up the centers of attention that language stabilizes and names in variously motivated ways.[62] This process that builds up the "logical scaffold known as common sense"[63] Langer calls "logical intuition." This scaffold supports human feeling and emotion in a unique or sui generis way, giving it its affective ground-tone. With respect to affectivity, Langer writes that "we have feelings toward the world—not transient excitements, but a permanent emotional attitude toward a permanent 'universe,'" giving us a coherent life of feeling[64]—or one marked by interlocked incoherences.

Langer's mature model of the symbolic mind, as worked out both in *Philosophical Sketches* and in the *Mind* trilogy, establishes that the originary threshold of human sense-making is "a matrix of vague, great significance, physical and emotional at once, felt rather than understood."[65] Felt significance is connected with symbolic seeing which Langer has shown "gives the world its fundamental unity, much deeper than the unity of its causal connectedness."[66] Felt significance permeates what Langer calls "common sense," a mélange of concepts which, being "rough but firm," is not disturbed in what Langer calls "settled" and "normal" society. Indeed, Langer thinks of common sense as "chiefly a fabric of images"[67] embedded in and guiding individually and socially intertwined affectively charged habits of attending, acting, and thinking which, Langer remarks, function in settled and tacitly operating ways. Rapid changes of the status quo push people out of these ways such that a kind of radical insecurity spreads over core domains of life—"providence and its plan, the credentials of human authorities, the validity of morals and institutions, the value or vanity of work and of life itself."[68]

Langer's descriptions are unsettling premonitions of our current situation, which is perhaps not as novel as we would like to think. The world we live in is anxious in many respects and what Langer says about the European world already in 1956, looking back over and beyond the twentieth century, has general import: its form is "broken. We feel ourselves swept along in violent passage, from a world we cannot salvage to one we cannot see; and most people are afraid"[69]—or deeply uneasy. What Langer sees, she claims, is a flood of novel experiences, states of nervous tension "verging on hysteria," a growing failure of words, and a spreading feeling of the inadequacy of our conceptual powers "in the face of the new world" that is marked by accelerated changes and expansion of thought.[70] Indeed, more fundamentally, "the world image has collapsed"[71]—and not just the world image of the European West. Tolerance of different world images and their diverse practical consequences has collapsed, too, both nationally and internationally.

There is a speeding up of things, with radical changes in the physical nature of places. There are fewer and fewer secluded places, with deep auratic or metaphorical character, where one can be alone or find oneself in attunement with the earth, although the tide is definitely turning. Old metaphors, Langer writes, "have lost their aptness" and old-world models broken or, one can say, brought into interminable warfare with one another. While it is true that whole segments of humanity have become bereft of their "mental orientation and moral certitude," Langer does not see the situation as hopeless but rather as a challenge for philosophy—and in particular for certain forms of philosophy that show symptoms of "intellectual collapse" by reason of their avoiding constructing a coherent way to counter the practical social effects, technical, economic, political, of the abstract world of mathematics which is "entirely alien to any metaphorical images we can muster,"[72] with the attendant loss of the power of universal ethical symbols.

Consequently, Langer argues, we need a philosophy such as the one she is proposing, that is "a critique of working concepts in all domains of life, especially those where old concepts are obsolete and new ones still incoherent and perhaps more than a little metaphorical."[73] Langer's primary working concept is symbolic transformation, the access point to a model of minding that defines its origins, branching powers, and results, both positive and negative.

VITAL FOUNDATIONS AND THE BREAKING OF SYMBOLIC ORDERS

Langer's conceptual construction, and reconstruction, of such a model and its bearing on human life culminates in her *Mind* trilogy, which incorporates and supplements in great detail the thematic and conceptual distinctions of the studies collected in *Philosophical Sketches*. Langer returns in the third volume of *Mind* to fundamental cultural consequences, both positive and negative, that we are faced with and offers "a way to a rational concept of human mentality"[74] that is able to throw more light on its irrational products and that at the same time recognizes the inevitability of their appearance, indeed, a permanent tendency to irrationality. What Langer calls its "vital foundations"[75] are the insuperable bio-psychological conditions embedded in the human mind's characteristic products: society, religion with its rituals and myths, conceptual thought, and personal intent and action. These products are themselves both the frameworks and the results of various forms of individuation and its core insight: death as the "inevitable finale of every life ... [the] implicit consequence of a basic evolutionary process, individuation."[76]

These products have histories that are subject to the conditions in which they have arisen. They arose, and continue to develop, in the intertwined processes of the individuation and involvement of associated individuals. Individuation is marked by what Langer calls "entrainment" and opens a "way for small impulses to find expression in the wake of a great and vigorous one" so that these small impulses "fit into the larger unity as elaborations of its passage."[77] In the third volume of *Mind* Langer foregrounds the human organism as a "vortex" of acts that is "functionally centralized and thereby divided from its environment."[78] The symbolic play of the brain is continuous and the self/body is an "action-built matrix" wherein, Langer fatefully points out, every act of the organism "inscribes itself on the cumulative formation of the historic individual. And so does every act of mentation."[79] These acts have deep linkages[80] to one another, and each individual body/self-life is marked by an inward rhythm. "Each life is a rhythmic structure ... The inward rhythms of each of different individuals vary widely and provide groundworks of their separate individualities," although each life exemplifies "one deeply felt rhythm" ending in the cadence, "the decline to death."[81]

The recognition of such a decline, and the sobering tragic vision of life that seemingly entails, would appear to induce a search for a defense against this decline, the original manifestation of which is religion and magic—and the subsequent development of mythic narratives and ritual practices occupied with dealing with a transcendent pole. The history of these efforts, however, is marked by progressive tension between the development of civilization, with its present world demands and achievements, and the resistances offered by systems of transcendence and their deep affective commitments. This tension leads to what Langer calls "the breaking" that civilization sets in motion. But it can also lead to their political fusion, as in Hitler's National Socialism.[82]

Civilization, the practical and technical organization of life in Langer's conception, is like the processes and products of symbolization on which it is dependent in that it grows by its edges, expanding and developing the distinctive spheres and powers of human life.

The process of civilization leads to conflict in that it involves complex forms of rational organization that upset what Langer calls the "ethnic balance"' or "vital balance" that is rooted in the types of symbolic thought and symbolic communication that shape what she calls "tribal" societies and their affective bonds. Such societies are not necessarily "primitive" in the usual sense of that term as "undeveloped." They have and do operate with high forms of symbolic structures, especially narrative mythic cosmologies, and elaborate rituals of death-avoiding or even death-seeking sacrificial practices. The tendency of tribal societies to be closed within a self-sufficient symbolic frame marked by a sense of a transcendent, transhuman, or transpersonal pole leads to resistance and elaborate attempts to maintain and defend in multiple ways the fulcrum of social equilibrium with its patterns and deep currents of feeling.

Langer recounts in the third volume of *Mind* a schematic history of a breaking that "a human world stunned by civilization"[83] underwent and is still undergoing. The rise of civilization, as she puts it, with its increasing language-based intellectuality and formulated gradations of value, "has inscribed itself objectively on the face of the earth," leaving grand scars on it which are deposited by "the life history of the mind."[84] Such an inscription is exemplified paradigmatically in the rise and development of cities and the intermixing in patterns of thought beyond trade in goods, inventions, and so forth.

The picture Langer paints of the linkages between the semiotic dynamics of the processes of civilization and cultural consequences of uprooted and transformed patterns of feeling and symbolic commitments and its lessons are not of something exotic and foreign. The breaking of old orders and of the factors leading to it that Langer described half a century ago is a continuing pressing worldwide phenomenon with a seemingly permanent conflict between highly articulated forms of continuing tribalism with battling closed worldviews and the feeling that they, too, are breaking. The "evolving mind," Langer argues, needs to purge speech, fantasy, and ritual from their exclusionary and totalistic tribal forms and contents.

BALANCE AND THE OPEN AMBIENT

Looking through Langer's analytical lenses at the development of civilization and its cultural consequences, both positive and negative, we see that they are the outcomes of the symbolic mind's open ambient, the dynamic matrix of entwined sign systems and the worlds of meaning embodied in them. They constitute ways of being in the world, governed by semiotic logics and contents particular to them. They are the ways that feeling takes on form. These systems of forms, including their technical outcomes and societal groupings, articulate and stabilize our patterns of relations to the world and to one another on affective, actional, and symbolic levels. Each of these levels has a complex array of what C. S. Peirce called "interpretants," or "proper significate effects," that shift the patterns of relations in the sensory continuum, giving a fundamental "tone" to our individual lives and its processes of individuation and to the shared contexts of involvement in which it occurs.

These processes and contexts that structure the unstable balance between individuation and involvement in different societies are both the sources of social and psychological conflict as well as the motivation to find ways of mediating between them. How is the creation and maintaining of such balance to be achieved? Langer writes that dialectic of thought alone, "without a social need to evade or resolve inconsistencies arising from different basic concepts, does not generate enough 'drive' to maintain a progressive mental life."[85] But resolving them is necessary for the "'moral advance' of society" and demands an advance on "the conceptual structure of the moral structure itself."[86] This is our task. Human culture, constituted by complex patterns and deep currents of feeling and affective attunements, undergoes qualitative shifts and its progression "requires human culture as a whole to keep a certain balance between its highest and lowest degree of change … entraining the countless psychical activities that make up mental life and an individual's or society's *Weltanschauung*."[87] Langer writes of finding a way to "uphold the drive of incompatibilities"[88] within a "complex yet balanced front" that makes up the process of civilization.

Langer sees her contribution to finding this way as a conceptual one: to supply a comprehensive model of minding that uncovers the roots and consequences of symbolic transformation. Langer's philosophical approach to these issues oscillates between diagnostic description and recognition of its sources and a clear model of what is possible, given human symbolic powers. She exhibits a kind of pessimism of the intellect accompanied by an optimism of the will. On the one hand, she claims that charity, nobility, honor, and even pity are "never in very generous supply among average people" and that there is an "inherent viciousness of human being,"[89] some rift or tear in human nature that is source of the distance between great individuals and their avowed standards—and not just "great" individuals. On the other hand, Langer's symbolic model of minding as a dynamic branching of feeling in all its dimensions and the pivotal achievement of the capacity for the free use of symbols for ideas traces how the specific symbolic power of ideation that marks humans both contributes to deleterious cultural consequences and at the same time offers a way of bringing them to mind in such a way that one can deal with them by presenting novel possibilities of living, utopian as they may seem.

This power of ideation is expressed or exemplified in the linkages between language, imagination, and speculation. Language institutes the greatest, pivotal change in the human mode of being in the world: it is the "increased scope of awareness in speech-gifted beings,"[90] where, as noted before, "our past is a story, our future a piece of imagination" and our ambient a "symbolically conceived place."[91] Langer shows in a powerful way that this symbolically conceived place, with its many levels and dynamic rhythms, is not just in our heads, but permeates every fiber of our being. We are faced, she writes, with a worldwide situation in which "head-on clashes of old faiths and new scientific and (especially) pseudoscientific persuasions are more often fateful encounters."[92]

The socio-economic consequences of these clashes are enormous with the whole human ambient, social as well as physical, being thrown into "convulsion." Langer sees the modern world, with the growth of mathematical sciences and their extension to the technological, economic, and political orders, as "two great streams of mental evolution."[93] While the formalization of the number sense was revolutionary, with "inestimable power for its purposes,"[94] when the range of its purposes has been broadened to an overarching attempt to lay a formal abstract grid of concepts, especially economic and technological concepts, over the human world, the result has been the building up of "world-wide waves of emotional conflict … in every society, savage, barbaric, or civilized."[95]

Langer calls the present age "a precipitous, heady transitional age, the Age of Science" with an unknown goal.[96] The "terrifying acceleration" of the past three centuries, she claims, with some solid justification, "blasted" our moral and mental balance and we have no way of "guessing whether or how we shall retrieve it."[97] Her reason forty years ago for thinking so is as pertinent now as it was then. It belongs to the very nature of the symbolic mind that defines us as human: "that newest of natural phenomena—Mind—still faces the mystery of all things young, the secret of vital potentiality."[98]

OUR NORMATIVE TASKS

Langer, reflecting on the general import of art in *Feeling and Form*, reveals the inner nature of this secret: "life is incoherent unless we give it form,"[99] and shows us the power and range of an indispensable working principle formulated in *Philosophy in a New Key*: "In the fundamental notion of symbolization ... we have the keynote of all humanistic problems."[100] Our lives are lived within a vast array of dynamically ordered and disordered civilizational and cultural contexts. The relations between them have also to be given form and ordered. Her model of symbolic transformation and of the symbolic mind seeks to do just that on the analytical level. It is up to us to make her analytical scheme, with its normative implications, inform our life practices as a guide and as a warning. It points the way to creating rich and diverse cultural worlds and their civilizing supports that are worthy of the types of beings she has shown us to be capable of being. It also describes the type of world we will continue to face if we do not attend to her whispers about the future.

NOTES

1. This complex notion is taken up from different angles in this volume by Lachmann, Dryden, Gaikis, Nocek, and appears, by reason of its theoretical centrality, in various ways in many other chapters.
2. Susanne K. Langer, *Feeling and Form: A Theory of Art Developed from Philosophy in a New Key* (New York: Charles Scribner's' Sons, 1953), 372. These themes are explored in wide contexts and with many examples, bringing Langer into relation with Peirce's semiotics, Dewey's experience-based aesthetics, and many others in my *Dimensions of Aesthetic Encounters: Perception, Interpretation, and the Signs of Art* (Albany: State University of New York Press, 2022). They are also discussed from different complementary angles foregrounding relations between body-based affective and perceptual processes and the potentiating role of signs and symbol systems in my *Consciousness and the Play of Signs* (Bloomington: Indiana University Press, 1994).
3. Langer, *Feeling and Form*, 372.
4. Ibid.
5. Ibid., 405.
6. Ibid., 409.
7. Susanne K. Langer, *Mind: An Essay on Human Feeling*, vol. 1 (Baltimore: Johns Hopkins University Press, 1967), 87.
8. Langer, *Mind*, vol. 1, 151.
9. Ibid., 354.
10. Susanne K. Langer, *Mind: An Essay on Human Feeling*, vol. 2 (Baltimore: Johns Hopkins University Press, 1972), 355.
11. Susanne K. Langer, *Mind: An Essay on Human Feeling*, vol. 3 (Baltimore: Johns Hopkins University Press, 1982), 125.
12. Langer, *Mind*, vol. 3, 194.
13. Langer, *Mind*, vol. 2, 255–8, 328–31; *Mind*, vol. 3, 213–17.
14. Susanne K. Langer, *Philosophy in a New Key* (1942; Cambridge, MA: Harvard University Press, 1957), 90.
15. Ibid., xiv.
16. Langer, *Mind*, vol. 1, 100.

17. Ibid., 59.
18. Susanne K. Langer, *Philosophy in a New Key*, 90.
19. Ibid., 201.
20. Ibid., 201.
21. Susanne K. Langer, *Philosophical Sketches*, 1962 (New York: Barnes and Noble, 2009), 89.
22. Ibid., 89.
23. Ibid., 90.
24. Ibid.
25. Ibid., 91.
26. Ibid., 92.
27. Ibid.
28. Ibid., 93.
29. Ibid., 93–4.
30. Ibid., 95.
31. Ibid., 96.
32. Ibid., 97.
33. Ibid., 98.
34. Ibid.
35. Ibid.
36. Ibid., see 93 and 99.
37. Ibid., 99.
38. Langer, *Philosophical Sketches*, 104.
39. Ibid., 106.
40. Ibid., 107.
41. Ibid., 108.
42. Ibid.
43. Ibid., 110.
44. Ibid.
45. Ibid., 111.
46. Ibid., 113.
47. Ibid., 114.
48. Ibid.
49. Ibid., 121.
50. Ibid., 124.
51. Ibid.
52. Ibid., 126.
53. Ibid.
54. Ibid., 128. See also Robert Innis, *Susanne Langer in Focus: The Symbolic Mind* (Bloomington: Indiana University Press, 2009); "Affective Semiosis: Philosophical Links to Cultural Psychology," in *Psychology as the Science of Human Being*, ed. Jaan Valsiner et al., 87–104 (Cham: Springer International, 2016); *Between Philosophy and Cultural Psychology* (Cham: Springer, 2020); and *Pragmatism and the Forms of Sense: Language, Perception, Technics* (University Park: Penn State University Press, 2002); Lloyd Sandelands, *Feeling and Form in Social Life* (Lanham, MD: Rowman & Littlefield, 1998); and Adrienne Dengerink Chaplin, *The Philosophy of Susanne Langer: Embodied Meaning in Logic, Art and Feeling* (London: Bloomsbury Academic, 2020), ch. 10.
55. Ibid., 129.
56. Ibid., 131.

57. Ibid., 132.
58. Ibid.
59. Ibid., 133.
60. Ibid., 137.
61. Ibid.
62. Langer, *Mind*, vol. 3, 203–19.
63. Langer, *Philosophical Sketches*, 142.
64. Ibid.
65. Ibid., 144.
66. Ibid., 145.
67. Ibid., 155.
68. Ibid., 140, 155.
69. Ibid., 157.
70. Ibid., 159.
71. Ibid., 142, 159.
72. Ibid., 161.
73. Ibid., 165.
74. Langer, *Mind*, vol. 3, 89.
75. Ibid., 90.
76. Ibid., 90–1.
77. Ibid., 91.
78. Ibid., 92.
79. Ibid., 94.
80. Cf., ibid., 98.
81. Ibid., 109.
82. Langer's *Philosophy in a New Key* (1942) comments on the effects of the starvation of the uprooted industrialized individual through modernization. It had created "A vague longing for the old tribal unity" that made "nationalism look like salvation, and arouses the most fantastic bursts of chauvinism and self-righteousness." Langer, *Philosophy in a New Key*, 292.
83. Ibid., 154.
84. Ibid., 155.
85. Langer, *Mind*, vol. 3, 194.
86. Ibid., 196.
87. Ibid.
88. Ibid.
89. Ibid.
90. Langer, *Philosophical Sketches*, 105.
91. Ibid., 106.
92. Langer, *Mind*, vol. 3, 218.
93. Ibid.
94. Ibid.
95. Ibid., 219.
96. Ibid.
97. Ibid.
98. Ibid.
99. Langer, *Feeling and Form*, 400.
100. Langer, *Philosophy in a New Key*, 25.

BIBLIOGRAPHY

Dengerink Chaplin, Adrienne. *The Philosophy of Susanne Langer: Embodied Meaning in Logic, Art and Feeling*. London: Bloomsbury Academic, 2020.

Innis, Robert. "Affective Semiosis: Philosophical Links to Cultural Psychology." In *Psychology as the Science of Human Being*, edited by Jaan Valsiner et al., 87–104. Cham: Springer International, 2016.

Innis, Robert. "Between Feeling and Symbolization: Philosophical Paths to Thinking about Oneself." In *Thinking about Oneself*, edited by Waldomiro Silva-Filho and Luca Tateo, 97–122. Cham: Springer, 2019.

Innis, Robert. *Between Philosophy and Cultural Psychology*, Cham: Springer, 2020.

Innis, Robert. *Consciousness and the Play of Signs*. Bloomington: Indiana University Press, 1994.

Innis, Robert. *Dimensions of Aesthetic Encounters: Perception, Interpretation, and the Signs of Art*. Albany: State University of New York Press, 2022.

Innis, Robert. *Pragmatism and the Forms of Sense: Language, Perception, Technics*. University Park: Penn State University Press, 2002.

Innis, Robert. *Susanne Langer in Focus: The Symbolic Mind*. Bloomington: Indiana University Press, 2009.

Langer, Susanne K. *Feeling and Form: A Theory of Art*. New York: Charles Scribner's Sons, 1953.

Langer, Susanne K. *Mind: An Essay on Human Feeling*, vols. 1–3. Baltimore: Johns Hopkins University Press, 1967–82.

Langer, Susanne K. *Philosophical Sketches*. Baltimore: Johns Hopkins University Press, 1962. Cited after reprint edn, New York: Barnes and Noble, 2009.

Langer, Susanne K. *Philosophy in a New Key*. 1942. 3rd edn. Cambridge, MA: Harvard University Press, 1957.

Sandelands, Lloyd. *Feeling and Form in Social Life*. Lanham, MD: Rowman & Littlefield, 1998.

CHAPTER ELEVEN

Music as the DNA of Feeling, and some Speculations on Whitehead's Influence on Susanne K. Langer's Philosophy

LONA GAIKIS

INTRODUCTION

Susanne K. Langer's thoughts on the allusive aesthetic agent *music*—in her words, "the purest of symbolic media"[1]—take this chapter to the musical core of Langer's philosophy.[2] The meaning of (and the ineffable in) music is in itself an auratic frontier that has occupied the minds of artists and philosophers ever since Plato eliminated the *unbecoming* harmonies from the metrics of myth. As a music enthusiast and lover of the arts, the young logician Langer could not desist from attempting to understand what might lie beyond the formal structures of semantics, a question that had occupied many great minds at the dawning *new age* of logical analysis in twentieth-century philosophy. However, intrigued by the propositions of an early Wittgenstein and Rudolf Carnap,[3] Langer continued Ernst Cassirer and Alfred North Whitehead's attempts to unveil the reach of symbolization processes in art, cultural forms, and ritual acts. To this triangle of thinkers, these human practices enrich and secure the mental substrata from which the matrices of meaning are woven. By analyzing art as a symbolic form, Langer was "scouting the possibility that *rationality arises as an elaboration of feeling*."[4] In the further development of her hypothesis, M/mind is brought back to the intricacy of bodily processes and their reciprocity with living and non-living entities. Art as *presentational form* is an analogue of embodied and embedded *feeling*, which is not to be confused with a theory of "emotion" or "sensitivities." Yet Langer's "morphological"[5] conception of form in music—and her general phenomenology of the arts—was largely misunderstood as an emotivist approach of *reading feelings into artistic expression*. This was simply not the case.

"'Feeling' is a verbal noun ... that psychologically makes an entity out of a process," Langer surmises. "To feel is to do something, not to have something."[6] In volume 1 of *Mind: An Essay on Human Feeling* (1967), Langer's semiology of artistic forms takes a rigorous empirical turn. From meeting the artist "halfway" by *feeling* in the art symbol,[7] she refocuses her terminology to a biologically grounded system that conceptualizes the emerging of mentality—and its diverse forms of expression in art, ritual, and science—from the concatenations and patterning of living form:

> In the first place, the phenomenon usually described as "a feeling" is really that an organism feels something, i.e., something is felt. What is felt is a process, perhaps a large complex of processes, within the organism. Some vital activities of great complexity and high intensity, usually (perhaps always) involving nervous tissue, are felt; being felt is a phase of the process itself. A phase is a mode of appearance, and not an added factor.[8]

Langer's magnum opus *Mind*, vols. 1–3 reveals, at last, how much her philosophy had been influenced by and was infused with process thought. The following reading of her "new key" in philosophy traces her concept of *feeling* as it burgeons early on, seeded by her "great Teacher and Friend" Alfred North Whitehead. This chapter analyzes its meaning for Langer, and serves as an antidote to the misunderstanding of a female logician who had seen more in music than a solipsistic theory of self-expression.[9] Her highly abstracted conception of musical form, essentially as "dynamic sound-patterns,"[10] offers a model towards rendering an ontology of visceral and intra-organic processes.

PRACTICAL UNREASONING: HOW FEELING BECOMES FORM

The foundations of Susanne K. Langer's philosophy are laid out in many ways in her first book, *The Practice of Philosophy* (1930), as the young logician reaches beyond the confining structures of propositional logic. In her chapter 7, on "Insight," she wittily refers to a seeming "decline and fall of the author's Pure Reason, and the advent of some Practical Unreason," as its title—to "hard-headed logicians"—might sound as if she were aiming to "show how, in sentimental and ethical matters, logic fails us."[11] The dominating philosophical school in the US at the time was indeed a clear-cut positivism that sought truth mainly in formal propositions and scientific proof. The division in Snow's famous "two cultures" was in full swing. Langer's approach, however, was just as hard-headed—perhaps even more so, as she aspired to include the *unlogicized*[12] areas of life in the systems of meaning by insinuating the possibility of an "'unconscious' symbolism."[13] She supports her argument by reasoning, "If we cannot account for the sort of knowledge that is called 'insight,' but must leave this to a special undiscovered faculty, then our theory is not an expression of universal equations."[14]

What seems to be a dig at logical positivism resonates with even more verve when Whitehead's overall praising of the book is reduced to one remark: that some of its "conclusions may be controversial."[15] Throughout her body of work, Langer boldly challenges the great minds of modern philosophy. At that time, she seemed to already have a strong intuition for what would become the foundation of her philosophical investigations. Since "Music is the purest of symbolic media," she speculates whether "the final object of musical expression is the endlessly intricate yet universal pattern of emotional life?"[16]

Scanning through Langer's major works, *Philosophy in a New Key* (1942), *Feeling and Form* (1953), and the three volumes of *Mind: An Essay on Human Feeling* (1967, 1972, 1982) for direct references to Whitehead, leaves one—at first glance—with the impression that there are more anecdotes from her doctoral advisor than any lasting impact of the process philosophy and metaphysics he was developing while Langer was his student at Harvard. However, the term "feeling" is consistent in her writings, and Whitehead's numerous discussions of "feeling" and "feelings" buzz throughout *Process and Reality* (1929). Although *feeling*, for Whitehead, had been "a mere technical term,"[17] reading through Langer, it actually proves central to his process ontology and particular notion of *prehensions*. With *feelings*, Whitehead referred to the elemental perceptual processes through which we as humans apprehend *reality* in its vibrancy and complexity.[18] Whitehead's widely known critique of the bifurcation of nature into two realities—"one is conjecture and the other is dream"[19]—shimmers through in this notion, as he imports a Lockean desertion from a representative theory of perception to incorporate *direct experience* as the basic vehicle with which we grasp the world.[20] For Whitehead, *feeling* is the very first—and most formative—impression with which our minds operate. *Feeling* accesses a relational connection that describes, in his words, "the basic generic operation of passing from the objectivity of the data to the subjectivity of the actual entity in question." He goes on to say, "Feelings are variously specialized operations, effecting *a transition into subjectivity*" (emphasis added).[21] In Whitehead's connective theory, feeling emerges from *actual entities* (or *actual occasions*) as "the final real things of which the world is made up."[22] These units form the complex of all living and non-living matter, yet not in a material sense, but transient, in passing "drops of experience, complex and interdependent."[23] They are the metaphysical entities that represent the microcosmic realities of process and induce creative advance. Confusing their aggregate, however, as a "final reality," or the actual real, would mean committing the Fallacy of Misplaced Concreteness,[24] and here Langer seems to take process thought into a direction of her own. Langer's idea of "feeling," whose virtual analogue was to be perceived in artistic expression, is rooted in the simplest of actual organic processes. In *Mind*, vol. 1, she highlights:

> Feeling stands, in fact, in the midst of that vast biological field which lies between the lowliest organic activities and the rise of mind. It is not an adjunct to natural events, but a *turning point* in them. *There must have been several such turning points in the evolution of our world*: the rise of life on earth, perhaps the beginning of irreversible speciation, the first true animal form, the first shadows of a 'psychical phase' in some very active animal, and the first genuinely symbolic utterances, speech, which marked the advent of [hu]man. It is with the dawn of feeling that the domain of biology yields the less extensive, but still inestimably great domain of psychology (emphases added).[25]

This evolutional branching out of feeling,[26] of which Langer's conception of mind is composed, resonates seemingly little with the semiology of art she developed throughout the first half of her philosophical career, as she substantiates her reasoning of nature, art, and form in *Mind*, vols. 1–3 with empirical data and scientific accounts. In tracing back its roots and connecting the morphology of her models and terminology, one can see to what extent her philosophy actually reached for a process-based ontology.

In *Susanne K. Langer. Die lebendige Form menschlichen Fühlens und Verstehens* (2000), German scholar Rolf Lachmann gives a most detailed and carefully put together account of the Whiteheadian influence that traverses Langer's philosophy, especially when it comes to her concept of *act*, which can be equated with Whitehead's conceptualization of *actual entities*.[27] This reading is shared by Donald Dryden, who emphasizes that both Langer and Whitehead sought to transcend the limits of scientific materialism.[28] Yet, was Langer truly committed to developing a metaphysics? Her early ideas on the possibility of rendering proto-linguistic symbolization do align with

Whitehead's paradigmatic shift towards a semiology of inter- and intra-organic percepts in *Symbolism: Its Meaning and Effect* (1927), and his particular view on the advance of form from aesthetic experience. According to Lachmann, Langer continues Whitehead's ideas in *The Practice of Philosophy* (1930), though "not in terms of the intended scope, but in terms of its analytical level."[29] This is particularly evident in Langer's specification of the relationship of the symbol and its discursive and presentational nature, which she unfolds in her second book *Philosophy in a New Key* (1942). The musical connotation—already so apparent in the book's title—returns when she lays out her view of how symbolic relations differ from signifying ones in *signs* (or signals).[30]

Key to Langer's semiology is her notion of meaning being determined by both logical and psychological aspects. To her, "Meaning is not a quality, but a *function* of a term," because logic, aiming to understand the very foundations of how meaning comes about, essentially deals with relations.[31] However, the problem is that these relations are not simple two-termed affairs: meaning in symbols is determined by a variety of terms. As Langer remarks, "different types of meaning consist of different types and degrees of relationship."[32] Her intention is to conceptualize how these symbolic relations, given, for example, in connotation and denotation, differ from signifying *sign* (or signal) relations, and depend on the recipient. "A function," Langer continues, "is a *pattern* viewed with reference to one special term round which it centers; this pattern emerges when we look at the given term *in its total relation to the other terms about it*."[33] She elucidates this rather mathematical scheme most tellingly with an example from old organ music, where a chord is written down as a pattern that embeds the note and its surrounding keys. This "meaning-pattern," as Langer calls it, expresses the chord in functional terms because it is represented differently depending on the point of view within the term.[34] This allows for the sounding pattern to be articulated in multiple ways, without distorting its meaning, or tone.[35] Whereas a *sign*-function remains triadic, Langer's model for symbols is a four-termed relation, which means that it is denotational and covers "subject, symbol, conception, and object."[36] She deduces this idea from her example in music. The "new key" in her theory of symbolization highlights the need for fitting terms to unlock the patterning of creative advance. Hence, "Symbols are not proxy for their objects, but are *vehicles for the conception of objects*."[37]

By viewing meaning through this general principle as a function, and not simply a property, Langer shows how two rather controversial kinds of meaning—the logical, which would take the symbol as its key, and the psychological, which refers to the individual reading in the subject—could be reconciled.[38] Symbol functions (both discursive and presentational) *resonate* with corresponding keys. This makes Langer's idea of *analogy* rather flexible, causing a transversal sense of the symbol and what it can affect. Langer's reasoning evoked criticism in her colleagues from logical positivism at that time,[39] yet she continued, and pushed forward a new general distinction between discursive forms of language, logical syntax and scientific symbols, and the full scope of presentational forms to be found in the arts.

The significance of music, in this respect, lies in its inability (!) to express permanent contexts because it lacks any assigned connotation, yet it is able to allude to an abundance of open and associated meanings. This gives musical form particular importance in Langer's semiology of the arts, as it makes explicit the purest form of *presentation*. Even though "Music has all the earmarks of a true symbolism," Langer speculates, it can only serve as an *unconsummated symbol*. "Articulation," she adds, "is its life, but not assertion: expressiveness, not expression."[40] This insight establishes her general notion of artistic import, as she will substantiate later on, in *Feeling and Form* (1953): "Music has *import*, and this import is the pattern of sentience—the pattern of life itself, as it is felt and directly known. Let us therefore call the significance of music its 'vital import.'"[41] This feature is given in any kind of artistic media, yet with varying characteristics.[42]

WHAT IS MUSIC?

Music's shape-shifting properties matter largely for Susanne K. Langer's conception of form. She asserts that "If [music] reveals the rationale of feelings, the rhythm, and pattern of their rise and decline and intertwining, to our minds, then it is a force in our mental life, our awareness, and understanding, and not only our affective experience."[43] Her view acknowledges an inescapable carnal and sensually infused aspect of knowing that inspires what she initially phrased a *vaguely contagious hunch* in both the sciences and the arts.[44] This presupposes a highly abstracted idea of what music actually is. In *Philosophy in a New Key*, Langer references musicologist Eduard Hanslick's non-representational understanding of music as "tönend bewegt formen," *dynamic sound-patterns*, later coining the expression "sounding forms in motion."[45] Yet she has strong reservations regarding his general determination of "meaning" to conventional denotation.[46] Leveraged by this formal apprehension of music, Langer achieves a purely logical standpoint of what music is and what it does. She explores this in more detail in her following book *Feeling and Form* (1953), where the abstraction of music helps her to deduce a general model of how art perception occurs.

It is notable that Langer is never interested in analyzing the content of an artwork, nor establishing tools of any kind to assess an art form. Her intention is to understand what distinguishes artistic expressiveness and its particular object within the universe of things, and although there are, at times, traces of aesthetic preferences, her theory in general acknowledges a plurality of artistic form.[47] The whole of her analysis has the aim of understanding what makes an object an object of artistic value. And this is its ability to convey a very particular sense of illusion, or in Langer's words, the "virtuality" of *living form*.[48] Her quite unique view of aesthetics triggers a whole new classification of the arts, which is determined by natural philosophical principles rather than linguistic ones. Langer's *Feeling and Form* is a very detailed discussion of all main artistic mediums from sculpture to prose, which significantly reorganizes the principles of art perception according to geometry, motion, and time. This seeds a new materiality of the arts, as the *illusion of living form* expresses itself in virtual space and is actuated in virtual time. Music, from this perspective, becomes the image—*qua* model—of subjective time.

Langer devotes a whole chapter in *Feeling and Form* to the discussion of the "image of time,"[49] in which she conceives of music as an art form that captures a sense of time in passage. "*Music makes time audible, and its form and continuity sensible*" (Langer's emphasis),[50] she states. Unexpectedly for an American logician, she also makes explicit the obvious links to Henri Bergson's conception of *la durée réelle,* or rather *la durée pure*, as the intuitive perception of time.[51] Both philosophers correspond in their insistence that there exists a first-hand experience of time that exceeds chronological measure.[52]

Although Langer agrees with Bergson's critique, she conceives of it as a challenge to her logical construction, rather than a defeat to symbolic expression as a "metaphysical pitfall."[53] In her view, "it is not the intervention of symbolism as such that balks our understanding of 'lived' time; it is the unsuitable and consequently barren structure of the literal symbol."[54] This motivates her search for a symbolism whereby the impression of "firsthand knowledge," or "intuition" as she claimed earlier, can be conceived of and expressed.[55] She does this by bringing musical duration into play as a formal equivalent of *la durée pure*, as the direct intuition of time.[56] Music now serves as a *dynamic image* that displays *all that is felt*: "The semblance of this vital, experiential time is the primary illusion of music," she asserts.[57] Whereas Bergson considered the conceptual move to spatialization a general betrayal of intuitive knowledge, and remained defiant towards any kind of scientific formalization, Langer took it as a means to expand the idea of music's "one-dimensional formless flow."[58] Spatial illusions in music, whether in volume, tone or movement, are an attribute of music's presentational form. This musical space enables a temporal auditive realm of several dimensions. Because perceived space in music is relative to its acoustic environment, it can only be considered a secondary illusion.[59]

Indeed, the two-dimensional mesh of a picture theory of meaning could not stop Langer from unleashing the potentiality of a spatiotemporal alternative that presents the ever-changing flux of felt states as they apply "*for* a mind as well as *of* a form."[60] *Feeling* in its purest form is occurrent in music.[61] By now it should be obvious that Langer's philosophy has nothing to do with understanding the happy and sad elements in artistic expression. She maps out a space-time that renders a general model for how feeling comes about, how impingements rise and fall in intensity, how their impulses fade out, or carry on. Music as a reflection of the continuous morphology of feeling is a model for how these intricate subliminal and subcutaneous movements are realized. Within this framework, it makes perfect sense for Langer to suggest:

> The essence of all composition—tonal or atonal, vocal or instrumental, even purely percussive, if you will—is the semblance of organic movement, the illusion of an indivisible whole. Vital organization is the frame of all feeling, because feeling exists only in living organisms; and the logic of all symbols that can express feeling is the logic of organic processes. The most characteristic principle of vital activity is rhythm.[62]

Langer has now outlined a scheme that situates meaning-making both in nature and culture, and the becoming of form interstitially between the subject and the corresponding object. The materials of music are composed of the impingements from virtual movement in virtual time that are felt as a succession of physical events in rhythmic sequences. The concatenations of bodily processes fuse with the virtuality of occurring forms in rhythm. This creates a matrix, a web of reciprocal relations of the internal and external world, which makes musical form a collective and sensually fecund experience. Rhythmic continuity epitomizes the endless process that perpetuates the preparation of new events with the ending of previous ones.[63] Rhythm commands its future. Harmonic progressions, the culminating of pressure in dissonance or the gliding of tones in melodies are "rhythmic agents," Langer remarks.[64] These patterning successions are an interweaving of tensions and resolutions, bound to somatic and metabolic processes common to all human bodies.

"What we call mind, soul, consciousness, or (in current vocabulary) experience," Langer surmises, "is an intensified vitality, a sort of distillate of all sensitive, teleological, organized functioning."[65] The foundations of M/mind lie in the dynamic matrices between the subjective and objective—of what is felt as action and what is felt as impact. When Langer evolves her generative idea of the "musical matrix" in *Feeling and Form*, she anticipates a structure for her biological study of process that will inform her later philosophical engagement. Here, the *act* model, which displays the embedded and relational character of events, is rendered in a curve of initiation, acceleration, consummation, and cadence.[66] Langer maintains her undulating model of occurring form. Hence, music becomes the DNA of feeling.

THE LURE FOR FEELING

Music's potential for a system based in natural philosophical principles resonates in several points with Whitehead's organicist theory of perception. Having zoomed through key passages that show the progression of Susanne K. Langer's thinking and conceptual development, we now look back to the young logician herself. What Langer arrived at by means of analyzing feeling in artistic forms seems close to what Whitehead coined "vector feeling" in *Process and Reality* (1929).

Although "feeling" had not been central to Whitehead's own process ontology, his coining of the term appears to have been fecund enough to inspire Langer's lasting "imaginative research."[67] As part of the vanguard of philosophers who began to excavate and contextualize Langer's full body of work in the late 1990s, the previously mentioned scholars Rolf Lachmann and Donald Dryden not only analyzed her work in relation to process philosophy, but were also involved in researching her estate and archive of notes, making parts of them accessible. One pack of notes, edited by Lachmann and published in a special issue on Langer in *Process Studies* (1997), taps into Whitehead's "Course on the Philosophy of Nature" (1927–8),[68] which Langer attended after completing her doctoral thesis in 1926. This documentation is of particular interest, for it sheds light on the intellectual soil that Langer's thinking emerged from. Most of all, it proves that Langer, though never explicit, could have been directly inspired by Whitehead's "lure for feeling."

In a nutshell, *feeling* for Langer is an intra-organic dynamic patterning of intensities that activates a knowing below the limen of discursive expression. Whitehead, in turn, describes in *Process and Reality* a "primitive type of physical experience"[69] that returns our direct experience of the world to a deeper, primordial ground of emotional impressions. According to him, we *sympathetically* feel in and with the "other," thereby externalizing the physical realm of our own body.[70] In fact, for Whitehead, this first stage of "vague" emotional perception[71] triggers what he calls "aesthetic feelings,"[72] or what we might call *affect*. His event ontology renders the impact and immediacy of this preconceptual "present" as *vector feeling*, for its movement—derived from the past and merging into the future—behaves like a stream or waves of occurring feeling between entities. "In this vector transmission of primitive feeling," he explains, "the primitive provision of width for contrast is secured by pulses of emotion, which in the coordinate division of occasions appear as wave-lengths and vibrations."[73] This very technical description "in the phraseology of physics," as Whitehead remarks, describes the initial process that seeds the underlying matrix of subjective visceral and vital import. He describes an amplitude of differing intensities (contrasts) that take effect at a lower level of perception, and it seems that this idea might have become a *leitmotiv* for Langer's subsequent process-oriented theory of the arts. Whereas most interpreters root Langer's process philosophical ties to her later treatises in *Mind: An Essay on Human Feeling I–III*, a deeper look into the origins, structure, and especially the virtuous play of her philosophical concepts insinuates that the very core of her concepts may have been shaped much earlier by this influence from Whitehead's lecture on process thought.

On December 3, 1927, the idea of *vector feelings* surfaces in Langer's notebook; Whitehead had elaborated on the modes and importance of *feeling* for process in his lectures. The accuracy of Langer's notes is remarkable, and although her record may omit some aspects, its content is largely complete and illuminating—not only to Langer-researchers but also to those wanting a better understanding of Whitehead. Without going too far into his labyrinthine conceptual universe, we read his account that introduces feeling as a shared organic quality:

> The word *concrete* is derived from *concrescence*, growing together. Only *satisfaction* is not a term in the concrescence; it is the *outcome* of the process ... The intensity of satisfaction is determined by the order of the concrescence as it arises and developes [*sic*]. Thus every element in the concrescence becomes a *value* with reference to the ideal order ... The tone of feeling of a satisfaction passes on into the world which is an objectification of it ... The ideal order is the lure of feeling, the final cause by which concrescence proceeds. This lure of feeling is Bergson's "*elan vital*." [74]

Whitehead's explanations open intriguing planes for philosophical adventures within and beyond Langer's continuation, as they draw hidden and cross-continental alliances between thinkers. While Bergson's theories are examined at large in Langer's philosophical discussions, the absence of direct accounts of a possible heritage of Whitehead's ontology moves the analysis of these notes to a rather hypothetical realm, yet the coherences are striking. What we find in Langer is a deepening and an advancement of Whitehead's *feelings* that entangle with the progression of—and bridge the vastness of—organic form. *Vector feeling* is crucial to Whitehead's conception of symbolic reference, for it represents one of the leading factors that determine his theory of perception, as Langer's notes outline.[75] The "vector-feeling-tone" that Whitehead concretizes in *Process and Reality* is "perception in the mode of causal efficacy"—the realm of purely visceral feeling—which is yet formative in creating objectified relations and therefore one aspect of the symbolization process.[76] Its second aspect, determined by the intuitive and demonstrative nature of sense data, is that of *presentational immediacy*. Both, in interplay, constitute Whitehead's idea of "symbolic reference."[77] His critique—that modern philosophy had partially gone blind through the bifurcation of nature—lies right here. For in Whitehead's words, "It is evident that 'perception in the mode of causal efficacy' is not that sort of perception which has chief attention in the philosophical tradition. Philosophers have disdained the information about the universe obtained through their visceral feelings, and have concentrated on visual feelings."[78] This notion of a submerged intellectual capability of understanding might have led Langer to divert from logical positivism, and to scout for a semiology that could render visible these invisible forms of insight. Like her teacher, she attempted to "fuse psychology with physics" in *feeling*,[79] and she found in the arts a material with which she could identify mind's obscured and subcutaneous reach—*meaning-making* free from questions about purpose, functionality, or significance, yet open to the hidden operations that define feeling's lure, as it passes as an objectification into the world. She sought to render the concrete in presentational form.

An exceptional glimpse into Langer's own thoughts on Whitehead's organicism and the aspects she considered critical is given by Donald Dryden in the aforementioned issue of *Process Studies*. His comparative analysis of Whitehead's process philosophy and Langer's application of it highlights not only the parallels in both—how *actual entities* are continued in Langer's *act* model—but remarks upon why she might have branched off from his universalizing approach of *feeling* as a cosmic principle. Referencing an unpublished manuscript on Whitehead, Dryden found that Langer, in fact, deemed his generalization of feeling "unfortunate," for it excluded the detailed study of material properties at its nexus of body and mind, the internal and external world.[80] She, in contrast, sought to actualize its occurrence by conceiving of feeling as an actual phenomenon—an entirely new component in a biologically based process-oriented philosophy—that emerged "from the evolutionary course of life."[81] Feeling remained the creative drive for life's advance, but Langer wanted to locate it in actual matter, and turned to medicine, chemistry, physics, and biology.

This departure from Whitehead's metaphysics is regarded as a radical turn that subordinated the possible metaphysical scope of symbolization to scientific scrutiny—something Whitehead, by all means, would have dismissed—unavoidably making Langer's great beliefs in science "guilty of misplaced concreteness," as philosopher Randall Auxier points out.[82] But this observation only manifests the impression that Langer might not have continued a metaphysical doctrine, but perhaps contended it. Referencing Dryden, Langer was indeed deeply inspired by "the 'mythical import' of Whitehead's metaphysical vision"[83] and understood its prophetic appeal. Yet being her teacher's critical friend, she was well aware of its potential weaknesses. In reviewing Langer's philosophical development, it seems that she wanted to be an actuator for the possibilities that process thought offered in conjecturing a new *real* from Whitehead's cosmological dream.[84]

Following the lure of feeling, Langer facilitated a way to fathom an ontology of living form that made music its dynamic agent and virtually rendered the concatenations and patterning of life's experience as a whole. But first she was looking for proofs in the actual world. Her notion of culture infused by nature gave her the tools to establish a philosophy of mind based in biological processes. Her conceptual investment in the field of arts during the first half of her philosophical venture nurtured a framework that encompasses artistic expression as well as biological progression. *Natura Naturans*—nature doing what nature does—is what drives creative advance on all levels. Langer's *act* model, as Dryden puts it, "enables the biological sciences to make connections with chemistry and physics and therefore allows the study of mind to articulate with the rest of the natural sciences."[85] Vice versa, her generative understanding of "the boundaries between sciences being always somewhat fluid"[86] as she maintained in her critique, resembles the shape-shifting properties in artistic media that enable a continuous and fertile crossing of primary and secondary features in forms of artistic expression.[87]

The essence of "living," Langer once held, "is a process, a continuous change; if it stands still the form disintegrates—for *the permanence [of form] is a pattern of changes.*"[88] This insight also holds true for the way she conceived of the *living* in philosophical thought. Regarding her flirtations with a possible metaphysical theory of symbolization, Langer remarked in her introduction to *Mind*, vol. 1, "I do not reject or deprecate metaphysics; only it seems to me to be the natural end, not the beginning, of philosophical work."[89] Whitehead's metaphysical speculations noticeably took root in Langer's philosophy of mind. But Langer had sought to build her synoptic view through a "generalization of systematic knowledge."[90] She strived for the particular in empirical research, before heading towards its generalization in metaphysical speculations. This was her approach to the *Mind* trilogy—however, when the time came to complete her research "in a well-constructed epistemological and possibly even metaphysical theory," as she aspired to in 1982, symptoms of aging interrupted its culmination.[91] The resonances with Whitehead remain one of the most intriguing aspects of Langer's work, and the unspoken myth of his influence continues to inspire future generations of philosophers.

NOTES

1. Susanne K. Langer, *The Practice of Philosophy* (New York: Henry Holt, 1930), 160.
2. On the meaning of music in Langer's philosophy, see Giulia Felappi's Chapter 4 in this volume, "Susanne K. Langer on Logic as the Study of Forms and Patterns of Any Sort,"

as well as Christian Grüny, Chapter 8, "The Systematic Position of Art in Susanne K. Langer's and Ernst Cassirer's Thinking," both in this volume.

3. Though Langer explicitly roots her philosophy in Wittgenstein, Russell, and Carnap, she expands the scope of possible symbolisms beyond language with presentational form. "I will go with the logisticians [*sic*] and linguists as far as they like, but do not promise to go no further. For there is an unexplored possibility of genuine semantic beyond the limits of discursive language," she writes (86). "The study of such products they relegate to *psychology*, not *semantics*. And here is the point of my radical divergence from them" (86; emphasis added). Langer criticizes logical positivism in condemning the artistic, or metaphysical realms as a pseudo-symbolistic phantom. Countering Carnap's epistemological perimeter that *I can ask whatever language will express; I can know whatever experiment will answer*, she replies, "the interest in the function and nature of symbolism … does not condemn philosophical inquiry as such." She calls for nothing less than "*every philosophical problem to be recast*, to be conceived in a different form" (85; emphasis in original). Langer will draw her attention to the subcutaneous realms of artistic expressiveness and its meaning. See *Philosophy in a New Key* (Cambridge, MA: Harvard University Press, 1993), 82–7.
4. Langer, "The Art Symbol and the Symbol in Art," in *Problems of Art: Ten Philosophical Lectures* (New York: Charles Scribner's Sons, 1957), 124.
5. Cf. Langer, *Philosophy in a New Key*, 238.
6. Langer, *Mind: An Essay on Human Feeling*, vol. 1 (Baltimore: Johns Hopkins University Press, 1967), 20.
7. Langer, *Feeling and Form*, 393, quoted in Iris van der Tuin, "Bergson before Bergsonism: Traversing 'Bergson's Failing' in Susanne K. Langer's Philosophy of Art," *Journal for French and Francophone Philosophy* 24, no. 2 (2016): 193.
8. Langer, *Mind*, vol. 1, 21.
9. "*Sheer self-expression requires no artistic form*" (216; emphasis in original), Langer states, dispelling the theory of self-expression in music as a wide spread misconception. She concludes, "[i]f music has any significance, it is semantic, not symptomatic" (218). It is worth noting that the examples by which Langer contrasts the concept and origin of "pure music" and "emotional catharsis" in self-expression are curiously compiled, and would deserve a more thorough discussion. Langer contrasts the abandon with which indigenous song and dance express feeling, organized in rhythmical patterns and thus distinct from mere emotional outcry, however, also distinct from the meticulous purity of a musical aesthetics. Today's discourses are only beginning to include indigenous music, artistry, and ritual in terms of epistemological import and worlding. Notably, their appreciation has never been questioned by artistic avant-gardes. Furthermore, Langer's first demarcation point from music oddly references "[a] lynching-party howling round the gallows-tree" (216). One can only speculate as to whether she intended to contrast the crudeness of white supremacy with the sophistication of states of expressive abandon, lament, or "the wild syncopated shouts of African tribesmen" as the origin of music. See *Philosophy in a New Key*, 214–18.
10. Ibid., 225.
11. Langer, *The Practice of Philosophy*, 152.
12. Susanne K. Langer, "The Art Symbol and the Symbol in Art," in *Problems of Art: Ten Philosophical Lectures* (New York: Charles Scribner's Sons, 1957), 125.
13. *The Practice of Philosophy*, 152.
14. Ibid., 153.
15. Ibid., prefatory note by Alfred N. Whitehead, dated September 1929.

16. Ibid., 161.
17. Donald W. Sherburne, *A Key to Whitehead's Process and Reality* (Chicago: University of Chicago Press, 1981), 224–5.
18. Cf. Alfred N. Whitehead, *Process and Reality* (New York: The Free Press 1985), 164.
19. Alfred N. Whitehead, *The Concept of Nature* (Ann Arbor, MI: University of Michigan Press, 1959), 30.
20. Cf. Sherburne, *A Key to Whitehead's Process and Reality*, 225.
21. Whitehead, *Process and Reality*, 40–1.
22. Ibid., 18.
23. Ibid., cited and contextualized in Sherburne, *A Key to Whitehead's Process and Reality*, 205. Whitehead specifies, "There is no going behind actual entities to find anything [28] more real. They differ among themselves: God is an actual entity, and so is the most trivial puff of existence in far-off empty space."
24. Cf. Sherburne, *A Key to Whitehead's Process and Reality*, 206.
25. Langer, *Mind*, vol. 1, 32.
26. Cf. Robert Innis, "Psychological Dimensions, Cultural Consequences, and their Breakings in Susanne K. Langer's Symbolic Mind," Chapter 10 in this volume, 152.
27. Cf. Rolf Lachmann, *Susanne K. Langer. Die lebendige Form menschlichen Fühlens und Verstehens* (Munich: Fink Verlag, 2000), 175.
28. Cf. Donald Dryden, "Whitehead's Influence on Susanne Langer's Conception of Living Form," *Process Studies*, no. 26 1/2 (1997): 62.
29. Cf. Rolf Lachmann, *Susanne K. Langer. Die lebendige Form menschlichen Fühlens und Verstehens* (Munich: Fink Verlag, 2000), 61 (translation by the author).
30. From 1951 on, Langer replaces her analysis of "The Logic of *Signs* and Symbols" with "The Logic of *Signals* and Symbols" (italics added) with reference to Charles Morris's terminology. "The term 'signal' is stretched," she explains, "to cover not only explicitly recognized signals—red lights, bells, et cetera—but also those phenomena which we tacitly respect as signals to our sense." The advantage she sees in this is that "sign" can now be used as a generic term to denote any vehicle of meaning, be it a signal or symbol. Since Langer only remarks this in the preface but does not replace "sign" with "signal" in her text, I use her original (pre-1951) terminology with the corrected term in brackets. See Langer's "Preface to the Edition of 1951" in *Philosophy in a New Key*, x.
31. Langer, *Philosophy in a New Key*, 55 (Langer's italics).
32. Ibid.
33. Ibid.
34. Ibid., 56.
35. Langer gives the example of the A-chord with the sixth, fourth, and third notes above A, which can be expressed either as a function of its lowest note (A), or can be treated with reference to the note upon which it harmonically builds. Therefore, a musician might call this chord "the second inversion of the seventh-chord on the dominant, in the key of G," which treats the function as a pattern of D. Both expressions (*qua* function), however, represent the very same tone (*qua* pattern). Ibid., 55–6.
36. Cf. ibid., 64.
37. Ibid., 60–1.
38. Ibid., 56.
39. This stir of her philosophy is discussed in Charles Varela and Laurence Ferrara, "The Nagel Critique and Langer's Critical Response," *Journal for the Anthropological Study of Human Movement* 2, no. 2 (1982): 99–111.

40. Langer, *Philosophy in a New Key*, 240.
41. Langer, *Feeling and Form*, 31–2. Langer revokes her earlier concept, and explains, "instead of being the 'unconsummated symbol' postulated in *Philosophy in a New Key*, it might have, indeed a single reference. I suspect that this is the case, and that the different emotional values ascribed to a work of art lie on a more intellectual plane than its essential import: for what a work of art sets forth—the course of sentience, feeling, emotion, and the *elan vital* itself—has no counterpart in any vocabulary." Ibid., 373–4.
42. "The primary illusion always determines the 'substance,' the real character of an art work, but the possibility of secondary illusions endows it with richness, elasticity, and wide freedom of creation that make real art so hard to hold in the meshes of theory," Langer explains in *Feeling and Form*, 118.
43. Langer, *Philosophy in a New Key*, 238–9.
44. Cf. Langer, *The Practice of Philosophy*, 166.
45. Langer, *Feeling and Form*, 107.
46. Cf. Langer, *Philosophy in a New Key*, 239.
47. On the topic of judging art, Langer writes, "there is no immutable law of artistic adequacy, because significance is always *for* a mind as well as *of* a form. But a form, a harmony, even a *timbre*, that is entirely unfamiliar is 'meaningless' ... for we must grasp a *Gestalt* quite definitely before we can perceive an implicit meaning ... Therefore the most original contemporary music in any period always troubles people's ears ... The more pronounced its new idiom, the less they can make of it, unless the impulse which drove the composer to this creation is something of a common experience, of a yet inarticulate *Zeitgeist,* which others, too, have felt. Then they, like him [and her], may be ready to experiment with new expressions, and meet with an open mind what even the best of them cannot really judge. Perhaps some very wonderful music is lost because it is too extraordinary. It may even be lost to its composer because he cannot really handle his forms, and abandons them as unsuccessful. But intimate acquaintance with all sorts of music does give some versatile minds a power of grasping new sounds; people so inclined and trained will have a 'hunch,' at least, that they are dealing with true 'significant form' though they still hear a good deal of it as noise, and will contemplate it until they comprehend it, for better or worse." Ibid., 263–4.
48. See in Part II "The Making of The Symbol," *Semblance*, in Langer, *Feeling and Form*, 45–68.
49. Ibid., 104–19.
50. Ibid., 110.
51. Ibid., 113–16.
52. This critique draws on the ancient dichotomy of time in its quantitative sense (χρόνος, *chronos*) and qualitative sense (καιρός, *kairos*). Time, only displayed in scientific measure, loses its quality as an entirely free and flexible entity, whose dream-like nature appears to be completely detached from actuality.
53. Langer, *Feeling and Form*, 114.
54. Ibid., 114.
55. Cf. ibid., 114.
56. Langer's solution to Bergson's predicament of symbolization is seeded by two essays that responded to his problem of finding an adequate approach to render music: French philosopher Gabriel Marcel's "Bergsonisme et musique," *La Revue Musicale* 6, no. 5 (March 1925): 215–29 (published in Langer, Susanne K. ed. *Reflections on Art: A Source Book of Writings by Artists, Critics, and Philosophers* [Baltimore: Johns Hopkins, 1958]), and composer Charles Koechlin's "Le temps et la musique," *La Revue Musicale* 7, no. 3 (January 1926): 45–62. See Langer, ibid., 115.
57. Ibid., 109.

58. Cf. ibid., 116.
59. Key to Langer's construction of symbols is her notion that each artistic medium is defined by a primary illusion, which determines its "substance," and a secondary illusion, endowing the substance with the "richness, elasticity, and wide freedom of creation that make real art so hard to hold in the meshes of theory." Ibid., 117–18.
60. Langer, *Philosophy in a New Key*, 263.
61. Langer, *Feeling and Form*, 121.
62. Ibid., 126.
63. Cf. ibid., 126.
64. Ibid., 129.
65. Ibid., 127.
66. Cf. Langer, *Mind*, vol. 1, 324.
67. This expression has been borrowed from an online conversation with Donald Dryden on Langer's ties to Whitehead (December 2021).
68. Rolf Lachmann, ed., "Susanne K. Langer's Notes on Whitehead's Course on Philosophy of Nature," *Process Studies* 26, no. 1/2 (1997): 126–50.
69. Whitehead, *Process and Reality*, 162.
70. Cf. ibid., 162.
71. Ibid., 163.
72. Cf. ibid., 162.
73. Ibid., 163.
74. November 1, 1927. "Susanne K. Langer's Notes on Whitehead's Course on Philosophy of Nature," 134–5.
75. See December 3, 1927. "Susanne K. Langer's Notes on Whitehead's Course on Philosophy of Nature," 144–5.
76. Cf. Whitehead, *Process and Reality*, 120–1.
77. Cf. ibid., 121.
78. Ibid.
79. December 6, 1927. "Susanne K. Langer's Notes on Whitehead's Course on Philosophy of Nature," 145.
80. Langer notes, "Whitehead's identification of 'positive prehension,' a cosmic principle of process as such, with 'feeling,' seems to me unfortunate, for it precludes any detailed study of that most interesting phenomenon which distinguishes psychology from physiology, just as the phenomena of organic functioning distinguish physiology from chemistry and physics (the boundaries between sciences being always somewhat fluid)." Dryden, "Whitehead's Influence on Susanne Langer's Conception of Living Form," 78, from "On Whitehead," unpublished manuscript, n.d., Susanne K. Langer Papers, Houghton Library, Harvard University.
81. Cf. Dryden, "Whitehead's Influence on Susanne Langer's Conception of Living Form," 78, quoting Langer in *Mind*, vol. 1, 444.
82. Randall E. Auxier, "Susanne Langer on Symbols and Analogy: A Case of Misplaced Concreteness?" *Process Studies* 26, no. 1/2 (1997): 96.
83. Dryden, "Whitehead's Influence on Susanne Langer's Conception of Living Form," 84.
84. Cf. ibid.
85. Cf. ibid., 76.
86. Langer, "On Whitehead," unpublished manuscript. Ibid., 78.
87. Langer remarks upon the possibility of echoing the primary illusion of one art form in the secondary illusion of others as an indication of a basic art community, and references this, in particular, in the perception of space in music. Cf. Langer, *Feeling and Form*, 118.
88. Ibid., 66.

89. Langer, *Mind*, vol. 1, xxii.
90. Cf. ibid.
91. Susanne K. Langer, *Mind: An Essay on Human Feeling*, vol. 3 (Baltimore: Johns Hopkins University Press: 1982), 201.

BIBLIOGRAPHY

Auxier, Randall E. "Susanne Langer on Symbols and Analogy: A Case of Misplaced Concreteness?" *Process Studies* 26, no. 1/2 (1997): 86–106. Available online: https://doi.org/10.2307/44798815.

Dryden, Donald. "Whitehead's Influence on Susanne Langer's Conception of Living Form." *Process Studies* 26, no. 1/2 (1997): 62–85. Available online: https://doi.org/10.2307/44798814.

Koechlin, Charles. "Le Temps et la musique." *La Revue Musicale* 7, no. 3 (January 1926): 45–62. Available online: https://lmhsbd.oicrm.org/media/public/documents/ART-KOC-1926-01.pdf (accessed February 7, 2022).

Lachmann, Rolf, ed. "Susanne K. Langer's Notes on Whitehead's Course on Philosophy of Nature." *Process Studies* 26, no. 1/2 (1997): 126–50. Available online: https://doi.org/10.2307/44798817.

Lachmann, Rolf. *Susanne K. Langer. Die lebendige Form menschlichen Fühlens und Verstehens* (German). Munich: Fink Verlag, 2000. Available online: http://digi20.digitale-sammlungen.de/de/fs1/object/display/bsb00043407_00001.html (accessed December 4, 2021).

Langer, Susanne K. *Feeling and Form: A Theory of Art Developed from Philosophy in a New Key*. New York: Charles Scribner's Sons, 1953.

Langer, Susanne K. "Insight." In *The Practice of Philosophy*, 152–66. New York: Henry Holt, 1930.

Langer, Susanne K. *Mind: An Essay on Human Feeling*, vol. 1. Baltimore: Johns Hopkins University Press, 1967.

Langer, Susanne K. *Mind: An Essay on Human Feeling*, vol. 3. Baltimore: Johns Hopkins University Press, 1982.

Langer, Susanne K. "On Whitehead," unpublished manuscript, n.d., Susanne K. Langer Papers, Houghton Library, Harvard University. Accessed through Donald Dryden.

Langer, Susanne K. *Philosophy in a New Key: A Study in the Symbolism of Reason, Rite and Art*. 3rd edn. Cambridge, MA: Harvard University Press, [1942] 1993.

Langer, Susanne K., ed. *Reflections on Art: A Source Book of Writings by Artists, Critics, and Philosophers*. Baltimore, USA: The Johns Hopkins Press, 1958.

Locke, John. *An Essay Concerning Human Understanding*, vol. 1. Oxford: Clarendon Press, 1894.

Marcel, Gabriel. "Bergsonisme et musique." *La Revue Musicale* 6, no. 5 (March 1925): 215–29. Available online: https://lmhsbd.oicrm.org/media/public/documents/ART-MAGe-1925-01.pdf (accessed February 7, 2022).

Sherburne, Donald W. *A Key to Whitehead's Process and Reality*. Chicago: University of Chicago Press, 1981.

van der Tuin, Iris. "Bergson before Bergsonism: Traversing 'Bergson's Failing' in Susanne K. Langer's Philosophy of Art." *Journal for French and Francophone Philosophy* 24, no. 2 (2016): 176–202. Available online: http://jffp.pitt.edu/ojs/index.php/jffp/article/view/776/738 (accessed February 7, 2022).

Varela, Charles, and Laurence Ferrara. "The Nagel Critique and Langer's Critical Response." *Journal for the Anthropological Study of Human Movement* 2, no. 2 (1982): 99–111. Available online: https://jashm.press.uillinois.edu/2.2/2-2TheNagel_Varela&Ferrara99-111.pdf (accessed December 15, 2022).

Whitehead, Alfred N. *Process and Reality: An Essay in Cosmology*. Corrected edn. New York: The Free Press, 1985.

Whitehead, Alfred N. *Symbolism: Its Meaning and Effect*. 3rd edn. New York: Macmillan, 1985. Available online: https://archive.org/details/in.ernet.dli.2015.166010/ (accessed January 11, 2022).

PART TWO

Generative Ideas

CHAPTER TWELVE

Susanne K. Langer and Philosophical Biology

ADAM NOCEK

Susanne K. Langer's work on the philosophy of science has received precious little attention. This is not to say that her work on aesthetics and symbolism, for which she is better known, has gained the recognition it deserves either, though she has a modest following in American pragmatism, the philosophy of art and aesthetics, and to a lesser extent process philosophy. Nevertheless, her contribution to the philosophy of science, and the philosophy of biology in particular, has been met with little to no recognition. And when it has been championed, for instance by Donald Dryden (1997),[1] this praise stands in stark contrast to the skepticism the *Mind* trilogy (especially volume 1) inspired in Peter A. Bertocci (1970),[2] Randall Auxier (1997),[3] and others, who accuse her work of committing a host of philosophical sins: among others, leaving the philosophy of Alfred North Whitehead behind without committing to the path laid down by Ernst Cassirer; overcommitting to scientific empiricism; and failing to overcome the dualism it desperately sought to. There has been important scholarship since seeking to bring the *Mind* trilogy into the conceptual orbit of her other writing, even suggesting that *Mind* is generally consistent with the aims of *Philosophy in a New Key* and *Feeling and Form*.[4]

This chapter is not an attempt to right the wrongs done to Langer's biology in the philosophical literature. What follows is less concerned with the critiques leveled against her ambitious project, and is more interested in finding those elements in her "philosophical biology" that are relevant to the biological sciences today. If Langer sets herself the goal of developing a concept of mind that can be understood in terms of the "highest physiological processes,"[5] then an essential part of this project, as we will see, is "rooting" feeling in organic activities[6] without resorting to reductive materialism or metaphysical dualism. What is remarkable, if widely unrecognized, is the empirical detail required for her account of biological organization to deliver on all that it promises—namely, to "bring mental phenomena into the compass of natural fact."[7] This should come as no surprise, according to Dryden, since her works rarely read like disciplinary "philosophy," and more like an empirically oriented philosophical investigation of a specific domain.[8]

In any case, this chapter argues for the usefulness of Langer's empirically rich philosophical biology for the twenty-first-century biosciences. It may seem that I am overstating the case a bit. This is not only because the first volume of *Mind* was written over fifty years ago and there have been considerable developments in the theoretical study of organisms since 1967. It is also because one of the most unique, if perplexing, features of Langer's biology is that the symbolic function of art is essential to building adequate concepts of vital processes and physiologically rooted feeling/mind in particular. While this makes for interesting natural philosophy and aesthetic theory, it hardly seems to speak to the most pressing concerns of theoretical biologists working on evolutionary and developmental change at the intersection of data science and complex dynamical systems.

And yet, that is precisely what this chapter proposes: Langer's art-based biology has an essential role to play in the data-intensive landscape of the biosciences. To make this case, I will look specifically at how Langer poses the problem of mind in her 1967 work, inquiring into why the art image is needed to resolve certain epistemological dead ends that have come to dominate the study of mind in empirical psychology, and how art lays the groundwork for a conception of biology that explains the emergence of feeling/mind without recourse to reductionism or metaphysical dualism. This discussion will set the stage for turning to developments in the study of biological complexity, and how Langer's arts-based method of biological research responds to problems in conceptualization of organismic development that theoretical biologists are just beginning to grapple with. In offering a fresh take on Langer's biology, this chapter serves as a kind of prolegomenon to art research in the biological sciences.

FROM MODEL TO IMAGE OF MENTATION

Langer's principal task in *Mind: An Essay on Human Feeling* is to show how "the entire psychological field is a vast and branching development of feeling."[9] And feeling, she also intends to show, is a physiological process that does not lend itself to the physicist's methods of analysis.[10] The trouble with modern psychological research is its failure to appreciate this latter point: despite feeling/mind's physiological basis, one cannot "transfer methods of observation simply and directly from physics to psychology," since "the intraorganic character of the material presents a special difficulty and does not lend itself to those methods."[11] On this point, Langer is clear: the physical sciences do not adequately attend to the specificity of mental phenomena,[12] and their privilege leads to what she calls the "Idols of the Laboratory": namely, "Physicalism, Methodology, Jargon, Objectivity, Mathematization."[13] Thus, although psychic phenomena are physiologically rooted, they are irreducible to physico-mathematical methods of analysis (propped up by the so-called Idols), a view that is largely consistent with a wider tendency in early to mid-twentieth-century natural philosophy to reject reductive materialism or anything that looks or smells like mechanism.[14]

And yet Langer's philosophy of mind does not resort to dualistic theses either. In fact, she is deeply critical of substance dualism and all forms of theistic teleology that posit the "soul," "mind," or *elan vital* as an animating force that acts on or beside the material world.[15] Here is Langer:

> Despite the vastness of time and change that must have prepared what we call "the Mind"' today, I hold that the elements of that marvelous structure may all be found in nature, and the principles of its formation are those of organic chemistry, electrochemical action, or whatever substitutes for such current concepts the progress of scientific thought may dictate in the future.[16]

This presents Langer with an especially difficult thesis to defend: on the one hand, her theory of mind, or what amounts to the same thing, her theory of "feeling" as the "mark of mentality"[17] must be a "biological theory of feeling,"[18] fully embedded within the organism's naturally occurring processes. And on the other hand, it is a theory whose contents, namely, certain intra-organismic processes that can be elaborated into complex forms of mentation (imagining, fantasizing, reasoning, etc.), must also be marked by a certain degree of autonomy from the unfelt physiological processes which give rise to them. If this were not the case, then psychic phenomena could be investigated using the methods of physical science.

Langer proposes to solve this problem with a non-dualistic and non-physicalist theory of mentality. Of course, the legitimacy of this proposition rests on how she defines "feeling." For if feeling is the mark of mentality, and "mental events" just are "felt impingements and activities,"[19] then all investigations into mind must start with feeling as a physiological process that requires its own method of analysis due to the peculiarity of felt activity. Such a conception of feeling is precisely what Langer provides: feelings, and thus mind, are not entities or things, for her, they are characterized temporally and not spatially (they do not occupy space), and are "produced by physical (especially nervous) activities," although they "themselves [are] not physical."[20] Feelings are "phases" of physiological process.[21] They are not static states but activities that emerge within intra-organismic processes, and they index that something is being done, something is being "felt," and not that something is possessed. Here is Langer:

> In the first place, the phenomenon usually described as "a feeling" is really that an organism feels something, i.e., something is felt. What is felt is a process, perhaps a large complex of processes, within the organism. Some vital activities of great complexity and high intensity, usually (perhaps always) involving nervous tissue, are felt; being felt is a phase of the process itself. A phase is a mode of appearance, and not an added factor ... When iron is heated to a critical degree it becomes red; yet its redness is not a new entity which may have gone somewhere else when it is no longer in the iron. It was a phase of the iron itself, at high temperature.[22]

Feelings, then, are phases of certain physiological processes themselves, which are rooted in "the fabric of totally unfelt activities," such as protein synthesis, leukocyte extravasation, and millions of other biochemical processes.[23] Some of these unfelt activities, which make up an intricate web of biomolecular activity in the living organism, will end up being felt by the organism in psychic phase, e.g., digestion, but the nature and range of these phasal events are vast—from sensory awareness to complex emotions and rationality—and transient.[24]

Put in the most general terms, feelings come in two varieties: activities can be "felt as impact," that is, originating "from outside"; or as action, that is, "felt as *autogenic* action."[25] However, in reality, this distinction is far from absolute: autogenesis and exogenesis are constantly interchanging. They are mixed modes of feeling that characterize the organism as an "open system"—a creative center of felt receptivity and activity.[26] This not only parallels many other (quasi-)naturalistic and process-oriented theories of biological systems (Whitehead, Bergson, James, and even Deleuze) but it also troubles hard divisions between subjectivity and objectivity since whatever is felt as action (subjectivity) and as impact (objectivity) are "functional properties" and not metaphysical givens.[27]

From this basic scaffolding, Langer claims to be able to account for all modes of mentation, from sensory awareness to mathematical abstraction, without importing "metaphysical assumptions of non-zoological factors."[28] This is because "feeling," writes Langer, "stands, in fact, in the midst of the vast biological field which lies between the lowliest organic activities and the rise of mind. It is not an adjunct to natural events, but a turning point in them." And "[t]his is why," she continues, "I make feeling the starting-point of a philosophy of mind. The study of feeling—its sources, its forms, its complexities—leads one down into biological structure and processes until its estimation becomes (for the time) impossible, and upward to the purely human sphere known as culture."[29] Feeling is the backbone of Langer's naturalist philosophy of mind: it unites the abstract spaces of rationality with biomolecular processes of living systems without introducing novel substances or reducing unique features to more basic physical "stuff."

Still, at this point in the trajectory of Langer's overall argument she has not given us much to work with. She has provided little more than a blueprint of what she intends to study. She has yet to detail the structure of feeling, how it emerges, what the varieties of it are, and how it fits into the study of biological systems more broadly. And furthermore, what specific methods does she use to study it? Langer notes, and somewhat mystically, that we all have "direct knowledge of feeling," and this "includes the sensibility of very low animals and the whole realm of human awareness and thought."[30] However, "the real patterns of feeling," Langer goes onto suggest, "which compose the fabric of mental life, usually pass unobserved, unrecorded and therefore essentially unknown to the average person."[31] This realm of largely unacknowledged mental activity, which lies below the threshold of cognition, is accessed through what she calls "prescientific knowledge," and it is the foundation of any scientific inquiry: "science arises from prescientific knowledge about its subject matter, empirical and haphazard, but developed by practice to considerable detail and precision."[32]

One of Langer's chief and most challenging insights about biological systems is that we require a pre-scientific *image* instead of a cognitive *model* of feeling. If the sciences use models, then it is because "a model … always illustrates a principle of construction or operation; it is a symbolic projection of its object which need not resemble it in appearance at all, but must permit one to match the factors of the model with respective factors of the object, according to some convention."[33] But Langer's argument is that there is no such model of feeling without an image of it first. An image, as opposed to a model, "does not exemplify the same principles of construction as the object it symbolizes"; on the contrary: it "abstracts its phenomenal character … it organizes and enhances the impression directly received."[34] In other words, human organisms navigate the world "felt as impact" and "felt as action" through images of it. They produce visual, auditory, movement, and a host of other images capable of registering changes in the welter of experience. Images are multiple, incomplete, and transitory; they allow an object to appear *as something*, which allows an organism to recognize and remember certain features in the sensory manifold.[35] The point is that these images furnish the raw material for a concept of mind:

> The trouble with modern psychology, and why its concepts of mind are inadequate, is that, before we [the psychologists] had any clear image of the phenomenon we call "mind," we committed ourselves to a model, the system of physical laws, whereby the material was automatically cut down to what the model could represent, and the very subject matter of psychology—the psychical phase of vital functions—was eliminated altogether, leaving only its overt record, behavior.[36]

Physical theories of psychic phenomena and other "Idols of the Laboratory" are largely a function of putting the cart before the horse—namely, developing a model of mind without an adequate image of it first. But if this is true, then what image gives us access to the "real patterns of feeling"? Where do we find an image that will be the basis for a concept of mind?

ART IMAGES AND THE SEMBLANCE OF FEELING

For readers of Langer, who will notice clear resonances with *Feeling and Form*, it should come as no surprise that art furnishes the image of feeling required to build an adequate model of mind. Here is Langer: "Who has a naïve but intimate and expert knowledge of feeling? Who knows what feeling is like? Above all, probably, the people who make its image—artists, whose entire work is the making of forms which express the nature of feeling."[37] Indeed, "[a]rtistic form is always the form of felt life, whether of impression, emotion, overt action, thought, dream, or even obscure organic process rising to a high level and going into psychical phase."[38] In other words, in the art image there is an expression, or rather a "semblance," of the inner life of feeling: perceptible form is given to the elusiveness of the forms of felt activity.[39] Langer puts it succinctly:

> Feeling is a dynamic pattern of tremendous complexity. Its whole relation to life, the fact that all sorts of processes may culminate in feeling with or without direct regard to each other, and that vital activity goes on at all levels continuously, make mental phenomena the most protean subject matter in the world. Our best identification of such a phenomenon is through images that hold and present them for our contemplation; and their images are works of art.[40]

While much still needs to be said about the nature of the artwork and whether it is capable of delivering on all it promises, it is worth noting that it is at this point Langer's argument becomes difficult to stomach for some. Even for the initiated, her project in *Mind* strikes them as misguided and even contradictory. Bertocci, for his part, sees a host of problems linking the aesthetic and the organic via the art image, and draws special attention to the fact that the aesthetic loses the autonomy she wishes to grant it when it is required to have a special "kinship" with organic processes.[41] And for readers looking to Langer's natural philosophy for a concept of felt activity in tune with the material and functional complexity of organisms, the turn to the art image found in ballet, poetry, and painting, will likely seem eccentric.

Although these may be fair criticisms, there is still much to salvage in her epistemology of feeling for contemporary readers. Essentially, she argues that science alone is not sufficient to construct its concepts. In the case of the biological and psychological sciences, art is in some sense necessary for building conceptual models. As she explains, her purpose in mixing the study of biology and art is not to "penetrate the mysteries of art with the help of … biological knowledge," which is not an uncommon method; rather, she hopes to "gain some biological and psychological insights through the suggestiveness of artistic forms." This means that "the theory of art is really a prolegomenon to the much greater undertaking of constructing a concept of mind adequate to the living actuality."[42]

To the contemporary ear, this should sound reasonable enough. In the last several decades BioArt has explicitly and provocatively troubled divisions between art and science. It is not uncommon now to see artists working alongside bench scientists. Artists from various backgrounds—from musicians to video and performance artists—are tinkering with genome sequences, turning stool samples purple, experimenting with transplantation, making kimchee sing, and so on.[43] But for our purposes, and as I have written about elsewhere,[44] it is important to recognize that the biosciences have always been materially, epistemologically, and culturally mixed. If BioArt experiments with and often troubles the relation between art and science, then it is because biology has never been an isolated activity. Life scientists are constantly borrowing methods, technologies, and media from the arts in order to "feel" their way toward reliable claims about messy, wet systems.[45]

The cinema, computer animation, graphic design, and even acoustics are part and parcel of doing research in the biosciences. As Natasha Myers explains, modeling biomolecular structures using sophisticated computer graphics software is an embodied practice; it demands a rigorous choreography of bodies, machines, and molecular worlds.[46]

One might read Langer as making epistemologically explicit what is practically implicit. Art plays an important role in the research activities of scientists, especially life scientists. In this frame, Langer's work in *Mind* paints a philosophical portrait of what is now well known in the history of science. The trouble with this view, even if it sits nicely alongside contemporary scholarship on science and technology, is that it tames the spirit of Langer's thesis. As the media historian Scott Curtis argues, it is not as if the arts (cinema, for example) are useful to the sciences in themselves; instead, they first must be made epistemologically compatible to the requirements of the scientific field.[47] In other words, even when there is clear overlap between the biosciences and the arts, epistemological and normative hierarchies often remain in check.

But Langer is not making this claim. Her thesis is much stronger than this: the art image does not need to represent the natural world, or even be about it in any explicit way, as the way in which early experiments in microcinematography record the development of bacterium. The artwork is autonomous in this respect, and does not need to adjust or alter what it does or how it frames its problems to be epistemologically valuable to the sciences. If anything, it is the other way around: science needs to bend itself to art. The biological and psychological sciences have failed to adequately theorize the emergence of centers of feeling in the physical world precisely because they have not directed their attention to non-discursive images in art.

For Langer, the art image, as opposed to the model, has no "preconceived standards."[48] The only "criteria" for art is that it has "vital import." The nature of this import does not lend itself to straightforward characterization, but for our purposes art's import is not determined by its subject matter (a painting does not have to be *about* an organism to be "organic"), only by its ability to create a semblance of "the inward tensions that compose our life of feeling."[49] Crucially, this "semblance"[50] of tensions that characterizes feeling is created through a process of abstraction. Abstraction in art is different from abstraction in science. Where the latter operates according to generalization, or "generalizing abstraction," which is how we tend to think about abstraction, in art, by contrast, abstraction functions by means of "presentational abstraction."[51] In the art image, there is an imaginative transformation of what is inwardly felt into what is outwardly recognized, without it being generalizable, since it is a "single expressive form."[52] Now, the transformation of inward feeling into outward semblance occurs by means of what Langer calls the "primary illusion." This illusion is what "makes the most direct sort of presentational abstraction,"[53] and is achieved through the manifestation of "tensions" and "resolutions" among the work's elements. These twin tendencies (of tension and resolution) are expressible in an incalculable number of ways, differing according to genre, medium, intensity, space, timing, gesture, and so on, but their existence is fundamental to the creation of the primary illusion in art—namely, that inward feeling can be expressed as an objective semblance. The primary illusion is realized, then, as a pattern of tensions among elements.

There a couple things to note here. First, the elements in a work are not the same as "materials." Elements are "virtual" and are a part of the "total form of the work," whereas materials are not.[54] It is precisely these tension-filled relations among elements in the work that express the restless inner life of feeling—its "life-giving tension."[55] The second thing to note is that these tensions must also come to some form of resolution. This is because, for Langer, tensions are always a part of larger patterns of relations that compose an "entire work" and deliver a "single impact."[56] The idea here, which is not too far removed from Whitehead's aesthetics,[57] is that the tensions among the work's elements are not divergent or contradictory. Quite the contrary: they create deeper patterns of relation responsible for what Langer calls an "isolating abstraction" that gives the work its form. This allows the work to appear as a "unity," a "gestalt," that ultimately constitutes the work's *rhythm*.[58]

Other factors, such as the work's "gradients,"[59] as well as its semblance of "growth,"[60] "motivation" as the Kantian "telic form without purpose,"[61] and "individuation,"[62] contribute to the uniqueness of the artwork's rhythmic pattern. "Secondary illusions" are key ingredients in the piece's ability to express the multidimensionality of feeling. These are illusions that help "widen the scope of its [the art work's] expression,"[63] and have the characteristic of suddenly coming from "nowhere … and fading again into nothing."[64] Such illusions tend to mark a significant phase transition, "from one order of existence to another,"[65] evidenced, for example, when the layering of surfaces in an architectural structure suddenly appear to be planes of an image in a painting or in the cinema.[66]

Still, Langer's chief concern in *Mind* is not aesthetics sensu stricto. Rather, it is demonstrating that the art image grants unprecedented access to the life of feeling in all its biochemical complexity. Somehow, the rhythmical pattern of elements in the work actually "reflects the dynamism that is constantly building up the life of feeling."[67] The "livingness" in the image,[68] achieved through a buildup of tensions, complex resolutions, and new tension-filled relations, is a symbolic projection of feeling in the biological world. And yet, at this stage the "kinship between organic and artistic forms" remains tenuous. Langer reassures us that in so far as "the work of art is a projection of feeling, that kinship with organic nature will emerge."[69] The question we are left wrestling with is: how? If it is true that the "theory of art is really a prolegomenon to the much greater undertaking of constructing a concept of mind adequate to the living actuality,"[70] then Langer needs to account for *how* the art image *qua* symbolic projection *actually* contributes to our understanding of the biology of feeling, and thus mind.

FROM IMAGE TO CONCEPT

The genesis of Langer's organically rooted concept of feeling becomes more intelligible once we take note of her characterization of "elements" in terms of "acts." She writes that "[e]lements in art have not the character of things, but of acts. They are 'active,' act-like, even when they are not 'acts' in the dramatic sense."[71] "The term [act]," she continues, "has an instrumental value for building up a coherent and adequate concept of mind, and on that pragmatic basis I use it in the broad sense here."[72] The equivalence struck between elements and acts is significant for Langer's epistemology: it gives the "element-act" the status of a bridge that connects art's projection of feeling to the empirical realities of biochemistry. For if elements in art provide a semblance of the tension-filled vital processes that "include all mental life,"[73] then their elaboration as "acts" is an important step on the path to formalizing a biologically rooted concept of mentation. Put another way: the act, born out of the element's projection of life, returns to life in order to lend it conceptual clarity.

The concept of act is detailed and complex, but with it, Langer derives an original theory of biological organization ranging over everything from the origins of life and the possibility of evolutionary and developmental change to complex forms of mentation, such as cognition. The empirical and theoretical detail of her act-based biology should have secured Langer a position among the most prominent philosophers of biology in the twentieth century. That it did not is not entirely surprising, for reasons I discuss below, but for our purposes several features of the act are worth noting. For one, acts, as opposed to cells, genes, or other material elements, are responsible for the various and complex forms of material and social organization living systems are capable of achieving. The act is not exclusive to living matter, and like the element, is not a material thing, a static entity, or a state. Rather, the act is a spaciotemporal "occurrence" or "event" that conditions "the singleness of an overall-tension."[74] For instance, the grimace of a canine, an animal's emotional transition, or the metabolization of toxins in multicellular organisms are all acts. Each of these acts also has an "abstractable form," which Langer calls an "action," and it permits the classification of acts according to their "causal patterns" and "operative principles," allowing us to claim, for instance, that "the same action may be performed by an animal's heart and by a pump."[75] Finally, Langer adds to her classificatory scheme by noting that acts tend to fall into patterns and predictable series—e.g., breathing, circulation, etc.—where "the same general form is discernable over and over again."[76] These act sequences are what Langer calls "activities."

Now, it is this taxonomy, which divides the act into actions and activities, that serves as the backbone for Langer's important discussion of "individuation" and "involvement" in *Mind*. Rather than speak of acts as "individuals" and risk introducing atomism into her natural philosophy, she speaks of "individuations" (at around the same time that Gilbert Simondon was working on a similar conception).[77] Acts are responsible for the individuating occurrences "exemplified everywhere in animate nature, in processes that eventuate in the existence of self-identical organisms"; they are what explain cell differentiation, the emergence of mind in the material world, and countless other forms of real difference.

Individuation emerges, then, from out of the "singleness of an overall-tension"[78] in matter, which is "motivated by a vital situation … uniquely given for each act."[79] But notice, if the "vital situation" lends each act its tensive quality, then the act is far from an isolated occurrence; on the contrary: the act is necessarily related to all those act-events that constitute its situation. Thus, each act is also *involved* with other acts, so much so that acts actually enter into the internal constitution of one another to structure motivation, which is not unlike the internal relatedness of occasions in Whitehead. The individuation of the epidermis in plants, for instance, is made possible by the involvement of uniquely motivated events with one another—e.g., the absorption of minerals and the secretion of metabolic compounds are act-events that work together to individuate epidermal cells. Hence, "the internal involvement of acts with each other, known as 'integration of functions,' is the most important factor in individuation, i.e., in the establishment of self-contained, stabile, vitally active systems."[80] Such integration among *organic* individuations is possible, however, only because of a more primary mode of involvement—namely, heredity. Here is Langer:

> Genetic inheritance makes the unity and continuity of a biotypic stock; it is every living being's indisoluble bond to its own ancestry … The gene complement is the one permanent factor in its lifelong, progressive situation. Everything else may change; everything else that motivates the advancing flow of its impulses and its realized acts plays on this anciently established core. Heredity is the primary involvement of every

organism with other organisms ... The stock is the largest natural unit of life. Its beginning goes back to the unknown days of biogenesis.[81]

With the twin forces of individuation and involvement outlined, Langer then returns to rhythm: "Individuation and involvement are the extremes of the great rhythm of evolution, which moves between them in a direction of its own, always toward more intense activity and gradually increasing ambients of the generic lines that survive."[82] Here, we get a sense of just how much Langer's biology is shaped by her aesthetics: the rhythmical patterns that give meaning to the tensions found among elements in art also give meaning to the evolution of acts in the biological world. Rhythm characterizes the ongoing drama of resolving tension-filled activities—in ecological niches, for example—only to engender new conflicts that create unforeseen opportunities for organismic interdependence. It is precisely this rhythmical unfolding of individuation and involvement of act-events in evolutionary history that accounts for the emergence of genuinely novel phenomena. The ontogenesis of feeling is explained, not through the introduction of vital substances or essential differences between living and physical matter, but through individuations and involvements in the material world "growing" in complexity.

From the rhythmical patterns of evolutionary acts, Langer derives the entire range of psychic phenomena—from sensation to rationality—without additions form extra-physiological "stuff." Acts enter into increasingly complex rhythms of involvement and individuation to the point where phasal shifts in matter pave the way for the emergence of sensation, emotion, and eventually rationality. What is striking is that her empirically and physically grounded conception of biological organization is derived from the abstract images present in the artwork: "Though we have no physical model of this endless rhythm of individuation and involvement," writes Langer, "we do have its image in the world of art."[83]

CONCLUSION: A NEW IMAGE FOR THEORETICAL BIOLOGY

It is no secret that Langer's theories of evolution and biologically rooted feeling did not make the splash in the philosophy of science or theoretical biology that one might have hoped. There are certainly historical reasons that may help account for this relative lack of interest. Langer's conception of mind presupposes nothing short of a reimagined theory of evolutionary and developmental change. It is not simply that her language was unrecognizable to most—e.g., rhythms, acts, impulses, individuations, and involvements—and that her concepts challenged the tenants of modern psychological research, but her conception of biological organization undermined the dominant theoretical paradigm of mid-to-late twentieth-century biology.

Recall that James Watson and Francis Crick's two seminal papers, which proposed the structure and function of the DNA molecule,[84] were published just over a decade before Langer's first volume of *Mind* appeared. While the cascading effects of Watson and Crick's discovery are hard to measure, at minimum they set the stage for the so-called "molecular revolution" in biology, and in doing so provided a material basis for a gene-centered theory of evolution and development that fed right into the hands of neo-Darwinians, who now had an empirical grounding for a fully mechanized theory of heredity and development that was perfectly compatible with the forms of causation operative in the rest of the material world.[85] Langer's conception of organismic change runs counter to this neo-Darwinian thesis in just about every way imaginable. Before I address this, it is worth noting that the theoretical biologist Conrad Hal Waddington, another close reader of Whitehead, proposed a theory of evolutionary and developmental organization that resonates with Langer's work on many levels,[86] but it, too, was met with harsh criticism and dismissed for decades.[87] Of course, the theory Waddington pioneered was "epigenetics" and this has proven to be a game-changer for theoretical and experimental biology, even if the full weight of the theory has yet to be appreciated.[88]

In any case, like Waddington, Langer does not dismiss the importance of genes in shaping the evolutionary and developmental trajectories of organisms: "the gene complement, with all its possibilities of combination and mutation," writes Langer, "is what holds such a mighty unit [of life] together, and sets the limits of what may happen within its scope."[89] Although the genome may function as the "unit of activity," it is not "an agent, an individual, or an organism; such entities arise only by processes of individuation."[90] For Langer, the individuation of living organisms via the concatenation of act-events is not determined by molecular machinery of the genome, as if the latter were the sole unit of selection. Phenotypic diversity is a function of genomic involvement working in concert with a host of other molecular, cellular, and social acts to frame the developmental possibilities of an organism. "No matter which or how many genes affect a particular process," writes Langer, "the final body expression of it is still open to other than genetic determinants."[91] Still, this "openness" does not simply shift the mechanisms of causality to extra-genetic factors, leaving evolution and development under the spell of mechanism. While organismic individuations are shaped through involvements with proximate environments (e.g., epigenetic), they cannot be reproduced in a generalizable model. The involvements of organismic acts are their own; they are singular and cannot be mapped onto others.

What is remarkable is that Langer's biology does not simply parallel Waddington's work on the epigenetic landscape in significant ways—and also anticipate insights made by Stephen Jay Gould and Elizabeth S. Vrba on "exaptation,"[92] as well as Stuart Kauffman's on the "adjacent possible"[93] (both of which state that you cannot predetermine function of a phenotype)—but it also points to some of the most difficult and unresolved questions about *method* in theoretical biology today. In the wake of the human genome project and the subsequent deluge of molecular and cellular data pouring into "-omics" databanks, a more holistic, systems-level understanding of how these data work together is essential to epistemology in the biosciences.[94] Building theoretical models, using the mathematics of complex dynamical systems, has significantly contributed to integrating the nonlinear causal mechanisms operative in the distributed landscapes of organismic development.[95]

Despite the explanatory affordances of complex systems, significant gaps are opened up when the limits of physico-mathematical modeling come into focus. As the computer scientist and theoretical biologist Giuseppe Longo has shown, complex adaptive systems apply the mathematical language of physics to patterns of spatial and temporal change irreducible to those found in physicochemical systems.[96] While using the tools of physics to say something about biological systems is not in itself a problem, and can be incredibly useful, it becomes one when those who use these tools do not register all they are abstracting from them to make predictions. Unfortunately, this failure to account for abstraction tends to be the rule rather than the exception among theoretical biologists. So, what is sold as a targeted modeling technique for complex biological phenomena is nothing more than the language of physics repackaged for biologists, who fall prey to the "Idols of the Laboratory." This means that biologists are not working with biology-specific modeling techniques. If they were, their models would be able to compute how a system generates its own conditions of emergence as it develops. In short: the model would be able to dynamize parameters that remain constant in dynamical systems modeling. But what Longo, Stuart Kauffman, the physicist Paul Davies, and an increasing number of scientists are realizing is that these conditions are not computable: more specifically, there is no way to compute how the possibilities in phase space for organismic development emerge alongside the material and energetic exchanges an embryo shares with its environment as it develops. The transcendental space of organismic development turns out to be more Deleuzian than Kantian, since potentials do not preexist the real material exchanges an organism shares with its environment.[97] While there are important uses for computational modeling in biology, especially in biomedical applications of molecular epigenetics,[98] ontogenesis is beyond the language of the computational sciences.[99]

As scientists and philosophers come to terms with the fact that ontogenesis is not circumscribed by the computational language of complexity, they are also having to grapple with the attendant problem that there are no research methods for theoretical biology not already determined by physico-mathematical descriptions.[100] For these researchers, this is not a call to separate the conceptual spaces of the biosciences from the investigation of other physiochemical processes (*à la* vitalist metaphysics), on the contrary: it is a plea to recognize that at a certain level of material organization phenomena emerge that can no longer be explained solely in terms of the previous level of organization, which means that new methods of research are required to investigate these phenomena. As the Russian biologist, Boris Zavadovsky wrote in 1931:

> The true task of scientific research, is not the violent identification of the biological and the physical, but the discovery of the qualitatively specific controlling principles which characterize the main features of every phenomenon, and the finding of methods of research appropriate to the phenomena studied ... But biological laws do not in the least lose thereby their material quality and cognisability, requiring only in each case methods of investigation appropriate to the phenomena studied.[101]

This brings us back to Langer. Her *Mind* trilogy displays an acute sensitivity to these shifts in material organization. If Langer uses the art image to investigate these shifts, then it is because she has discovered a research method appropriate to the phenomena she is investigating: that is to say, modes of relation that do not admit of general description via physical models. What emerges out of her "image of feeling" is not only a concept of physiologically rooted feeling that does not resort to psychic additions or mechanistic explanations but also a method for analyzing phenomena that is specifically tailored to the object of investigation.

As I conclude this chapter, I think it is worth underscoring just how close we remain to Langer's concerns in the twenty-first-century biosciences. In the rush to model physical and biological worlds on the same mathematical template, and offer causally robust explanations of the material world, distinct levels of organization are cast aside on the false promise of explanatory success. This situation, which more or less characterizes the landscape of theoretical biology today, has left some groping for methods and conceptual tools similar to the ones Langer was in search of over a half-century ago. If Langer's organic theory of feeling is born out of an experiment in crafting these methods and tools, then it stands as an important one. This is not simply because she gets the biology "right"—although I think it is worth entertaining a biology composed of rhythmical acts evolving according to varying degrees of involvement and individuation—but it is because her biology is conceived from a domain of research-practice—art—that creates semblances of the tensions and resolutions in the natural world that cannot be directly modeled. In this way, art, for Langer, is a method for researching changes and shifts in living matter that do not admit of general description, and it is ultimately this method that feeds and nourishes her philosophical concept of mind. What is striking, in this regard, is that it is precisely these unforeseeable shifts in matter that raise a host of questions for biologists today, and are beginning to convince some that the physical sciences are not up to the task of characterizing such novelties.[102]

Langer's research method might offer, in the end, some important lessons for theoretical biologists today. What would it mean for biologists searching for adequate methods to consult presentational abstractions in art before settling on generalizing abstractions in science? Would it mean that, before trying to code, biologists learn to paint? Or would it be that scientists head to an opera before attempting to sonify molecular interactions using elaborate sonification software? Likely, the range of relevant art practice would be more expansive than we find in Langer's own work (her taste in art was rather elitist and limited), and inclusive of film, computer animation, and myriad other contemporary forms of art research,[103] but the sentiment of her work would remain intact. This final set of provocations is not intended to sound flippant; quite the contrary, it is intended as a serious proposition for arts-based research in the biosciences. In the end, I think taking the proposal seriously would have less to do with a strict deployment of Langer's concepts like "acts," and more to do with the aesthetic genesis of new concepts in biology that address phasal transitions in excess of the finite modes of description used in the computational sciences.

NOTES

1. Donald Dryden, "Susanne K. Langer and American Philosophic Naturalism in the Twentieth Century," *Transactions of the Charles S. Peirce Society* 33, no. 1 (1997): 161–82.
2. Peter A. Bertocci, "Susanne K. Langer's Theory of Feeling and Mind," *A Review of Metaphysics* 23, no. 3 (1970): 527–51.
3. Randall Auxier, "Susanne Langer on Symbols and Analogy: A Case of Misplaced Concreteness?" *Process Studies* 26, no. 1–2 (1997): 86–106.
4. See Robert E. Innis, *Susanne Langer in Focus* (Bloomington: Indiana University Press, 2009).
5. Susanne K. Langer, *Mind: An Essay on Human Feeling*, vol. 1 (Baltimore: Johns Hopkins University Press, 1967), 29.
6. Langer, *Mind*, vol. 1, 22–3, 29.
7. Susanne K. Langer, *Philosophical Sketches* (Baltimore: Johns Hopkins University Press, 1962), 25.
8. Donald Dryden, "Susanne K. Langer and American Philosophic Naturalism in the Twentieth Century," *Transactions of the Charles S. Peirce Society* 33, no. (1997): 165–6.
9. Langer, *Mind*, vol. 1, 23.
10. Ibid., 22–3.
11. Ibid., 65.
12. "The reason for the failure of this well-equipped expedition into the mathematical realm [a la Rashevsky] is that abstract concepts borrowed from physics, such as units of matter—even with the adjective 'living' to qualify them—and their motions, do not lend themselves readily to the expression of psychologically important problems." Ibid., 43.

13. Ibid., 47.
14. Dryden, "Susanne K. Langer and American Philosophic Naturalism in the Twentieth Century," 163.
15. Cf. Langer, *Mind*, vol. 1, 11.
16. Susanne K. Langer, *Mind: An Essay on Human Feeling*, vol. 3 (Baltimore: Johns Hopkins University Press, 1982), 89.
17. Langer, *Mind*, vol. 1, 4.
18. Ibid., xviii.
19. Ibid., 9.
20. Ibid., 4.
21. Ibid., 20–1.
22. Ibid., 21.
23. Langer writes that, "[i]t is this transiency and general liability of the psychical phase that accounts for the importance of preconscious processes in the construction of such elaborate phenomena as ideas, intentions, images, and fantasies, and makes it not only reasonable but obvious that they are rooted in the fabric of totally unfelt activities which Freud reified with the substantive term, 'the Unconscious.'" Ibid., 22.
24. Cf. ibid.
25. Ibid., 23.
26. Cf. ibid.
27. Ibid., 31.
28. Ibid., xvii.
29. Ibid., 32.
30. Ibid., 55.
31. Ibid., 57.
32. Ibid., 56.
33. Ibid., 59.
34. Ibid.
35. Cf. ibid., 59–60.
36. Ibid., 59.
37. Langer, *Mind*, vol. 1, 64.
38. Ibid.
39. Ibid., 67.
40. Ibid.
41. Bertocci, "Susanne K. Langer's Theory of Feeling and Mind," 543.
42. Langer, *Mind*, vol. 1, 244.
43. See Robert Mitchell, *Bioart and the Vitality of Media* (Seattle: University of Washington Press, 2010).
44. See Adam Nocek, *Molecular Capture: The Animation of Biology* (Minneapolis: University of Minnesota Press, 2021).
45. On this topic see Phillip Thurtle, *Biology in the Grid: Graphic Design and the Envisioning of Life* (Minneapolis: University of Minnesota Press, 2018); Hannah Landecker, *Culturing Life: How Cells Became Technologies* (Cambridge, MA: Harvard University Press, 2007), or Natasha Myers, *Rendering Life Molecular: Models, Modelers, and Excitable Matter* (Durham, NC: Duke University Press, 2015).
46. See Myers, *Rendering Life Molecular*.
47. See Scott Curtis, *The Shape of Spectatorship: Art, Science, and Early Cinema in Germany* (New York: Columbia University Press, 2015), 4–5.
48. Langer, *Mind*, vol. 1, 140.

49. Ibid., 146, 96–7.
50. For a more detailed discussion of semblance, see Langer's *Feeling and Form* (1953), ch. 4.
51. Langer, *Mind*, vol. 1, 156.
52. Ibid.
53. Ibid.
54. See Langer, *Feeling and Form*, 125.
55. Langer, *Mind*, vol. 1, 161.
56. Ibid., 169.
57. See Alfred N. Whitehead, *Process and Reality*, especially 279–80. See also Steven Shaviro's, *The Universe of Things: On Speculative Realism* (Minneapolis: University of Minnesota Press, 2014), for a detailed account Whitehead's aesthetic theory.
58. See Langer, *Mind*, vol. 1, 172.
59. Ibid., 211.
60. Ibid., 199, 213.
61. Ibid., 220.
62. Ibid., 222–3.
63. Ibid., 238.
64. Ibid., 240.
65. Ibid., 239.
66. See Bernard Cache, *Earth Moves: The Furnishing of Territories* (Cambridge, MA: MIT Press, 1995).
67. Langer, *Mind*, vol. 1, 199.
68. Ibid., 199–253.
69. Ibid., 244.
70. Ibid.
71. Langer, *Mind*, vol. 1, 202.
72. Ibid.
73. Ibid., 257.
74. Ibid., 294, 304.
75. Ibid., 304.
76. Ibid., 305.
77. See Gilbert Simondon, *L'Individu et sa genèse physico-biologique: L'individuation à la lumière des notions de forme et d'information* (Paris: Presses universitaires de France, 1964).
78. Langer, *Mind*, vol. 1, 294.
79. Ibid., 311.
80. Ibid., 342.
81. Ibid., 336.
82. Ibid., 355.
83. Ibid.
84. Shown in both papers by James Watson and Francis Crick, "A Structure for Deoxyribonucleic Acid," and "Genetical Implications of the Structure of Deoxyribonucleic Acid," published in *Nature* 171 (1953): 737–8; 964–7.
85. See Dennis Noble, "Neo-Darwinism, the Modern Synthesis and Selfish Genes: Are They of Use in Physiology?" *Journal of Physiology* 589, Pt 5 (2011): 1007–15.
86. See Conrad H. Waddington, *The Strategy of Genes* (New York: Routledge, 1957).
87. As in Erik L. Peterson, "The Excluded Philosophy of Evo-Devo? Revisiting C. H. Waddington's Failed Attempt to Embed Alfred N. Whitehead's 'Organicism' in

Evolutionary Biology," *History and Philosophy of the Life Sciences* 33, no. 3 (2011): 301–20.

88. See Adam Nocek, "Transcendental Biology," *Philosophy Today* 63, no. 4 (2019): 1153–78; "The Epimedial Landscape," *Angelaki: Journal for the Theoretical Humanities* 25, no. 3 (2020): 24–40; and the book *Molecular Capture: The Animation of Biology* (Minneapolis: University of Minnesota Press, 2021).
89. Langer, *Mind*, vol. 1, 336.
90. Ibid.
91. Ibid., 335.
92. See Stephen J. Gould and Elisabeth S. Vrba, "Exaptation—A Missing Term in the Science of Form," *Paleobiology* 8 (1982): 4–15.
93. See Stuart Kauffman, *Humanity in a Creative Universe* (Oxford: Oxford University Press, 2016).
94. I have pointed to this in the publications from 2019, 2020, and 2021.
95. See also Leonie Ringrose, ed., *Epigenetics and Systems Biology* (Amsterdam: Elsevier, 2017).
96. See Giuseppe Longo and Maël Montévil, *Perspectives on Organisms: Biological Time, Symmetries and Singularities* (Berlin: Springer, 2014).
97. See Nocek, "Transcendental Biology," as well as Gilles Deleuze, *Difference and Repetition* (London: Continuum, 1994).
98. Cf. Ringrose, *Epigenetics and Systems Biology*.
99. See Cary Wolfe and Adam Nocek, "Ontogenesis beyond Complexity: The Work of the Ontogenetics Process Group," *Angelaki: Journal for the Theoretical Humanities* 25, no. 3 (2020): 3–8.
100. Cf. Longo and Montévil, *Perspectives on Organisms*.
101. Zavadovsky cited in Joseph Needham, "A Biologist's View of Whitehead's Philosophy," in *The Philosophy of Alfred N. Whitehead* (New York: Tudor Publishing Company, 1951), 254.
102. There is an open question concerning whether the phase shifts that interest Langer are the ones that interest theoretical biologists. It is clear, for instance, that Langer is not promoting Whitehead's pan-experientialism, and neither are theoretical biologists like Longo. The latter would likely locate important phase shifts at the first sign of life, but this does not seem to be Langer's assertion. Although, she does make it clear that living individuations determine the meaning and scope of their involvements. What is more, recent developments in biosemiotics, using C. S. Pierce, would likely bring Langer in even closer proximity to the biosciences today. See Jesper Hoffmeyer, "Why Do We Need a Semiotic Understanding of Life?," in *Beyond Mechanism: Putting Life Back into Biology*, ed. Brian G. Henning and Adam C. Scarfe (Plymouth, UK: Lexington Books, 2013), 147–68.
103. See Adam Nocek, *Molecular Capture: The Animation of Biology* (Minneapolis: University of Minnesota Press, 2021), 326–30.

BIBLIOGRAPHY

Auxier, Randall. "Susanne Langer on Symbols and Analogy: A Case of Misplaced Concreteness?" *Process Studies* 26, no. 1–2 (1997): 86–106. Available online: https://doi.org/10.2307/44798815.

Bertocci, Peter A. "Susanne K. Langer's Theory of Feeling and Mind." *A Review of Metaphysics* 23, no. 3 (1970): 527–51. Available online: https://doi.org/10.2307/3332574.

Cache, Bernard. *Earth Moves: The Furnishing of Territories*. Translated by Anne Boyman. Cambridge, MA: MIT Press, 1995.

Curtis, Scott. *The Shape of Spectatorship: Art, Science, and Early Cinema in Germany*. New York: Columbia University Press, 2015.

Davies, Paul. "The Epigenome and Top-Down Causation." *Interface Focus* 2, no. 1 (February 2011): 42–8. Available online: https://doi.org/10.1098/rsfs.2011.0070.

Deleuze, Gilles. *Difference and Repetition*. 1968. Translated by Paul Patton. London: Continuum, 1994.

Dryden, Donald. "Susanne K. Langer and American Philosophic Naturalism in the Twentieth Century." *Transactions of the Charles S. Peirce Society* 33, no. 1 (1997): 161–82. Available online: https://www.jstor.org/stable/40320624 (accessed February 18, 2022).

Gould, Stephen Jay, and Elisabeth S. Vrba. "Exaptation—A Missing Term in the Science of Form." *Paleobiology* 8 (1982): 4–15. Available online: https://www.jstor.org/stable/2400563 (accessed February 18, 2022).

Hoffmeyer, Jesper. "Why Do We Need a Semiotic Understanding of Life?" In *Beyond Mechanism: Putting Life Back into Biology*, edited by Brian G. Henning and Adam C. Scarfe, 147–68. Plymouth, UK: Lexington Books, 2013.

Innis, Robert E. *Susanne Langer in Focus*. Bloomington: Indiana University Press, 2009.

Kauffman, Stuart. *Humanity in a Creative Universe*. Oxford: Oxford University Press, 2016.

Koutroufinis, Spyridon. "Teleodynamics: A Neo-Naturalistic Conception of Organismic Teleology." In *Beyond Mechanism: Putting Life Back into Biology*, edited by Brian G. Henning and Adam C. Scarfe, 309–44. Plymouth, UK: Lexington Books, 2013.

Landecker, Hannah. *Culturing Life: How Cells Became Technologies*. Cambridge, MA: Harvard University Press, 2007.

Langer, Susanne K. *Feeling and Form: A Theory of Art Developed from Philosophy in a New Key*. New York: Charles Scribner's Sons, 1953.

Langer, Susanne K. *Mind: An Essay on Human Feeling*, vol. 1. Baltimore: Johns Hopkins University Press, 1967.

Langer, Susanne K. *Mind: An Essay on Human Feeling*, vol. 3. Baltimore: Johns Hopkins University Press, 1982.

Langer, Susanne K. *Philosophical Sketches*. Baltimore: Johns Hopkins University Press, 1962.

Langer, Susanne K. *Philosophy in a New Key: A Study in the Symbolism of Reason, Rite and Art*. Cambridge, MA: Harvard University Press, [1941] 1957.

Longo, Giuseppe, and Maël Montévil. *Perspectives on Organisms: Biological Time, Symmetries and Singularities*. Berlin: Springer, 2014.

Longo, Giuseppe, Maël Montévil, and Stuart Kauffman. "No Entailing Laws, but Enablement in the Evolution of the Biosphere." ArXiv:1201.2069, January 2012. Available online: http://arxiv.org/abs/1201.2069 (accessed February 18, 2022).

Mitchell, Robert. *Bioart and the Vitality of Media*. Seattle: University of Washington Press, 2010.

Myers, Natasha. *Rendering Life Molecular: Models, Modelers, and Excitable Matter*. Durham, NC: Duke University Press, 2015.

Needham, Joseph. "A Biologist's View of Whitehead's Philosophy." In *The Philosophy of Alfred North Whitehead*, edited by Paul Arthur Schlipp, 241–71. New York: Tudor Publishing Company, 1951.

Noble, Dennis. "Neo-Darwinism, the Modern Synthesis and Selfish Genes: Are They of Use in Physiology?" *Journal of Physiology* 589, Pt 5 (2011): 1007–15. Available online: https://doi.org/10.1113/jphysiol.2010.201384.

Nocek, Adam. "The Epimedial Landscape." *Angelaki: Journal for the Theoretical Humanities* 25, no. 3 (2020): 24–40. Available online: https://doi.org/10.1080/0969725X.2020.1754017.

Nocek, Adam. *Molecular Capture: The Animation of Biology*. Minneapolis: University of Minnesota Press, 2021.

Nocek, Adam. "Transcendental Biology." *Philosophy Today* 63, no. 4 (2019): 1153–78. Available online: https://doi.org/10.5840/philtoday2020128317.

Peterson, Erik L. "The Excluded Philosophy of Evo-Devo? Revisiting C.H. Waddington's Failed Attempt to Embed Alfred North Whitehead's 'Organicism' in Evolutionary Biology." *History and Philosophy of the Life Sciences* 33, no. 3 (2011): 301–20. Available online: https://www.jstor.org/stable/23335143 (accessed February 18, 2022).

Ringrose, Leonie, ed. *Epigenetics and Systems Biology*. Amsterdam: Elsevier, 2017.

Scarfe, Adam C. "Epigenetics, Soft Inheritance, Mechanistic Metaphysics, and Bioethics." In *Beyond Mechanism: Putting Life Back into Biology*, edited by Brian G. Henning and Adam C. Scarfe, 369–408. Plymouth, UK: Lexington Books, 2013.

Shaviro, Steven. *The Universe of Things: On Speculative Realism*. Minneapolis: University of Minnesota Press, 2014.

Simondon, Gilbert. *L'Individu et sa genèse physico-biologique: L'individuation à la lumière des notions de forme et d'information*. Paris: Presses universitaires de France, 1964.

Thurtle, Phillip. *Biology in the Grid: Graphic Design and the Envisioning of Life*. Minneapolis: University of Minnesota Press, 2018.

Waddington, Conrad Hal. *The Strategy of Genes*. New York: Routledge, 1957.

Watson, James, and Francis Crick. "A Structure for Deoxyribonucleic Acid." *Nature* 171 (1953): 737–8. Available online: https://doi.org/10.1038/171737a0.

Watson, James, and Francis Crick. "Genetical Implications of the Structure of Deoxyribonucleic Acid." *Nature* 171 (1953): 964–7. Available online: https://doi.org/10.1038/171964b0.

Whitehead, Alfred North. *Process and Reality: An Essay in Cosmology*. 1929. Corrected edn. Edited by David Ray Griffin and Donald W. Sherburne. New York: Free Press, 1978.

Wolfe, Cary, and Adam Nocek. "Ontogenesis Beyond Complexity: The Work of the Ontogenetics Process Group." *Angelaki: Journal for the Theoretical Humanities* 25, no. 3 (2020): 3–8. Available online: https://doi.org/10.1080/0969725X.2020.1754005.

CHAPTER THIRTEEN

Thinking Non/Humanly with Susanne K. Langer

ELDRITCH PRIEST

Whenever Langer's name is invoked it is often with respect to her theory of symbolism and philosophy of art. And not without good reason. *Philosophy in a New Key* (1942) and *Feeling and Form* (1953) are impressive works. They show an erudite and original thinker carrying out an assault on logocentrism in a way that not only presages the concerns that would preoccupy poststructuralist thinkers during the second half of the twentieth century, but also anticipates certain feminist interests in the importance of experience, the polyvocality of meaning, and the centrality of feeling in thought and life. But Langer's philosophy has always been a philosophy of mind. In fact, she herself made this clear when she undertook her decades-long project to write the multi-volume *Mind: An Essay on Human Feeling*. In this paper, I want to consider Langer's characterization as crucial to our understanding of her work, but not for the reasons one might think. As typically understood, Langer's concern for the status of mind is a concern for the status of human being. For her, "mind" is the exclusive remit of homo sapiens. But to arrive at this position Langer has to conscript the non-human animal in a way that makes the latter integral to her thinking about mind. In other words, animals are foundational to Langer's philosophy of mind, and in this respect, there is something to be said about her untold contribution to thinking non-human mentality. However, her contribution has its limits and it is in the way Langer construes animal mentality as largely the feeling of actionable lures with practical effects that she overlooks the way non-human animals might be said to think when they play, when they exhibit an "impractical enthusiasm." But to grasp this it is first necessary to lay out Langer's reasoning that shows mind as an aspect of feeling to which human beings have developed a species defining attitude.

FEELING?

The somewhat counterintuitive image of mind as feeling that Langer develops comes most explicitly at the end of her career and reflects not only her commitments to philosophical naturalism but an ongoing (if somewhat obscured) debt to her teacher Alfred Whitehead's metaphysics. For Whitehead, feeling is a technical term that describes the way entities in the process of actualizing themselves respond to what precedes them, including their own previous state of being and, in principle, everything within the entire world.[1] But for Langer feeling, which is also a technical term, is a strictly intra-organic affair. That is, whereas Whitehead admits feeling into any and all processes, Langer admits them only into those processes that count as organisms. Feeling is, therefore, something that only pertains to living beings; only "vital activities of great complexity and high intensity, usually (perhaps always) involving nervous tissue, are felt."[2] But that is not all, for as Langer has it, feeling is properly understood as an activity, not a thing. "To feel is to do something, not to have something," she writes.[3] Thus, feeling is a doing, a doing taking account of its own goings-on. For Langer, feeling is the mark of mentality. However, while feeling marks mentality, only those vital activities that are felt purely for their expression, for their sheer appearance, can be felt as mind. In other words, not all doings that are felt do the same thing, and only some doings do "mind." So how, then, is mind done? Which is to say: how is mind felt? Langer's position, which never really changes over the course of her career, is that mind is done by (felt as) a process of symbolization, a process that she also identifies as a specific activity of abstraction. And while abstraction is a wholly natural process, one that extends from microorganisms to great apes, it is only when abstraction leads to an "attitude toward objects *in absentia*"[4] that it begins to be felt as "mind." Abstractions that do not promote such an attitude still constitute a form of mentality, but this mentality does not make a mind.

Thus, while we can say that insects, earthworms, and simians, for example, have a mentality, they do not have a "mind." This is not, however, because their abstractions, their feelings lack complexity. It is because the attitudes that animals develop towards what William James famously called the "blooming, buzzing confusion"[5] of the world only ever enter into their existential situation as a stream of ongoing signals for immediate action. Langer writes, "If we would speculate on what an animal sees or fails to see in its environment" (or, I would add, what it feels abstractly) "we must start from what it is doing; for it sees [feels] whatever will implement or frustrate its acts."[6] As such, the mentality of animals—their ways of feeling patterns of impingements pass into signals and indications to act—is largely oriented by practical demands.[7] Said another way, what animals feel, what they "think" are not forms or concepts, but values that extend their creaturely behavior into a vital milieu.

But the "minded" creature, the human being, develops a rather different attitude towards the welter of impingements and goings-on that compose what, as Massumi notes in this volume, become its *forms* of feeling. Its attitude to the perplexity of feelings from which it formulates or abstracts a world of ideas develops out of what Langer sees as an unprecedented *need* to symbolize, a need to transform the valences of experience into devices for making an abstraction. Unlike animals, which she argues only need the flow of experience to yield actionable values, humans seem to require that it yield expressive ones, too. That is, for humans, something of the blooming, buzzing confusion needs to make an appearance apart from its sheer carrying on. The feeling called "mind" can be understood, then, as an activity whereby experience is felt abstractly as it is transformed into something that stands for something else. Mind, therefore, distinguishes itself *formally* from simple (animal) mentality by the way humans treat sense data as abstractions that *represent* experience and thereby satisfy the distinctive human need for symbolic expression.

However, there is another aspect to this symbolic transformation that is obscured by the formalization of experience. Given that mind, like all forms of feeling, is wrought by the vital rhythms and motivations that underwrite organic existence, it is also distinguished *practically* (and ironically) by its impractical enthusiasms. As Langer puts it:

> Only a part—howbeit a very important part—of our behavior is practical. Only some of our expressions are *signs* [by which Langer means signals], indicative or mnemonic,

> and belong to the heightened animal wisdom called common sense; and only a small and relatively unimportant part are immediate *signs of feeling.* The remainder serve simply to express ideas that the organism yearns to express, i.e., to act upon, without practical purpose, without any view to satisfying other needs than the need of completing in overt action the brain's symbolic process.[8]

If it is right, as Langer argues, that the mind is "an organ in the service of primary [human] needs"[9]—including this need to symbolize—and is distinguished from creaturely mentality by the production of acts that are expressive rather than exclusively practical or even communicative, then the satisfaction of *this* need has to be found in an order of activity that is somehow set off from or *exceeds* the practicalities of biological existence. That is, while the need for expression is determined (as all organismic processes are) by physiological processes, its fulfillment is affirmed by activities that yield something other than practical effects. Said another way, the symbolic transformation of experience satisfies the need for activities that are marked precisely, if not entirely, by their impractical enthusiasms.

AN IMPRACTICAL ENTHUSIASM

But what exactly marks an enthusiasm as impractical? In Langer's model nothing is impractical—even expression—so long as it satisfies a biogenetic need. And that is exactly what abstraction does for humans—it satisfies a need. This need, however, manifests in a particular way. As Langer writes, "It is only natural that a typically human function [the symbolic transformation of experience] should require a typically human form of overt activity," and this activity "is just what we find in *the sheer expression of ideas.*"[10] (The phrase "overt activity" is confusing here, because what Langer means by overt in this case is not an outwardly observable behavior but simply the *explicit* activity of ideating. Although this activity is, practically speaking, covert, since it is felt rather than unfelt by the organism, it is, strictly speaking, an overt activity.)

So, the chatter in our head, or the daydreams we have, or the intractable scrap of melody we notice only after we have been virtually auditioning it for hours, days, or weeks on end is practical to the extent that it satisfies an organic craving to abstract, to symbolize. But given that what is practical is a matter of degree, as Gilles Deleuze and Félix Guattari illustrate in their analysis of Rube Goldberg and stone-sucking machines,[11] we could argue that inner chatter, earworms, reveries, and so on, which exemplify the sheer expression of ideas, are *functionally* impractical enthusiasms. These, however, are not the only activities that are functionally impractical. To this list of private passions, we can add, as Langer does, the more public enthusiasms of myth, ritual, and art. For Langer these activities are marked as impractical enthusiasms to the extent that their very occasion coincides with their expression, which is to say (somewhat tautologically), that what is impractical is expressive, and what is expressive is impractical.

But what does it mean for an activity to be impractical? In short, it means to be "empirically senseless."[12] However, more broadly it means being symbolic. Over the course of her career, Langer offers a number of ways to consider what a symbol might be, and directly writes in *Feeling and Form* that a symbol is "any device whereby we are enabled to make an abstraction."[13] But a clearer way to understand a symbol's impractical enthusiasm can be found in Langer's discussion of Freud's take on ritual activities. Freud, she writes, "recognized that ritual acts are not genuine instrumental acts, but are motivated primarily *a tergo*, and carry with them, consequently, a feeling not of purpose, but of compulsion."[14] Ritual acts are not carried out for utilitarian ends so much as they extend from "a sheer inward need," and "are best interpreted as *expressive* behavior."[15] Thus, from Freud's insight she concludes that "human behavior is not only a food-getting strategy"[16] but also a performance that entails the production of feelings that feel their own abstraction from the push and pull of life's practical relays. This means, "Every *move* is [for human beings] at the same time a *gesture*"[17]—namely, an expressive form. In Langer's estimation, a gesture is expressive before it is communicative, an impractical enthusiasm felt abstractly before it is felt functionally. An act becomes symbolic, then, to the extent that it is felt as a "quality without practical significance," a quality that, incidentally, "makes the forms of things … present themselves *in abstracto*."[18] In a sense, symbolization is a process of self-abstraction, a process that, given Langer's emphasis on its human estate, surprisingly finds expression in the non-human kingdom.

THE PARADOX OF PLAY

It was British anthropologist and early cyberneticist Gregory Bateson who, unlike Langer, understood that self-abstraction is not exclusively a human affair. Abstraction, for him, is the direct perception of a difference, and this difference is itself expressive. In his theory of play and fantasy, Bateson suggests that an activity which draws value from its own performance stages a paradox that establishes the conditions by which, as he says, map and territory—symbol and symbolized—can be discriminated.[19] Bateson's example is the case of play-fighting monkeys whose actions, which he observed while visiting a zoo, did not denote what the actions for which they stand would have denoted. In their mock contest, the simians, Bateson argued, express a disparity between their performance and actual combat such that one and the same gesture—a "bite" for instance—comes to be (logically) distinguished from another—a "nip." Yet, the expression of this disparity—the activity of abstraction—is not exactly experienced as a symbol proper. It is recognized in the execution as an immanent value. To use Langer's words, in play abstraction is "felt as a quality rather than recognized as a function."[20]

The paradox of play—play is something that is not what it would be—marks, for Bateson, an evolutionary leap in communication. But I would suggest that the abstraction that is play also marks a step in the evolution of a what is a human need—symbolization. The execution of a fight that is not a fight has a value that satisfies a need which is chiefly expressive. That is, before play-fighting serves as an instructional model of how to wage war, play's abstraction is felt as a vitalizing force. But if play's abstract vitalization gives something to the animal, then Bateson's theory would seem to undermine Langer's thesis regarding symbolic transformation as an exclusively human need. To the extent that the paradox of play creates a semiotic situation, it would seem, then, that animals do have some inkling of mind, for play's paradox shows in its peculiar form of abstraction an impractical enthusiasm—sheer symbolic, or as Langer puts it, "symbolific," activity.

***

That abstraction matters to animals is not so outrageous. Langer readily admits that animals respond to and orient their actions around elements *drawn from* their immediately lived situation. Indeed, she writes: "The interpretation of signs is the basis of animal intelligence."[21] But for her, "directly felt inward and outward acts, springing from impulse and ambient pressions and opportunities, are sufficient for all animal needs."[22] However, the fact that animals play—that at times their actions do not denote what those actions for which they stand would denote—suggests there is something about play that satisfies what should also probably be called an expressive need.

This is something that Brian Massumi (2014) takes up in his work on animal instinct and sympathy. Building on Bateson's insights, and making a refrain of the idea of the "ludic gesture," Massumi creates a conceptual territory that gives animal mentality its expressive due and restores an instinctive edge to the human mind. For Massumi, a ludic gesture is, very simply, a form of abstraction. More precisely, it is a behavioral form that "performs an abstraction on its action" to yield an expressive excess.[23] The *manner* in which an activity is carried out, Massumi says, is what gives certain animal behaviors their "-esqueness," their pure expressive value.[24] The ludic gesture is, in other words, a feeling. But its feeling is an abstraction that makes it a *thinking*-feeling absolutely one with doing.

In this respect, the ludic gesture has the same function as Langer's symbolic transformation, for both concepts point to a process by which an "excess is felt as a palpable enthusiasm."[25] For Langer this enthusiasm finds its way into distinctly human acts such as ritual and art but also, more mundanely, as the endless chatter in our head. However, from Massumi's point of view, this transformation, this abstraction need not be construed as outright symbolic activity. An animal, he says, "performs something extra to the functions of its behaviors" through the -esqueness of its act—a flourish of the tail, a flash in the eye, a bounce in the step all satisfy an expressive need.[26] The status of the transformation as "symbolic" is, for the animal, not as important as the doing of this something extra.

***

Now, it is quite possible that I myself am reading something extra into this idea of expressive need that Langer did not intend. And perhaps I am even saying something heretical by virtue of the way my extrapolation converts her concept of symbolic transformation into an abstract force of impractical enthusiasm. But given that I am human, and humans have a need to transform experience into forms that are purely expressive, or interpretively creative, my heresy is not only condoned by Langer's notion of symbolic transformation, it is conceptually pre-endorsed. Furthermore, humans, like all animals, like to play. In fact, "When we humans say 'this is play,' we are assuming our animality."[27] And when we assume our animality, we are composing ourselves around "the -esqueness of the analog animal."[28] Becoming-animal is, then, a way of reading something extra into vital activities *with vital activities themselves*. But this is also how animals experience their becoming-human.

If that is not reason enough to take up the conceptual overspill of expressive need into a creative interpretation, then consider this: the symbolic transformation of experience that makes a bite a nip is an animal capacity. Yet, if it is an animal capacity, then just as equally, "when animals play, they are preparatorily enacting human capacities."[29] Thus, "Play dramatizes the reciprocal participation of the human and the animal, from both sides"[30] to effect an effective paradox that makes it possible to ask: do animals have wistful fancies? Do they have a need to express the extra-being that belongs to their form-of-life signature style? Do they have impractical enthusiasms? If so, is the manner in which they express these enthusiasms felt as thought, or enacted as style?

EPILOGUE

Now, although it is possible to ask these questions, it is not really possible to answer them in any straightforward manner, for all that would be given, would, from the perspective of an impractical enthusiasm, be something that *could be* answered. This is to say that because *play* and the symbolic transformation of experience in *feeling* surpass the given, they traffic in possibilities rather than actualities. As such their purview is at heart speculative, and therefore these words which you are reading do not denote what those words for which they stand would denote.

This chapter is published as "What It's Like to Think What It's Like to Think," in Eldritch Priest, *Earworm and Event* (Durham, NC: Duke University Press, 2022).

NOTES

1. Alfred N. Whitehead, *Process and Reality* (New York: The Free Press, 1978), 219–35.
2. Susanne K. Langer, *Mind: An Essay on Human Feeling*, vol. 1 (Baltimore: Johns Hopkins University Press, 1967).
3. Langer, *Mind*, vol. 1, 20.
4. Susanne K. Langer, *Philosophy in a New Key: A Study in the Symbolism of Reason, Rite, and Art* (Cambridge, MA: Harvard University Press, 1942), 31.
5. William James, *The Principles of Psychology*, vol. 1(New York: Dover Publications, [1890] 1950), 488.
6. Susanne K. Langer, *Mind: An Essay on Human Feeling*, vol. 2 (Baltimore: Johns Hopkins University Press, 1972), 77.
7. More accurately, Langer would say that animals are not motivated by a series of conceptually framed purposes and means as much as they are motivated by a desire to consummate the overall tension of acts that arise intra-organically from situation to situation. Langer, *Mind*, vol. 2, 45–102.
8. Langer, *Philosophy in a New Key*, 43.
9. Ibid., 38–9.
10. Ibid., 43.
11. C.f. Gilles Deleuze and Félix Guattari, *Anti-Oedipus: Capitalism and Schizophrenia* (Minneapolis: University of Minnesota Press, 1983), 3.
12. Langer, *Philosophy in a New Key*, 51.
13. Susanne K. Langer, *Feeling and Form* (New York: Charles Scribner's Sons, 1953), xi.
14. Langer, *Philosophy in a New Key*, 50.
15. Ibid., 50–1.
16. Ibid., 51.
17. Ibid.
18. Langer, *Feeling and Form*, 51.
19. Gregory Bateson, *Steps to an Ecology of Mind* (Northvale, NJ: Jason Aronson, 1972), 183–98.
20. Langer, *Feeling and Form*, 32.
21. Langer, *Philosophy in a New Key*, 59.
22. Langer, *Mind*, vol. 2, 138.
23. Brian Massumi, *What Animals Teach Us about Politics* (Durham, NC: Duke University Press, 2014), 5.
24. Ibid., 10.

25. Ibid., 9.
26. Ibid., 59.
27. Ibid.
28. Ibid., 59.
29. Ibid., 8.
30. Ibid.

BIBLIOGRAPHY

Bateson, Gregory. *Steps to an Ecology of Mind*. Northvale, NJ: Jason Aronson, 1972.

Deleuze, Gilles, and Félix Guattari. *Anti-Oedipus: Capitalism and Schizophrenia*. Minneapolis: University of Minnesota Press, 1983.

James, William. *The Principles of Psychology*, vol. 1. 1890. New York: Dover Publications, 1950.

Langer, Susanne K. *Feeling and Form*. New York: Charles Scribner's Sons, 1953.

Langer, Susanne K. *Mind: An Essay on Human Feeling*, vols. 1–2. Baltimore: Johns Hopkins University Press, 1967–72.

Langer, Susanne K. *Philosophy in a New Key: A Study in the Symbolism of Reason, Rite, and Art*. 1957. 3rd edn. Cambridge, MA: Harvard University Press, 1942.

Massumi, Brian. *What Animals Teach Us about Politics*. Durham, NC: Duke University Press, 2014.

Whitehead, Alfred North. *Process and Reality*. New York: The Free Press, 1978.

CHAPTER FOURTEEN

From Aesthetic Frights to the Politics of Unspeakable Thought with Susanne K. Langer

BRIAN MASSUMI

Susanne Langer's philosophy hinges on the analysis of the relation between discursive and non-discursive expression, language-borne and non-verbal thought. This essay works transversally across her three major philosophical statements, *Philosophy in a New Key*, *Feeling and Form*, and *Mind* (vol. 1), to investigate the imbrication of these two dimensions of expression. It examines Langer's distinction between "vital import" and signification as it plays out between sensuous and non-sensuous modes of perception. This interplay raises the question of the literal and the figurative, reworked by Langer in counterintuitive, but importantly suggestive, ways. It is in the passage from non-discursive thought to discursive expression, conditioned by the interplay of the literal and the figurative, that error arises. Expanding upon Langer's theory of error, and extrapolating from it, the essay ends with a brief speculative foray into the contemporary reign of error inaugurated by the politics of Trumpism and the concomitant rise of conspiracy thinking.

WHAT THE TOADSTOOL DID

A report of curious behavior in apes plays a pivotal role in Susanne K. Langer's discussion in *Philosophy in a New Key* of the conditions of possibility of language.[1] The report is from Wolfgang Köhler, in his work on the mentality of apes.[2] The chimpanzees he studied had a predilection for novel objects proposed for play and exploration. Every once in a while, however, a new object would elicit an immediate reaction of horror. For several of the chimpanzees, it was toadstools. Pinecones, just as novel, were fine. Thinking it might be an instinctive reaction to the possibility that the toadstool could be poisonous, Köhler presented it wrapped in cloth, so the odor would come through, but the offending fungus was hidden from view. It would be reasonable to assume that if an instinctive aversion were involved the smell of a potentially poisonous substance would elicit an avoidance response. It did not. Furthermore, the toadstool was not alone. It was just one in a motley series of inexplicably frightful, non-toxic, things: a pair of blue pants, leather gloves, a flat and rusty tin can, a burlap bag filled with hay, and primitive dolls with buttons for eyes. The last was the scariest of all, sending the full cohort of chimpanzees stampeding to the farthest corner of the compound, each riotously competing to bury its head deepest in the scrum.

What do these objects have in common? Nothing that I, as a non-chimpanzee ape, can discern. Humanly speaking, the response to the doll is the easiest to relate to. A doll can easily enter the "uncanny valley" where it carries an intimation of animation that clashes with the deadness of its button eyes. The contrast of qualities it carries places it in an indeterminate category. Everybody, human and presumably chimpanzee, knows that it is just a raggedy thing. Chimpanzees are not stupid. But its aspect connotes more—too much. An "over-determination":[3] a participation in too many qualities, of the sort that do not and should not go together. We *know* that it is just a thing, and we can easily see what its pragmatic parameters of action are. But this knowledge is overridden by the *feeling* of other, unplaceable possibilities, of scarily uncertain tenor, enveloped in the ambivalence of the overall quality through which and with which the thing appears. We know one way, and feel, qualitatively, another. That qualitative feeling is just as immediately a thinking. It conveys alternate possibilities, and alternate possibilities are only-thought. The feeling is a *thinking-feeling*. The doll's uncanny quality *connotes* something that cannot be fully expressed in words, even among those apes, such as we, who have a facility with them, because the felt quality signals an over-determination of meaning, a surplus of significance. Connotation is a linguistic operation. Here, it is functioning in pre-linguistic animals, enveloped in a directly experienced, complex quality. The fact that it is operative among our fellow apes bears witness to their possessing the primary conditions for language (making it more just to call them proto-linguistic rather than pre-linguistic).

Langer terms the thought-felt fears of the chimpanzees "aesthetic frights."[4] She as much as admits that chimpanzees also have aesthetic joys when she notes that some objects had the opposite of the toadstool effect, eliciting an immediate affection. These were adopted as companion objects, and treasured. A smooth, round stone is an example. Clearly, the fetish-readiness of the stone resides in its qualities of smoothness and roundness. These are qualities that co-compose without friction or ambivalence, unlike those of the object of aesthetic fright. The roundedness takes the smoothness on a roll. The coming-together of these qualities connotes the possibilities the stone carries, enveloped in the overall quality they co-compose: the possibility of continuous, rhythmic caress; the possibility of being pleasantly sucked; the possibility of being carried like a mascot in the crease between the lower abdomen and the thigh. These are soothing actions, connoting peace and comforting.

Although more harmonious, these qualities are no less uncontainable in language. Their "import," to use a key Langerian term, is not reducible to the signification of the words that denote them. The signification is "clothed"[5] in an affective tone that makes the meaning directly felt, in a qualitative more-than the semantic way. The fullness of the *situation*[6] with allure, tendency, incipient possibility, is swaddled into expression. The situation's charge of meaningful *activity*—the immanent swarm of "proto-acts"[7] it carries—comes to expression. It is literally seen in stone shape, without actually appearing. The shape of the stone in the hand of the ape is the direct expression of import.

The direct expression of import in a perceptual event is what Langer means by "symbolism." What the symbol presents, without it actually appearing, is the composition of contrasting proto-acts immanent to the situation. Langer calls this virtual activity complex the situation's "pattern of sentience."[8] The pattern of sentience is the "play of impulses form[ing] the dynamic matrix of life"[9] in the situation, at that moment, for this event. It is better to call it a "plexus"[10] than a complex, because the proto-activities come folded together in the clothing of the experience, the activity of each suffusing the activity of all, modulating the affective tonality felt overall while retaining its own sub-quality as an "*element*"[11] of the event. Each such element is a "microsituation":[12] a patch of relation, to others in the fold and to the "total relation"[13] that is their confluent situation and joint resultant. Relations within relation.

Chimpanzees certainly are not stupid. If they are particularly susceptible to aesthetic frights, and more resistant to being wheedled out of them, it is perhaps because they think-feel just as intensely as we humans but lack the palliative of full-fledged language. Language can exert the stabilizing influence of denotation. Denotation establishes "a 'short-cut' relation between names and things," with the result that "the relations in them are turned into something like objects."[14] The reified relations appear, actually this time, as the perceptible properties of the object. The clothed complexity of the pattern of sentience that is import appeared *non-sensuously*.[15] It was thought-felt, suggestively more than surface, and more than what was actually present: potential. This vital matrix now rises to the surface of the object, where it actually appears, transmuted into a pattern of actual sensuous qualities. The relations within relation of import are "projected" onto the surface like a "flattened map" of the situation.[16]

Elements of import, as agitating impulses, are "act-like."[17] Their proto-activity agitates. Properties of an object, appearing as sensuous qualities, on the other hand, present themselves for passive reception. They display themselves. Language, in its denotative functioning, indexes itself to the display. Words affix to points on the map, in one-to-one correspondence. Words are discrete, recombinable units of language: *parts* of language, as opposed to "elements." Their application to the map construes the points on the map as similarly discrete and recombinable, parts of the object. Parts interacting discretely, one beside the other, form external relation. They are "*in* relation"[18] to each other, extrinsically. Their interrelating spreads itself extensively, along a line, over a surface, or through a volume.

Elements, for their part, *are* relations. Beings of relation. They are intrinsically related. They are indivisibly involved in one another, the activity of one arcing into or out of another, tendential, in potential. They call for and call forth each other. They are indicators and bearers of each other. Rather than forming an extensive spread, they flex and plex together in a dimension of intensity. They form involutions, "further and further acts subsumed under almost any act."[19] Activity all the way down: an "immense involution pervading the realm of life."[20] They constitute "internal relations."[21] This does not mean that they are contained in an interior. It means that the activity of the one is in the others, mutually inflecting, modulating, resonating, rhythming,[22] in any number of potential ways, harmonious or discordant, or both at the same time, expressible in aesthetic frights or joys.

Elements are "indivisible and inalienable from the whole."[23] But their over-determination means that "no analysis can uncover a simple and direct relation of one element to another or of an element to ... the whole."[24] The elements fall in with each other, all together differentially. They present unity only in the expressed form of their concerted affective tonality or non-sensuous quality. Their falling-in with each other is transformational (eventful), the expressed quality a lived token of that transformation. "Suddenly, [an element's] initial character changes, or some other element has taken its place, not dislodged it but swallowed it so that the old element is somehow 'in' the new but is transformed by figuring in another functional pattern altogether."[25]

Language splays the involution of the elements out into an extensive order, linear in nature. It "string[s] out our ideas even though their objects [the elements and their qualitative expressions] rest one within the other; as pieces of clothing that are actually worn one over the other have to be strung side by side on the clothesline. This property of verbal symbolism is known as discursiveness."[26] Language's discursive order hangs the intensity of symbolic expression out to dry. "Language in the strict sense is essentially discursive; it has permanent units of meaning which are combinable into larger units; it has fixed equivalences that make definition and translation possible; its connotations are general, so that it requires non-verbal acts, like pointing, looking, or emphatic voice-inflections, to assign specific denotations to its terms."[27] Language transposes the "symbolic" thought-feeling of aesthetic frights and joys and other vital imports into a register of general ideas. This, in Langer's vocabulary, is the register of "conception," in contradistinction to "*concepts*," or the thinking in the feeling of the event of symbolic expression.[28] "Wordless symbolism, which is non-discursive and untranslatable, does not allow of definitions within its own system, and cannot directly convey generalities."[29]

Conceptions have signification, of a general order. Concepts have singular significance: vital import. Each appearance of a concept is tinged with an affective tonality all its own. They are "embodied—sometimes rather too much":[30] over-determination. The order of concepts is continually self-transforming in a way that generates an ever-more than linearly articulable meaning. Concepts are qualitatively expressed. They are by nature expressive, while the conceptions of discursive language have a vocation to be demonstrative and governmental function.

Discursive language's demonstrative vocation is to some degree stymied by the generality of the ideas it forms. That is why it has to be supplemented by "non-verbal acts like pointing."[31] However demonstrative its gesturing, it cannot recover the singularity of the thinking-feeling that accompanies its stringing out, along an involutionary course co-involved with that linear progression. Demonstration drapes a general over a particular. The fit is loose. A great deal of complexity sinks into the folds of the experiential plexus. "An idea that contains too many minute yet closely related parts, too many relations within relations, cannot be 'projected' into discursive form; it is too subtle for speech."[32]

Every concept is such an idea. The granular level of feeling that informs import is lost in discursive language. Conceptions represent a loss of conceptual power, at the same time as they raise thought to a higher power. What discursive language loses in intensity, texture, and nuance of feeling, is counterbalanced by a gain in general powers of extensive ordering, of the kind undergirding scientific and technological achievements and governmental function.

It is important not to treat Langer's distinction between discursive language and non-discursive expression as a rigid binary. Discourse's running line of words is footloose and lossy, but "our conceptual activity seems to flow through them."[33] Discursive and non-discursive thought are imbricated. There is "an unconscious, spontaneous process of abstraction, which goes on all the time." This is the "process of recognizing the concept in any configuration given to experience, and forming a conception accordingly." Langer calls this "abstractive seeing," and credits it with being the "foundation of our rationality."[34]

In short: no concept, no conception; conception, lossiness of concept. There is the rub in rationality.

Our human rationality is what the chimpanzee beholding a toadstool lacks. It is not at all clear that it is to their general loss. For the rub of abstractive seeing is also the foundation of humankind's advanced and seemingly endless capacity for error.

LANGUAGE AND THE USELESS

Something funny happens on the way to rationality. The literal is turned on its head. The notion of abstractive seeing makes the point that the literal *is* a form of abstraction: "literal knowledge, the abstracted conception of things."[35] "Our first understanding of forms is normally a literal comprehension of them as *typical things* or *such-and-such events* … which may be called *practical vision*."[36] Practical vision hinges on words having attained "fixed, general connotations, so that they may serve in a conventional, literal fashion."[37] "Literal" here means reliably pointing to an object or conception in a conventionally actionable way, recruiting typical things for such-and-such events. Only certain practically significant aspects of the relations within relation composing the total relation of the situation as directly thought-felt are taken up, and these are taken to be "the" meaning. Units of literalized meaning can be repeated and recombined, in relative independence of any given context, enabling a systemization of practices. This selective systemization comes to be taken for the full reality. "Speech becomes increasingly discursive, practical, prosaic, until human beings can actually believe that it was invented as a utility."[38]

The becoming-utilitarian of language works to neutralize the vagaries of context by placing the emphasis on repeatability. Once language has become maximally discursive, practical, and prosaic, contexts can be treated from the bias of what they can be abstractively seen to have in common. When new contexts are encountered, they are assimilated to the normative frame of such-and-such events featuring typical things. Connotation is pacified, the agitative relational activity of its elements muffled. Like objects, it becomes reified relation. This is in fact exactly what is meant by objective: the rendering normatively pre-recognizable of contexts, pointed to for their readiness to host useful activity. Connotation becomes limited to the background indication of contexts, tamed. All of this mobilizes tremendous powers of organization, but by rounding off the singularity of the situation and losing the granular detail of its relational complexion, it also introduces the possibility of error. It is the case in point of Whitehead's "fallacy of misplaced concreteness."[39]

Useful language neutralizes context—but context was what empowered the generalizing abstraction at the basis of it to begin with. Without something more ample in its embrace of complexity and potential to depart from, the selective generalization of discursive language becoming-useful would have nowhere to stand. Its taming of context produces a faded image of context in its more embracing shape, as full of vital import, and more than utilitarian affordance.

This is where context is turned on its head. Context in a more expansive, primary sense is what harbors vital import, as directly perceived in and as the singular affective tonality of aesthetic frights and joys. Connotation, pre-practical vocation, is the proto-linguistic mode in which vital import comes to conceptual expression. The primacy of context in this sense is obvious if you consider that the context of an expression "is what tells us what is its sense—whether we should take it literally or figuratively, and how, in the latter case, it is to be interpreted—it follows that *the context itself must always be expressed literally*, because it has not, in turn, a context to supplement and define its sense" (emphasis added).[40] The literal in the useful-language, denotative sense is a figuration (projective mapping) of the literal as it directly conveys import in symbolic expression. The "literal" is figurative. The figurative in the everyday sense, coming out the far end of utilitarian language use, is a second-order figuration that partially recovers the in-each-otherness of import's relations within relation. This takes the form of metaphor.[41]

A mode of expression that lacks conventional words "expresses what is unspeakable in verbal speech."[42] This is the case of course for artistic modes of expression (none more powerfully than music in Langer's estimation). Artistic expression "can reveal the nature of feelings with a detail and truth that language cannot approach," albeit "untranslatably."[43] This "revelation" is its power of thinking—in concepts, directly presented, with non-sensuous intensity.

But as we have seen, so mundane an event as the sudden appearance of a toadstool in your compound can carry an aesthetic force. Non-discursive conceptual activity accompanies everyday perceptual modes, just as it "runs through discourse," in an intensive dimension. Every experience is tinged with an involutionary more than a sentient patterning. Every act is run through with the conceptual force of the aesthetic, defined as the feeling of the thinking-feeling under way; the feeling of potential *more thans*, as vital matrix. This is the revelatory mode of thinking that we term "insight" or "intuition."[44] What Langer terms symbolism brings insight into the "literally unspeakable realities" of the vital matrix.[45] It brings what William James terms "pure experience" to effective expression. "The limits of language are not the last limits of experience, and things inaccessible to language may have their forms of conception, that is to say, their own symbolic devices. Such nondiscursive forms [are] charged with logical possibilities of meaning."[46]

Yes, we actually come to believe that language was invented as a utility. Langer dedicates many pages in *Philosophy in a New Key* to demonstrating the opposite.[47] The utility widely theorized at the proto-linguistic level of non-human animals and neonate humans, before denotative language develops, is the communication of basic needs. The signs by which proto-linguistic beings express their needs constitute a subset of what Langer calls "natural signs," which are "symptoms,"[48] as opposed to both the signifying signs of denotative language and the thinking-feeling of symbols. Observation reveals that language does not dawn in the communication of needs. "All attempts to teach apes or the speechless 'wild children' to talk, by the method of making them ask for something, have failed; whereas all cases where the use of language has dawned on an individual, simian or human, under such difficult circumstances, have been independent of the practical use of the word at the moment."[49] Accession to language comes in the form of ejaculations, particularly of joy, without any utilitarian intent—purely expressive. They are particularly of joy because, although outcries of pain or fear have an expressive aspect, they are symptomatic in a way that calls for practical response. This places them more in the needful category of natural signs. The expression of joy, on the other hand, is redundant in relation to practical matters. It doubles the practicality of the situation with a pure expression of the situation's affective tonality, attached to a word that will later come to have a denotative relation to the salient features of the situation, and will develop over time a utilitarian vocation. In the emergent, expressive, use of the word, the vocalization punctuates or culminates the situation with an immanent report of its non-sensuous quality, more than it designates any thing. It is less communicative than participatory. The case is similar for human babies in the babbling stage. They possess "a great sensitivity to 'expressiveness' of any sort," and are exuberant in letting us know that in their vocalizations.[50] "Young children learn to speak … by constantly using words to bring things into their minds, not into their hands."[51]

Substitute the "activity of thinking-feeling" for "mind." Langer, of course, is far too sophisticated a philosopher to take up the notion of "mind" as a subjective container of images or concepts. Like Whitehead, she is interested in understanding mentality as a full-body mode of activity that is found in various degrees along a continuum of life. She defines the first degree of mentality as "felt passage,"[52] echoing Spinoza's definition of affect. First words embody mentality to the first degree in bubbling expression. "It is best to admit that ... the purely communicative aspect of language has been exaggerated,"[53] along with its fundamental usefulness.

> The transformation of experience into concepts, not the elaboration of signals and symptoms, is the motive of language. Speech is through and through symbolic; and only sometimes signific. Any attempt to trace it back entirely to the need of communication, neglecting the abstractive experience at the root of it, must land us in the sort of enigma that the problem of linguistic origins has long presented.[54]

It is easy to see that at this emergent threshold, the "concepts" into which experience is transformed are "lived abstractions"[55]—thinking-feelings participatory in the event they bring to expression.

THE GENEALOGY OF ERROR

Context is literal in Langer's sense because it is thought-felt in direct perception, and direct perception, as Whitehead argues, is fundamentally without error. It is exactly as it happens. "A thing is what it seems."[56] But its seeming, as it is, nevertheless carries the seeds of error.

Whitehead uses the example of a mirror.[57] Under certain conditions, you might look in the mirror and take what is reflected behind you for something in front of you, where the surface of the mirror lies. You may even reach out your hand to grasp an object and knock your knuckles on the mirror. Whitehead's point is that this result was built into the structure of the situation, which was objectively illusionary. Or better, it featured an objective figuration: the mirror itself accomplished a projection, flattening the three-dimensional span of the situation and mapping it onto a two-dimensional surface display occupying the center of attention. The gesture of grasping expressed an active understanding of the context, as it really appeared, in its objective figuration.

The positioning and posturing of the body, with its grasp-happy habits of perception, were constituent elements of the context. The embedded, active understanding expressed by the hand gesture shows that the direct perception was directly a thinking. The movement of the hand was a performative deduction of the location of the object reached for and the possibilities for manipulation it affords, accomplished without an act of cogitation separate from the perceiving (in Peirce's terminology, it was an abduction). There is a direct conceptual element in the make-up of the act. In other words, the act of grasping qualifies as a symbolic act, performed under conditions where the seeming of what the thing is, was objectively "dissembl[ing]."[58]

The taking of the object behind for one before was a mistake—a mis-take—in the sense that it did not land felicitously; it did not complete its tendential arc. Langer characterizes such *mis-takes* as proto-errors.[59] *Proto-errors* are abductive short-circuiting of action. They are corrected pragmatically, by the unhappy result. Full-fledged errors are adventitious outgrowths of proto-errors. They thrive under conditions in which signifying gestures are sheltered from such immediate, pragmatic tests as knuckle-banging. For example, rather than going to grasp the object, you might point to it while describing it as lying before you. Error occurs in the passage from direct perception to denotation, particularly as it is supplemented, as Langer says it always must be, by demonstrative gestures. "Error arises only on the higher level of 'intellect' (discursive thinking)."[60] It consists in mis-taking the symbolism of the situation—that is, the expression of the literal context in Langer's sense—as "literal" in the everyday sense. Error is misplaced literality.

If there is a thinking in the perceiving that is one with the act, might the situation have potentially played out differently, avoiding the unhappy consequences of the proto-error? Yes, if, for example, your quality of attention had been different in the situation. In other words, had the context been different, because of a variation in that constituent element of it that is your self-positioning, self-posturing body and the habits and tendencies it carries. But it was not. The context was what it was. So, this is a counterfactual argument. The thing is, *the context as it was, included counterfactuality, in potential.* Had your quality of attention been different, you might have had an increased sensitivity to the periphery of your perception that might have clued you into the location of the object, altering your abductive thinking of the situation. Or subtle variations in light and shadow might have taken on abductive values as cues. A context is complexly primed, full of texturings and backgrounding-foregroundings that are susceptible of making a difference, but are not necessarily destined to.[61] These are contributory elements of the situation that include what, looking back at the genesis of any particular act that happened to be performed within it, were seeds of alternative outcomes. These seeds of potential are really, literally in the situation, co-constitutive of the context, co-constructive of any symbolization of it. They are no less there when they fail to bear fruit. They are part of the concept (and are the reason why the concept is always more inclusive than any given conception bringing it to actual expression).

"Wherever an actualization occurs there has been an option which the actualization has decided ... options belong to the very nature of acts."[62] "Every act within an individual has to get out of the way of other acts which, nevertheless, are making its situation and perhaps implementing its advance to consummation."[63] If an act misfires, "its abortive dynamism adds itself to the unanalyzable matrix of the agent."[64] "Out of the flood of unfelt options arise the larger ones that resolve themselves in behavioral acts."[65]

"Resolve *themselves.*" An agent is not a subject, or a deciding mind. "An agent is a complex of actions"[66] reaching its own resolve. The complex includes actions that remain in potential and actions that consummate, proto-acts that are effectively felt and those passing unperceived. All of these belong to the texture of the situation, as real elements in its constitution.

UNFELT FEELINGS: TOWARDS A POLITICS OF UNSPEAKABLE THOUGHT

Acts emerge from a flood of unfelt options. The way the flood resolves itself into an effectively channeled, individualized flow of action[67] constitutes who (what) the subject will have been at each step of the way as it continues across the felt passages of a life. "Life is the self-expression of impulses,"[68] or elemental tendencies, each with its own affective sub-flavor. The behavioral acts composing a life are over-acts. They roil over, and channel out, their outcome selectively resolving an energizing tension among a crowd of "act-like" agitations. These proto-acts do not appear in their individuality, but neither are they erased. They are felt, in effect, in the way feelings felt and unfelt issue in a joint result. In the proto-commotion immanent to the behavioral act's outcome, the act-like impulses are "internally involved with each other."[69] They are mutually included in each other's ferment.

> Emotional opposites—joy and grief, desire and fear, and so forth—are often very similar in their dynamic structure, and reminiscent of each other. Small shifts of expression can bring them together, and show their intimate relations to each other, whereas literal description can only emphasize their separateness. Where there is no exclusion of opposites, there is also, strictly speaking, no negative.[70]

No exclusion, no negative: positive plenum of entangled tendential potential. Plexus. Roiling over and channeling out to constitute the complex of over-acts that is the agent, subject of a life as it continues, stringing itself out as it gathers into itself the beaded episodes of activity through which it passes.

"Small shifts of expression" can bring the elements together—and bring their immanent coming-together over and out in a new expressive variation on their involvement with each other, like a kaleidoscope ever presenting a new facet. The practice of small shifts bringing out varieties of expression is the specialty of art: "many aspects of life that never rise to feeling may appear in the art symbol; and they appear there as though they could be felt."[71]

But it is clear, from Langer's discussion of aesthetic joys and frights, as well as many other aspects of her thinking, that while art is the consummate expression of unfelt proto-acts "as if they could be felt," this symbolic function is not confined solely to the realm of artistic activity. It is present in all acts, as an aspect of their complexion. It is expressed ephemerally in the overall affective tonality, or non-sensuous quality, of each and every event.[72] Artistic events are those that conserve the plexus of unfelt feelings, "embedded and inherent in objects [or performances] from which they can not be actually separated, but only distinguished by abstraction."[73] These objects or performances, which cannot be actually separated from the generative plexus of potential proto-acts, "share in the non-sensory character of [the] relational forms"[74] characterizing the plexus. They constitute the paradoxical category of "*objective feelings*"[75]—one of the most original contributions of Langer's thought.

Whereas the plexus immanently exceeds the agent, in the usual sense of an enduring subject of experience, objective feelings transcend it. They carry the relational forms of the plexus forward, into a *beyond* of the individual life, prolonging its vitality, or patterns of sentience, into other lives and other futures. In both dimensions, infra-(micro, molecular) and over-(macro, molar),[76] symbolic expression exceeds the personal. The hallmark of Langer's thought is the concern for *impersonal intensities* of experience.

The phrase "objective feelings" should not be allowed to overshadow the contribution to the plexus of what were termed above "positionings" or "posturings" that prime the expression by texturing its proto-active field of emergence into a contrasting pattern of backgroundings and foregroundings, felt and unfelt. The terms *positionings* and *posturings* are not ideal. They imply a preexisting agent as their enduring subject, when for Langer, the agent is emergent in and through its acts of expression, and is serially regenerated by each iteration of them, leaving (in the case of the artist) objective feelings behind as "monuments" of their passing.[77] A better term than positionings or posturings would be "tendencies."[78]

Another word, suggested by Whitehead's philosophy (which was so central to Langer's intellectual development) would be "valuations." The orienting of proto-expression inclines the coming event of expression toward certain tendential "options" at the expense of others, making their rise into feeling more probable. It may also reconcile contrasting tendencies to each other in a new self-catalyzing accommodation, thus inventing new patterns of sentience that add potential to the life of expression. Small, act-like "shifts in expression" occurring immanently to the act, in ways that do not actually appear, can inflect the course of expression. They can make a difference in the outcome that can be felt, in affective tonality, even if they cannot be consciously thought individually or ordered discursively. Even shifts that remain at the level of non-discursive thought will have contributed to inflecting the course of discursive thinking. Unspeakable thought would then express in language (in it, through it, but not of it).

Can this non-discursive inflection of discursive thinking be made a practice, or even a politics? If small, act-like shifts can be contrived to inflect discourse, the resulting course variations in discursive thinking could be observed, then fed back into the emergent level of proto-expression. Iterated, this process would enlarge the parameters of discourse, bringing different facets of the plexus to non-sensuous expression in turn. The selective complex of actions constituting the agent would be brought into greater overlap with the plenum of its generative plexus. Across the series of small, act-like shifts, impersonal intensities of feeling would resolve to express *themselves* more fully, through repetition and variation. The repetition and variation would cumulatively enrich the vital import carried through the iterative process.

It is just such a practice of small shifts in expression that Erin Manning analyzes under the concept of the "minor gesture,"[79] situating its politics in relation to forms of activity that straddle art, philosophy, and activism that had not yet come into their own in Langer's time, and which go by such contemporary names as "research-creation."[80]

The political import of pragmatic practices for the collective (impersonal) exploration of unspeakable thought becomes evident if we return to Langer's account of the genealogy of error. The derangements in present-day discourse, which have risen to plague proportions since the successful presidential campaign of Donald J. Trump (to limit the discussion arbitrarily to the paradigmatic American case), can be seen as errors of projective mapping. Mis-takes that have come to be systemically embedded in social discourse distort the texture of the plexus of potential. They foster dissemblings of causal relation that stand up to subsequent experience no better than trying to grasp an object behind by darting a hand out in front.

Length limitations forbid a fully developed analysis here. A brief allusion to one prominent example will have to do. Think whiteness as the surface of a systemic mirror. The projection of the diversity of tendencies plying the social field onto that normative surface produces a misleading mapping. The texture of the social field's constituent tendencies is blurred, the effective complexion of the contrasts between them muted. The full spectrum of options, and "optional feelings," that are proto-active in potential is relegated to unfeeling. The thinking-feeling of potential is thus limited, in ways that selectively channel over-acts of behavior into predictable patterns that fall into a precast mold, reinforcing existing structural inequalities observable on the macro level.

Combined with the loose-fitting clothing of the general ideas enshrined in dominant discourse, this structural dissembling of felt potential spawns egregious intellectual errors. Systemic problems with complex relational etiologies are misattributed simple causes mapped onto discrete agents, or parts of society, such as members of racialized groups or immigrants. Everything is flattened onto a macro-surface of expression that occults the generative infra-activity of the social elements of all categories. This is a case of misplaced literality, taking form in a mis-mapping of the societal context. On its basis, slippage easily occurs. The fatally loose fit of the general ideas afoot allows the causal misattributions to skid further and further off the tracks. Associative links, not far from free association, madly proliferate, gaining range and speed through social media dissemination. They self-express in the mode of indirect discourse, free from anchoring in particular attributions ("they" say …; I heard …; didn't you hear?). A machine of collective enunciation spins out of control, as the general social media agent "they" says more and more, more and more associatively. General ideas progressively lose what definition they had. Elisions occur, producing monstrous amalgamations (such as "liberal communists"). Conspiracy thinking is now all over the map. Conspiracy thinking: the discursive industry mass-producing metonymic assoications and metaphorical substitutions stemming from and disseminating errors in causal connection. With conspiracy thinking comes a change in the register of the overall affective tonality, into one of paranoia, hate, and fear. The utility of language dangerously misfires.

In Langerian terms, the mis-takes embedded in direct perception by the valuations of whiteness are amplified by misplaced literality, leading to intellectual errors enshrined in general ideas. The violent swings of conspiracy thinking let these errors loose. A malady of literality develops into a metonymic/metaphorical plague. Conspiracy thinking is a discursive will to truth gone tragically, farcically wrong. All of this is symptomatic of a deficit of symbolic thinking-feeling, the hallmark of which is the extreme personalization of the cultural and the political that we see today.

Extrapolating from Langer's work, an understanding of what she analyzes under the now archaic-sounding rubric of "symbolism" can take on urgent actuality. It can potentially inform a collective counter-politics of expression drawing on impersonal intensities of thinking-feeling, in ways aesthetically more embracing of the texture of the social plexus, and less mis-takenly flush with vital import. A fuller account of both the maladies and potential counter-practical salves will have to wait for another opportunity.[81]

NOTES

1. Susanne K. Langer, *Philosophy in a New Key: A Study in the Symbolism of Reason, Rite, and Art* (New York: New American Library, 1948), 89–93.
2. Wolfgang Köhler, *The Mentality of Apes* (New York: Harcourt, Brace & Co., 1925).
3. Susanne K. Langer, *Feeling and Form: A Theory of Art* (New York: Charles Scribner's and Sons, 1953), 242.
4. Langer, *Philosophy in a New Key*, 91.
5. Langer, *Philosophy in a New Key*, 58. Langer's use of the word "clothed" recalls Whitehead's in *Process and Reality* (New York: Free Press, 1978), 85–6.
6. Susanne K. Langer, *Mind: An Essay on Human Feeling*, vol. 1 (Baltimore: Johns Hopkins University Press, 1967), 281.
7. Langer, *Mind*, vol. 1, 274, 436.
8. Langer, *Feeling and Form*, 31.
9. Langer, *Mind*, vol. 1, 304.
10. Ibid., 304.
11. Ibid., 292, 428.
12. Ibid., 303.
13. Langer, *Philosophy in a New Key*, 44.
14. Ibid, 63–4.
15. Cf. Langer, *Feeling and Form*, 21–3.

16. Langer, *Philosophy in a New Key*, 64. This is an allied concept to Whitehead's "transmutation" (*Process and Reality* [New York: Free Press, 1978], 27).
17. Langer, *Mind*, vol. 1, 202, 273.
18. Langer, *Feeling and Form*, 21.
19. Langer, *Mind*, vol. 1, 273.
20. Ibid., 275.
21. Ibid., 204.
22. The generative role of rhythm in compositions of feeling is a central concern of Langer's; see, for example, *Mind*, vol. 1, 213.
23. Ibid., 200.
24. Ibid.
25. Ibid.
26. Langer, *Philosophy in a New Key*, 66.
27. Ibid., 78.
28. Ibid., 48–60. "Concepts" in Langer can be compared to Deleuze's "ideas" as they are defined in *Logic of Sense* and *Difference and Repetition.*
29. Ibid., 78.
30. Ibid., 49n.
31. Whitehead similarly emphasizes the "elliptical" nature of linguistic propositions and their necessary supplementation by demonstratives in *Concept of Nature* (Cambridge: Cambridge University Press, 1964), 7–12.
32. Langer, *Philosophy in a New Key*, 75.
33. Ibid., 61.
34. Ibid., 58. The predominance of vision here should not be taken as limitative: other senses also perform abstractions. Of particular importance are the kinesthetic senses, in particular proprioception.
35. Ibid., 216.
36. Ibid., 216–17.
37. Ibid., 114.
38. Ibid., 115.
39. Whitehead, *Process and Reality*, 7–8.
40. Langer, *Philosophy in a New Key*, 114.
41. Ibid., 112–14.
42. Ibid., 190.
43. Ibid.,
44. Langer, *Mind*, vol. 1, 129–30.
45. Langer, *Philosophy in a New Key*, 211.
46. Ibid., 215.
47. Ibid., 83–116.
48. Ibid., 46–7.
49. Ibid., 96.
50. Ibid., 99.
51. Ibid., 98.
52. Langer, *Mind*, vol. 1, 426.
53. Langer, *Philosophy in a New Key*, 89.
54. Ibid., 103.
55. Massumi, *Semblance and Event Activist Philosophy and the Occurrent Arts* (Cambridge, MA: MIT Press, 2011), 17–18, 28.

56. Langer, *Feeling and Form*, 49.
57. Whitehead, *The Concept of Nature*, 147, 151–3, and *Process and Reality*, 126–7.
58. Langer, *Feeling and Form*, 49.
59. "Prototypes of error": Langer, *Philosophy in a New Key*, 23.
60. Langer, *Feeling and Form*, 381.
61. Erin Manning, *Pragmatics of the Useless* (Durham, NC: Duke University Press, 2020), 103–14.
62. Langer, *Mind*, vol. 1, 436.
63. Ibid., 429.
64. Ibid., 377. This is Whitehead's "negative prehension"; *Process and Reality*, 41.
65. Ibid., 436.
66. Ibid., 314.
67. For Langer's definition of action, in contradistinction to the allied terms activity, act, and potential act (or act-likes), see *Mind*, vol. 1, 299–305.
68. Ibid., 376.
69. Ibid., 340. This is what I term "bare activity" in my own work.
70. Langer, *Feeling and Form*, 242.
71. Langer, *Mind*, vol. 1, 199.
72. My reading of Langer as asserting an aesthetic aspect to all activity and all expression motivates my method in this chapter of roving transversally across her three core works, often combining citations that in their original emplacement were embedded in an analysis of a particular domain of expression, from art (*Philosophy in a New Key*, *Feeling and Form*) to biology (*Mind*, vol. 1). This approach rests on the assumption of an operative analogy between domains and strata, such that the same functors recur in each, under continuous transformation (as is the case with the conceptual primitives of Whitehead's metaphysics).
73. Langer, *Feeling and Form*, 20.
74. Ibid., 20.
75. Ibid., 20.
76. Langer, *Mind*, vol. 1, 288, 305.
77. Gilles Deleuze and Félix Guattari, *What is Philosophy?* (London: Verso, 1994), 164–8. Chapter 7 of this work ("Percept, Affect, and Concept," 163–99) develops a theory of the work of art as a "block of sensations" that is startlingly reminiscent of Langer's "objective feelings." They argue, as Langer does in *Feeling and Form*, that the expressive power of art cannot be understood either in terms of the intentions of the artist or in terms of the reception of the observer, but that the work of art captures in its material form an impersonal force that enables it to "stand on its own."
78. Langer, *Mind*, vol. 1, 304.
79. Erin Manning, *The Minor Gesture* (Durham, NC: Duke University Press, 2016).
80. Erin Manning and Brian Massumi, *Thought in the Act: Passages in the Ecology of Experience* (Minneapolis: University of Minnesota Press, 2014), 83–152.
81. Brian Massumi, The Personality of Power: A Theory of Fascism for Anti-Fascist Life (London: Verso, 2024).

BIBLIOGRAPHY

Deleuze, Gilles. *Difference and Repetition*. Translated by Paul Patton. New York: Columbia University Press, 1994.

Deleuze, Gilles. *Logic of Sense*. Translated by Mark Lester with Charles Stivale and edited by Constantin V. Boundas. New York: Columbia University Press, 1990.

Deleuze, Gilles, and Félix Guattari. *What Is Philosophy?* Translated by Graham Burchell and Hugh Tomlinson. London: Verso, 1994.

Köhler, Wolfgang. *The Mentality of Apes*. New York: Harcourt, Brace & Co., 1925.

Langer, Susanne K. *Feeling and Form: A Theory of Art*. New York: Charles Scribner's and Sons, 1953.

Langer, Susanne K. *Mind: An Essay on Human Feeling*, vol. 1. Baltimore: Johns Hopkins University Press, 1967.

Langer, Susanne K. *Philosophy in a New Key: A Study in the Symbolism of Reason, Rite, and Art*. New York: New American Library, 1948.

Manning, Erin. *The Minor Gesture*. Durham, NC: Duke University Press, 2016.

Manning, Erin. *For a Pragmatics of the Useless*. Durham, NC: Duke University Press, 2020.

Manning, Erin, and Brian Massumi. *Thought in the Act: Passages in the Ecology of Experience*. Minneapolis: University of Minnesota Press, 2014.

Massumi, Brian. *Semblance and Event: Activist Philosophy and the Occurrent Arts*. Cambridge, MA: MIT Press, 2011.

Massumi, Brian. The Personality of Power: A Theory of Fascism for Anti-Fascist Life. London: Verso, 2024.

Whitehead, Alfred N. *The Concept of Nature*. Cambridge: Cambridge University Press, 1964.

Whitehead, Alfred N. *Process and Reality*. New York: Free Press, 1978.

CHAPTER FIFTEEN

Towards Vitality Semiotics and a New Understanding of the *Conditio Humana* in Susanne K. Langer

MARTINA SAUER

INTRODUCTION

In hindsight, it is primarily Susanne K. Langer's theory of *act*, and only secondarily her theory of art, that is central to the conception of Vitality Semiotics. It focuses on affective, semiotically relevant forms that constitute our world experience, human social interaction, and ultimately art experience. Thus, this somewhat unusual distinction between these two aspects of Langer's work is not only important for art and our understanding of the world, but can also be seen as fundamental to social interaction and, furthermore, to the universal formation of culture(s).[1] Underlying this broad perspective is the question (already shaped by early semiotic approaches such as those of Ernst Cassirer and Alfred N. Whitehead) of how we, as humans, come to make sense of our experiences with the world and to adjust our drive and determination accordingly. Everything we become attentive to only becomes significant through a process of symbolization that develops "images" (German, *Vorstellungsbilder*),[2] whose "forces"[3] determine the immediate future. According to this tradition, we can identify two levels of orientation regarding how our experiences take on meaning: 1) the process of meaning-making is general, distanced and objective (epistemological term); 2) meaning-making caters to decisions and actions (action-relevant term). But how do both relate, and what is their common ground? Both must be embodied, as Langer suggests with her theory of act. Strictly speaking, this precedes cognition and action altogether.

In her later writings in *Mind*, vol. 1 (1967) and vol. 2 (1972), Langer hints at this but does not elaborate on its significance for social life, and beyond that, for cultural development.[4] A closer look, however, reveals that these two dimensions are already inherent in Langer's *act* model. She argues that with respect to acts, understood as central units of experience, when the question of motivation—central to any theory regarding action—is raised, it is an "*effect of decision.*"[5] In this way, the "active" part of acts relative to experience becomes obvious. From this she concludes that acts themselves become the actors.[6] *Their* "*decisions*" influence others: they result in processes of involvement and individuation—in Langer's words, "That structure is the agent's body."[7] The act model developed by Langer is not only a vital process of concrete sense-making (epistemological term), but leads to conscious decisions and volitional actions which form the basis for social structures (communicative term).[8] An expanded view beyond epistemology to a theory of communicative (and thus socially and culturally relevant) action is fundamental to Vitality Semiotics. On this basis, it can be presented as a new understanding of the *conditio humana*.

Essentially, taking this step entails the assumption that experience and therefore *all* act-relevant processes are based on this common foundation, namely the effect of decision, and must also concern the perception of artistic artifacts. Langer, however, did not draw this conclusion from her approach. Yet this insight is essential to Vitality Semiotics, for if it is followed through, the dominant understanding of art perception as being irrelevant for decisions and actions—valid for Langer, and commonly held to this day—has to be revised.

Subsequently, the question arises as to whether the assumptions upon which aesthetic theory is based are correct at all. This needs to be considered carefully, especially since according to Langer there is no difference between the experience of the world and the experience of art. She states that art "presents a form which is subtly but entirely congruent with forms of mentality and vital experience." These are all based on an intricate network of dynamic acts. In the particular case of art, they are triggered by artistic means.[9] Why does Langer nevertheless tie in with aesthetic theory?

A closer look shows that Langer speaks of the fact that in art, in contrast to design, there is no pursuit of an objective.[10] This assumption of Langer, however, can only be defended if artists—in contrast to designers—are committed to this noble goal. No other explanation can be claimed upon considering her theory of experience based on analogies. This leads one to question, however, what the artistic means of artists (which correspond to those of designers) are used for, if not to fulfill an objective? The distinction drawn by Langer between art and design is rooted in her idealistic assumption that artists do not pursue personal, group, or third-party intentions in creating. Rather, they take their inspiration from the vividly felt forms or "living forms" of nature ("internal forms") and implement these by artistic means ("external forms"). In this way, artists convey an *impartial image of feeling* or of "the movement of emotive and perceptive processes" of nature.[11] For Langer (and for Ernst Cassirer, as we will soon see), this implies that an artist is able to transform the perceptual values felt in natural forms into an artwork. He or she objectifies with the work what has been experienced. He or she thus *re-expresses* it with artistic means. The motif thus undergoes a process of subjectification, the expression of which can be perceived by the viewer. Hence, Langer concludes: "Art is the objectification of feeling, and the subjectification of nature."[12]

Langer's understanding of art as neutral is precisely the weakness of her reasoning, especially since Western artistic works up to modern art at the turn of the twentieth century were usually commissioned works. From this point on, it depends more or less on the artist which motif is chosen and why. The consequent expression essentially depends on which motifs are taken up and what possible purpose they serve. This is all the more significant because motifs can all be subjectivized by artistic means and their expression can be felt by the viewer. After all, it is the motif as a "motive"—in the double sense of subject and intention—that is fundamental to both art and design.

If we are to accept this assumption, it is no longer possible to speak of the "innocence" of art. What art presents must be critically reexamined to see whether it is actually free from self-interest and the interests of others. This is all the more important because art in particular seems to have an unnoticed influence on others. However, it is Langer's consistently functional understanding of experience and meaning-making processes (as seen in her act theory) that can be used as a solid basis to describe the underlying social (and therefore cultural) effective processes. Thus, it is also true that the expression triggered via the artistic means in interaction with the motif not only affects the body, but also conveys a specific image (in German, *Vorstellungsbild*) of the motif. This means that only if this image is recognized as more than one of feeling, but also as a statement about something, the beholder can concretely refer to it and respond. He or she can accept or reject the presented viewpoint. If this recognition does not exist, however, the image, because of its double mode of operation, can subliminally influence the beholder's opinion and thus also their future decisions and actions, which are always part of social and cultural development. If this approach, presented with the concept of Vitality Semiotics, is taken seriously, it implies that both the creator and recipient of art should take responsibility for their actions. It also means that the habitual attitude of beholders to turn to art only as a *bon vivant* is questionable, for they should be aware of art's effects.[13]

LANGER'S (IMAGE-)ACT THEORY: IMPLICATIONS FOR VITALITY SEMIOTICS

It is subsequent (and contemporary) semiotically oriented, cultural-anthropological empirical research that confirms Susanne K. Langer's *embodied act* model. Her conception can be seen as a basis for the immediate evaluation of and response to everything we encounter. This encompasses our encounters with the world in all its diversity, from animate to inanimate nature, and also the human-made. This is why art should also be included. However, only in the extension beyond epistemology to a theory relevant to communication can Langer's conception also form the basis for any social and cultural development. Only in this respect can her (image-)act theory function as a solid foundation for a new empirical-psychologically-based conception of the *conditio humana*.

However, the extension of perceptual functions in the sense of a *conditio humana*, as proposed here, turns out to be possible only if one takes into account that *feelings* or *vital affects and vital effects* play a central role in the evaluation of all that we experience. This reveals an understanding of perception and its experience close to Langer's thinking. Remarkably, this approach is already found in Whitehead and Cassirer, as well as in empirical research in developmental psychology and neuroscience that runs in parallel, and picks up in the 1970s. For Vitality Semiotics (originally building on research from Visual Culture) empirical psychology is crucial,[14] as it studies vital affects and effects, significant factors for evaluation and determination processes, as well as selection and (re)cognition, and, essentially, emerging decisions and actions. Accordingly, the first thing that constitutes an evaluation or judgment is not, as one might suppose, based on a logical weighing of the pros and cons that ultimately speaks in favor of one action over another. The type of research discussed here refers unanimously to the non-discursive aspects that provide the initial impetus for an evaluation of experience. According to this approach, any engagement will already contain a primary reaction. The subsequent action or "decision" is an effect, not a judgment as such. This means *a world experienced and felt as alive, as vital affect, is followed by vital effective responses.*

Although Langer's focus lies primarily in the epistemologically relevant process of meaning-making itself, she has developed with her act theory a model that simultaneously provides the basis for an extension as a communication theory. This is because it is characterized by concrete, dialectical, and logical reactions towards what we encounter. That which is to be understood as reaction in perception turns out to be the product of a dynamic process of "tensions and resolutions."[15] This process involves numerous acts, whose effects are based on "decisions." In order to demonstrate this connection, Langer's first priority is to prove it with empirical research.[16] This is all the more necessary because in everyday life, as she puts it, signs and discursive symbols serve as foundations for *conscious* "intelligent" adaptation to the world:

> A form that is both sign and symbol ties action and insight together for us; it plays a part in a momentary situation and also in the 'science" we constantly, if tacitly, assume ... Yet all our signs and symbols were gathered from sensuous and emotional experience and bear the marks of their origin—perhaps a remote historical origin.[17]

However, the relevance of Langer's act theory for decisions and actions, and thus also for social and cultural processes, takes a back seat to her self-imposed task of uncovering the non-discursive foundations of experience as the ground for meaning-making. With this in mind, she focuses on showing that the living processes of superordination and subordination of single acts form the basis for the comparability of artistic and biological acts.[18] For Langer, the starting point is the fact that even the simplest interaction (comparable to artistic acts) is characterized by rhythms and a dialectical exchange of energies, forms, and qualities, and—to that extent—by non-discursive aspects. Accordingly, it is this structure of the acts that can be considered form-giving. It is characterized by an initial starting phase (impulse or energy change/discharge, also "potential act" or "event") followed by an increase in complexity, guided by "tensions and resolutions," to a turning point and a concluding phase (cadence).[19] Each interaction has its own rhythmic pattern. The process, structured in this way, can be described as a sequence of events leading to a "transposition of matter in space."[20] In this respect, the act can be regarded as the central functional unit of living form, which is vital to plants and animals, as well as to humans.[21] Furthermore, Langer holds that it plays a crucial role for perception in general, and to art in particular:

> Dialectic rhythms ... play such a major role in vital functions that their importance in the activity and even the physical existence of organisms makes them an essential mark of living form in nature, as their virtual images is of "living form" in art.[22]

LANGER'S PREDECESSORS: IMPACT ON VITALITY SEMIOTICS

Langer follows an understanding of cultural form that was already introduced in the 1920s, in the early semiologies of process philosopher and mathematician Alfred N. Whitehead and neo-Kantian philosopher and cultural anthropologist Ernst Cassirer.[23] Both held the view that cognition does not emerge from a distanced relationship with the world, but from a sensually endowed world perception that is determined by expressivity and vitality, rather than pure factuality.[24]

In Whitehead's terms, human perception is essentially composed of sense data or *qualities*, such as colors, sounds, taste sensations, tactile sensations, and bodily sensations, that have spatial, temporal, and pragmatic terms, and are accordingly pursued with "vivid apprehension."[25] Langer shares the view of "retreating from" and "expanding to" impulses that determine our perception,[26] and in this way find their emotional equivalent in—and secure power over—other impulses.[27] Again, it is the structure of the "acts of experience"[28] that can be seen as an "*effect of decision*" as Langer calls it, or as "*causal efficacy*" in Whitehead's words. For both, this is characterized by affective-emotionality that becomes central for the symbolic interpretation and modification of the environment, and therefore also for cultural development.[29] In their view, however, the original emotional stimuli are only attuned to an aesthetically pleasing, harmonious whole by means of the arts,[30] and insofar have no relevance for action.

Langer was influenced not only by her doctoral supervisor Whitehead but also by Cassirer, who she met during his last exile years in the United States in 1941. Cassirer, too, assumed that the human ability to grasp the world symbolically must lie in a particular mode of perception that is not merely objective. It is based on a strong instinctive substrate,[31] which he introduces as "expressive perception" (in German, *Ausdrucks-Wahrnehmung*).[32] According to this assumption, Cassirer states that the human being is basically not oriented towards factually differentiable forms. Instead, he emphasizes, following the research of two psychologists, Ludwig Klages and his colleague from Hamburg, Heinz Werner, that human perception is primarily oriented towards forms of motion and spatial forms. In this respect, they are not perceived by us as static, but as living. That which is perceived through expressive perception therefore turns out not to be a factual but a living form. This corresponds to a mode of perception that is consequently non-discursive.[33] Cassirer's proximity to Langer is unmistakable. These experiences of the perception (the forms of motion and spatial forms) of the world only become significant because they are characterized by a striving for action. Again, the so-called "effect of decision," which, according to Langer and Whitehead characterizes the acts of experience, is also described in Cassirer's philosophy. However, according to Cassirer, the first impressions of the expressive perception are replaced in a second step by symbol-forming acts that are shaped by will. It is redeemed by a mythical, descriptive, and cognizant symbolic formation.[34] It is essential that this happens in a gradual process of externalization that eventually breaks with the original expressive world.[35]

Eventually, both Cassirer and Langer develop an awareness of this unconscious primordial experience that materializes in art. Langer emphasizes this in her book *Philosophy in a New Key* (1942), which Cassirer seems to pick up in 1944.[36] Yet for him, as for Langer, art has no relevance to action, since only the expressive form itself is experienced and felt as its "image." Cassirer proposes a transformation process to have taken place, from originally non-discursive forms into artistic forms whose vital effects are consciously felt. Cassirer illustrates these effects with an example by describing his feelings upon viewing a landscape painting. Instead of grasping "*living things*" he engages with "*living forms*," which are consciously experienced. Cassirer notes:

> No longer in the immediate reality of things, I live now in the rhythm of spatial forms, in the harmony and contrast of colors, in the balance of light and shadow. In such absorption in the dynamic aspect of forms consists the aesthetic experience.[37]

It is expressive perception as such that comes to light through art. Not only does the doing as an expressive function become conscious and comprehensible, so too does the process of the formation of symbolic meaning. At the same time, considered as an anthropological condition, this shows that "in the work of the artist the power of passion itself has been made a formative power."[38] Here, Cassirer again coincides with Langer.

Whatever is experienced as expressive form undergoes a specific intensification with art: Cassirer therefore claims that art is an "intensification of reality." It is a process of concretization.[39] Although Cassirer is convinced that the artist does not arbitrarily invent the form of things, he is in accord with Langer in that there is no connection to purpose.[40] "Since the art symbol is not a discourse, the word 'message' is misleading," as Langer concludes.[41] Instead, both draw a connection between truth (the *appearance* of nature in feeling) and beauty (the *experience* of nature in feeling).[42]

With the founding of Hamburg University in 1919, Ernst Cassirer and psychologist Heinz Werner met for the first time and shared an office. With Cassirer and other researchers united in the so-called *Hamburg Circle*, Werner already specialized on issues of aesthetic and psychophysiological phenomena shared an interest in the extended processes of perception.[43] By comparing human development not only from an ontogenetic (individual) but also from a phylogenetic (mankind) point of view, he comes to a similar conclusion as his colleagues, which he published in *Einführung in die Entwicklungspsychologie* in 1926. It became a reference book for developmental psychology and was translated into English in 1933, after Werner's immigration to the United States. He assumes that human perception—or rather, experience—can primordially be regarded as a syncretic and indivisible unity; a distinction between a sensorimotor, perceptual, and affective organization is not possible. Instead, he characterizes it as a *dynamic and physiognomic* apprehension of things.[44] The world is experienced as a vital network of actions. Consequently, a clear separation between object and subject, object and state, feeling and action does not occur.[45] Rather, it reflects an original view of the world that children hold and artists retain, but which the average adult lacks.[46]

A specific context of how Werner's primordial and direct way of accessing the world is relevant in social contexts was only elaborated in the 1980s by child psychologist Daniel N. Stern. His research in particular bridges the gap from a merely symbolically relevant interpretation of the world to one that concretely includes decisions and actions, and elevates their relevance for social interaction and thus for cultural development. He incorporates not only Werner's psychological considerations on the development of human perception but also Langer's act model and theory of art.

What is remarkable about Stern's findings is that an organizing principle of perception can be observed in infants as young as two months old, inasmuch that "their social capacities are operating with vigorous goal-directedness to assure social interactions." [47] It is these interactions that give rise to affects, perceptions, sensorimotor events, memories, and other cognitions. At the core of his research, as emphasized by his predecessors and now confirmed by Stern, stands the infant's ability to develop abstract representations of perceptual properties at the earliest stage.

> These abstract representations that the infant experiences are not sights and sounds and touches and nameable objects, but rather shapes, intensities, and temporal patterns—the more "global" qualities of experience.[48]

According to this, it is abstract representations, concrete forms, degrees of intensity and time patterns, which are considered essential for perception as such, as well as for the perception of art, as Stern himself later points out. Extending his approach, the researcher also refers (as Werner did earlier) to the specific quality of this experience, which tends to translate perceptual qualities into emotional ones. Stern characterizes these as *vitality affects*. These are best described in dynamic, kinetic terms such as "surging," "fading away," "fleeting," "explosive," "crescendo," "decrescendo," "bursting," "drawn out," and so on.[49] It is self-reflection and language that lead to forgetting or detachment from this global mode of experience between the fifteenth and eighteenth month of human life.[50]

This kind of experience should be regarded as fundamental not only in the field of human development but also in the understanding of the arts. Essential to this assumption is the question—first sown by Langer and later continued by Stern—as to whether a reflection of this external experience is possible within. In this context, Stern formulates a promising research question that was specifically inspired by Langer's research on art.

> How, then, do we get from intensity, timing, and shape to "forcefulness"? This is the question that lies at the heart of understanding one aspect of how art works, and perhaps a look at how the question has been approached in the domain of art may be helpful in understanding it in the domain of behavior ... She [Langer] suggests that, in works of art, the organization of elements seems to present an aspect of life. The feeling that is presented is in fact an apparition, an illusion, a *virtual* feeling. [emphasis added][51]

Following Langer, Stern points to the possible virtual feeling of three-dimensional space in painting, the virtual feeling of kinetic volume in sculpture, that of virtual time in music, and the virtual realms of power in movement and gesture in dance. Particularly his conception of activation contours (intensity in time), perceived in the overt behavior of another, and becoming a *virtual vitality affect* when experienced by an observer (within the self), coincides with the research questions of a group of Italian neuroscientists led by Giacomo Rizzolatti, who in 1996 linked their discovery of mirror neurons to the human capacity for empathy.[52] In a joint research group with Stern, they pursued this question and published their evidence in 2013.[53] More recent research from 2020 and 2021 links again to Stern and coincides with the core hypothesis of Vitality Semiotics.[54] These modern researchers assume that vitality forms, mediated not only by gestures and actions but also by words, "characterize social interactions by providing information about affective states of the actors involved" and in this way communicate their mood.[55] These thoughts are currently being further developed in a joint research project on "art and multimodality" and on "art and atmosphere and mood."[56] The starting point for this is again the connection to Stern and Langer and to their relevance regarding research on communicative processes by means of art. In agreement with Langer's and Stern's views that a painter's style—that is, the way he or she deals with forms—corresponds to vitality affects:

> The translation, then, from perception to feeling in the case of style in art involves the transmutation from "vertical" perceptions (color harmonies, linear solutions, and the like) into such virtual forms of feeling as calmness. The analogous translation from perception of another person's behavior to feelings involves the transmutation from the perception of timing, intensity, and shape via cross-modal fluency into felt vitality affects in ourselves.[57]

The difference between grasping vitality affects in social behavior and in art lies, as Stern puts it, in the contemplation and therefore awareness that the art experience unfolds what is usually considered impossible in daily life due to confinements and conventions. What consequences can be drawn from this for the question of the relevance of art for communication, and furthermore, for action?

LANGER'S CONCEPT OF ART VERSUS THAT OF VITALITY SEMIOTICS

The ability to become aware of one's own reactions, and thus of what is being reacted to and how, distinguishes humans from other living beings. The arts assist in making us aware of our reactions and what we react to. Engaging art does not necessarily trigger instant reactions, i.e., concrete decisions and actions, which makes it possible to free the mind and consciously experience the input. In this way, we feel the experience, reflect on it, and become aware of what has triggered it. In this sense, the duality of perception is revealed to us through art: on the one hand, there is our self-forgetfulness in experiencing the work of art, and, on the other, our experience of living physical activity. The former has no practical relevance to us since it is not noticed. In contrast, the latter becomes evident to us in the conscious living experience, e.g., in the tone and scale of music, the visual play of colors and forms, the movements in dance, the sweet, sour, or bitter taste in foods, or the tone of voice in language.[58] This is where the fundamental elements of the *conditio humana* manifest in artistic form.

Remarkably, however, our self-forgetfulness is also significant, because it highlights the blindness of the beholder to possible intentions of the work of art or the artist. This manifests itself in the general rejection of the possibility of the artist desiring to have an intentional influence on the beholder and, accordingly, in the rejection of possible reactions and actions of the beholder to what is conveyed by the work. Therein lies the crux of art as such and the conceptions surrounding it. It ultimately rests in the false assumption that art is *free of purpose*, while this is not the case for design, advertising, or propaganda.[59] But this is precisely where the contradiction lies, for each mode of creative expression—whether in art or design—is based on the same principle of appealing to feeling. In the same vein, *Gestalt*, as created or intended by artists and designers, pursues a specific purpose—whether to render a motif, address a theme, express a taste, or formulate a message. It is not possible to claim neutrality towards content or intent. A conception of art falls short if it assumes that only pleasure, displeasure, and some possible insights are conveyed.

Here, a re-evaluation seems appropriate. We can neither claim that art is just a human approach to the world that is based in feeling, nor can art be understood as having its value solely in the conscious feeling of its expression. Moreover, what the artist chooses as a motif is not limited to nature as a model, but can be chosen at will. Whatever becomes the motif, with the new conditions indicated, is not accessed by the beholder via an *external world*, but an *internal one* created by the artist. The artist's image (in German, *Vorstellung*) can be experienced by the viewer through the perception of the "living forms" of the artistic means. Against the backdrop of Langer's general act theory, this experience is characterized by "decision effects." In the case of art, it depends on the artist's "decision acts" for a particular gestalt. In this way, art conveys a specific view about something that is fixed or determined by the artist. However, the artist's viewpoint, which is subliminally communicated to viewers, can be obscured by traditionally false assumptions about its presuppositions. Its true function as a means of communication then remains hidden.

Langer herself pointed out that the product of art "is more than a vital expression, it is the expression of his [the artist's] idea, his personal conception of the ways of feeling."[60] It is consequently the artist's—or his or her client's—conception of or intention for a subject that is realized in the viewer's perception. In this sense, art can very well have an effect on revelations of our inner life. But Langer's statement, "that it shapes our imagination of external reality according to the rhythmic forms of life and sentience, and so impregnates the world with aesthetic value"[61] falls short. In essence, this is contradicted by her approach, according to which all products produced and thus shaped by us are based on the same premises. For they cannot be understood as actual organic tensions of internal forms of nature, but as virtual organic tensions of perceptually generated tensions of artistic or external forms.[62] Art, then, is not only the objectification of feeling but also the embodiment of the feeling as a motif. And because this motif is mediated by feelings, the subjectivation Langer speaks of is not necessarily that of an external motif (such as nature), but can be anything the producers want to convey.

In fact, Langer hints at the possibility of manipulation not only in design and advertising but also in art. This becomes clear in her remarks on the analysis of the perception of images in art. To this end, she says that form is not an actual representation of organic tensions in nature, but *virtually generated tensions of artistic or outer forms.*[63] Her reference to the fact that the viewer experiences these feelings consciously in their encounter with the arts simultaneously explains the effects it can have accordingly. That which is conveyed thus recedes into the background, and the experience itself becomes determining. When Langer speaks of a subjectification of the motif itself, this also becomes apparent. Unlike Langer, however, I claim that this motif need not be nature, but whatever the producer shows us. This means that what we are shown and what we subsequently feel depends on the producer's intentions and will. It is not neutral at all, as Langer sees it when she makes the connection to nature.[64]

> The secret of the "fusion" is the fact that the artist's eye sees in the nature, and even in human nature betraying itself in action, an inexhaustible wealth of tensions, rhythms, continuities and contrasts which can be rendered in line and color; and those are the "internal forms" which the "external forms"—paintings, musical or poetic compositions or any other works of art—express for us.[65]

Against this backdrop, for example, a late landscape painting of the Montagne Sainte-Victoire by Cézanne, seems actually, as Langer suggested, oriented to the external model. A love song or the *pas de deux* danced by a pair of lovers can also be judged as comparatively harmless with regard to a possible influence on opinion formation, since these do not convey personal, group, or client intentions, but address universal feelings. Less chaste are sales intentions or ideologically driven views in the arts, whether they are *weltanschaulich*, political, or religious. Some want to draw us into buying, by using slogans and stimulating colors and forms; others want to win us over ideologically, with electrifying speeches, gestures, and marches. As is known in research oriented towards the philosophy of art, it was already the idealists Plato and Kant who criticized these abilities of the arts, and called for their orientation towards higher ideas (in German, *Vorstellungen*); the true, the beautiful, and the good. In contrast, it was the cultural critic Walter Benjamin who was less idealistic and looked to *models in art itself*, such as DADA, to break free from this. Only in this way does he see the possibility of those affected by art to become aware of its influence and able to critically engage with its content. Only if this succeeds, Benjamin continues, it is possible to *liquidate* the "traditional value of the cultural heritage" that constitutes art.[66] After centuries of dependence on art in the service of the patron, Benjamin's hope seems to have been realized. This is clearly recognizable among modern artists, who today are independent of commissions and thus have and use the opportunity to discuss potentially controversial social and environmental issues.

LANGER'S BASICS FOR A NEW UNDERSTANDING OF THE *CONDITIO HUMANA*

The human condition at its core implies that our sense of feeling permanently responds to the stimuli that impinge on us. What emerges as an *image of feeling* or *image of mood* (generated in accordance with the stimuli, and in interaction with our own culturally shaped being) forms the foundation for our decisions and actions. From here, we can draw further conclusions on the phenomenon of the human urge to express and lend *form* to what is felt internally. If one extracts all positive or negative evaluations and considers this phenomenon independently, it represents the core aspect of being human. This is true if every form of artistic expression, be it language, picture, sound, or even smell, is understood not only as an epistemological but also as a communicative term. Independent of possible insights that the formative process can produce, there is an equally vital connection to communication aspects. Communication—redefined as vital forms that express one's own feelings and intentions—therefore consists of the transmission of self-images: our views or opinions about something. These take shape with the words, pictures, tones, tastes, and fragrances, etc., we create. In this way, they can emerge from us and can simultaneously be perceived and understood by others. In continuation of Langer's extended theory of *acts* as the *conditio humana*, neither images, nor language, sound, or smell are neutral.[67]

The remarkable thing about the functioning of the process of perception and evaluation is that what is intended by art's presentational ability—through the organization of forms and their subliminally vital affect and vital effect—is hardly perceived by us. According to our everyday habits of perception, we can only be certain of what has been said and represented. This corresponds to what Langer called the "infallible, all-supporting primary illusion"[68] that dominates our perception from the age of fifteen to eighteen months, as Stern showed. This means that in view of what we believe to be true because it is felt bodily, and due to its effects on our actions, it follows that basically everyone who creates (and we do that in every moment of our lives) bears *responsibility* for their creations, as does the one who perceives what has been created. In this respect, both the "speaker" (the creator) and the "receiver" (the beholder) should be alert that they are active in the process of forming and being formed. This requires awareness of the mutual and subliminal processes of creation and perception, and the knowledge that in these processes there is no such thing as neutrality.

NOTES

1. For the influence of the arts on social life, see Martina Sauer, "Visualität und Geschichte. Bilder als historische Akteure im Anschluss an Verkörperungstheorien," in *Jenseits des Illustrativen. Visuelle Medien und Strategien politischer Kommunikation*, ed. Niels Grüne and Christian Oberhauser (Göttingen: V&R unipress, 2015), 39–60.
2. Symbolization or the formation of "images" (German, *Vorstellungsbilder*) as described by Ernst Cassirer in *The Philosophy of Symbolic Forms*, vol. 3: *The Phenomenology of Knowledge*, 1929 (New Haven, CT: Yale University Press, 1976), 162–90, here 181–4. See also Martina Sauer, "Bildkraft und Tatkraft: Zum Verhältnis von ästhetischer Erfahrung und Technik im Anschluss an Cassirer, Langer und Krois," *Techne—poiesis—aisthesis*, ed. Birgit Recki: *Kongress-Akten, Deutsche Gesellschaft für Ästhetik* 3 (2018): 1–25.

3. It is Whitehead who describes the power of symbolization in this way. See Alfred N. Whitehead, *Symbolism: Its Meaning and Effect* (New York: Little and Ives Company, 1927), 57.
4. Instead, Langer focuses on a different aspect of sociocultural life, as Anne Pollok and Robert Innis argue here in their chapters. It starts from an understanding of the sociocultural as a specific realm of "representative symbols" (or "presentational symbolization"), which originate in mythic and ritual consciousness. (See Anne Pollok's Chapter 7 in this volume, "Susanne K. Langer's Theory of Self-Liberation Through Culture," and Robert Innis, Chapter 10 in this volume, "Psychological Dimensions, Cultural Consequences, and their Breakings in Susanne K. Langer's Symbolic Mind".
5. See Susanne K. Langer, *Mind: An Essay on Human Feeling*, vol. 1, 1967 (Baltimore/London: Johns Hopkins University Press, 1985), 257–306, 275–99, expanded on 307.
6. Ibid., 314.
7. Ibid., 307–59, 329. A further discussion on Langer's non-discursive forms is found in Martina Sauer, "Ästhetik und Pragmatismus. Zur funktionalen Relevanz einer nicht-diskursiven Formauffassung bei Cassirer, Langer und Krois," *Image* 20 (2014): 49–69.
8. Susanne, K. Langer, *Mind: An Essay on Human Feeling*, vol. 2 (Baltimore and London: Johns Hopkins University Press, 1972), 103–40, 137–3, 265–316, 301–12). In contrast to human interaction, exchanges between animals cannot be understood as a social process (ibid., 200–14). This is a matter of instinct-driven action without the formation of ideas or object awareness (ibid., 45–101, 55, 62).
9. Langer, *Mind*, vol. 1, 55–106, 67, for reference to nature, see 87.
10. Susanne K. Langer, *Feeling and Form: A Theory of Art*, 1953 (London: Routledge & Kegan, 1967), 319–25; cf. Langer, *Mind*, vol. 1, 127.
11. Langer, *Mind*, vol. 1, 67.
12. Ibid., 87. See here Anne Pollok and Robert Innis, Chapters 7 and 10. Both agree that with the formative power of the human being an artistic-symbolic form is created, through which we become aware of the emotional level. "Art is 'about' emotions ... Art unfolds these feelings, offering us a semblance of the world" (Pollok, "Susanne K. Langer's Theory of Self-Liberation," this volume, Chapter 7, 108). In this context, Innis also refers to the normative nature of Langer's approach, which is concerned with fostering "a joyous pulse of life ... to create a 'defense against outer and inner chaos'" (Innis, "Psychological Dimensions, Cultural Consequences, and their Breakings," this volume, Chapter 10, 150).
13. Cf. Martina Sauer, "From Aesthetics to Vitality Semiotics—From l'art pour l'art to Responsibility. Historical Change of Perspective Exemplified on Josef Albers," in *BildGestalten. Topographien medialer Visualität*, ed. Lars C. Grabbe, Patrick Rupert-Kruse and Norbert M. Schmitz (Marburg: Büchner, 2020), 209–10.
14. See Martina Sauer, *Faszination und Schrecken, Zur Handlungsrelevanz ästhetischer Erfahrung anhand Anselm Kiefers Deutschlandbilder*, 2012 (Heidelberg: arthistoricum.net, 2018), 35–106, 191–266.
15. Langer, *Mind*, vol. 1, 206–53, 329, cf. (with reference to the arts) Langer, *Feeling and Form*, 207, and 47–59 and 372, respectively.
16. Langer seeks evidence for this approach in psychology, developmental psychology, physiology, neurology, biology, and zoology, the results of which she presents most notably in vol. 2 of *Mind*.
17. Susanne K. Langer, *Philosophy in a New Key: A Study in Symbolism of Reason, Rite, and Art* (1942; New York: American Library, 1954), 230–1.

18. Langer, *Mind*, vol. 1, 261.
19. Ibid.
20. Ibid., 263.
21. Ibid., 292; for elaboration of the connection, see 272–99.
22. Ibid., 324.
23. Similarly to Langer, Whitehead and Cassirer referenced concrete empirical research and mathematical considerations in their philosophy. All three were concerned with proving that, on a molecular level, there are interactions, which are "controlled" by physical and chemical processes, and rise up as more complex organic, psychic, and eventually artistic processes. See Whitehead, *Symbolism,* 64–5; and Cassirer's correspondences with neurologist Kurt Goldstein (Frankfurt a.M), who did research comparable to the biosemiotics of Jacob von Uexküll. In *John M. Krois: Bildkörper und Körperbilder. Schriften zur Verkörperungstheorie ikonischer Formen* (Berlin: Akademie Verlag, 2011), 53–62, 176–93. Cassirer's connections to mathematics and physics are apparent in Cassirer, *The Philosophy of Symbolic Forms*, vol. 3, 279–447.
24. See Martina Sauer, "Ästhetik und Pragmatismus." In this context, the approach of Cassirer's colleague from the Hamburg Circle (the art historian and cultural scientist Aby M. Warburg) is also interesting. However, his concept of "pathos formulas" will not be specifically discussed here, since there is only an indirect connection to Langer. For this, see: Sauer, *Faszination und Schrecken*, 213–25.
25. Whitehead, *Symbolism*, 26–59, 45, and for more on "pathos," 47.
26. At the same time, Jacob von Uexküll developed his bio-semiotically oriented theory of "Umwelt" (environment) by observing the behavior of animals. See Jacob von Uexküll, *Umwelt und Innenwelt der Tiere* (Berlin: J. Springer, 1909).
27. Whitehead, *Symbolism*, 74–88; see also 45, 48–9, 54.
28. Ibid., 51.
29. Ibid., 60–88, here 66, cf. also 78–88.
30. Ibid., 85.
31. Cassirer, *The Philosophy of Symbolic Forms*, vol. 3, 66.
32. Ibid., 73.
33. Ibid., 80.
34. Ibid., 162–90, 181–4, cf. with reference to the process of symbolic formation with consideration of Max Scheler's philosophy: 99–107, here 87–8.
35. Ibid., 99–107, 84.
36. Cassirer never realized his planned fourth volume of the philosophy of symbolic forms in art. This can be explained, among other things, as an after-effect of Cassirer's immigration in 1933 via detours to the USA. It may therefore have been the correspondences with Langer since 1941, and her book from 1942, which provided him with points of departure for his approach to the function of the arts in his last publication *An Essay on Man: An Introduction to a Philosophy of Human Culture*, 1944 (New York: Doubleday & Company, 1953). This book was intended as a summary of his previous three volumes for the English-speaking world. However, considering the key points laid out in *An Essay on Man*, it can also be understood as an extension and an actual application of the approach he advocates beyond language, myth, and knowledge to science, art, and history.
37. Cassirer, *An Essay on Man*, 176–217, 194.
38. Ibid., 190.
39. Ibid., 186–91, 184,
40. Ibid., 178–94, 187.
41. Langer, *Feeling and Form*, 393.

42. Ibid., 395. Cassirer, *An Essay on Man*, 186–91, 184.
43. See Ulrich Müller, "The Context of the formation of Heinz Werner's ideas," in *Heinz Werner and Developmental Science: Path in Psychology*, ed. Jaan Valsiner (New York: Kluwer Academic, 2005), 25–54.
44. See Heinz Werner, *Comparative Psychology of Mental Development*, rev. edn, 1940 (New York: International University Press, 1957), ch. 2, part "primitive perception as dynamic: physiognomic perception": 59–142, and especially 67–103.
45. Ibid., 59–67.
46. Ibid., 71–2. This was elaborated by Werner in collaboration with the Bauhaus master Gertrud Grunow, using Kandinsky as an example. See Heinz Werner, *Einführung in die Entwicklungspsychologie*, 1926 (Munich: Barth, 1959), 46–7, and 66–7.
47. Daniel N. Stern, *The Interpersonal World of the Infant: A View from Psychoanalysis and Developmental Psychology* (New York: Basic Books, 1985), 28.
48. Ibid., 47–68, 51.
49. Ibid., 53–61, 54.
50. Ibid., 27–8, elaborated on 162–82.
51. Ibid., 157–61, 158, See also Stern's second book and his conclusions regarding the performing arts such as music, dance, theater, and film: Daniel N. Stern, *Forms of Vitality: Exploring Dynamic Experience in Psychology, the Arts, Psychotherapy, and Development* (Oxford: University Press, 2010), 75–101.
52. Giacomo Rizzolatti et al., "Action Recognition in the Premotor Cortex," *Brain* 119 (1996): 593–609. Following this research, concrete references to art were being made by a member of this group, Vittorio Gallese, together with the American art historian David Freedberg (2007). This research has been constantly deepening ever since. See David Freedberg and Vittorio Gallese, "Motion, Emotion and Empathy in Esthetic Experience," *Trends in Cognitive Sciences* 11, no. 5 (2007): 197–203, cf. Vittorio Gallese and Cinzia Di Dio, "Neuroesthetics: The Body in Esthetic Experience," in *Encyclopedia of Human Behavior*, ed. V. S. Ramachandran (Amsterdam: Elsevier, 2012), 1–7. On embodied simulation: Vittorio Gallese, "Embodied Simulation: Its Bearing on Aesthetic Experience and the Dialogue Between Neuroscience and the Humanities," *Gestalt Theory* 41, no. 2 (2019): 113–28.
53. Cf. Giacomo Rizzolatti et al., "The Neural Correlates of 'Vitality form' Recognition: An fMRI Study." *Social Cognitive and Affective Neuroscience* 9 (2013): 951–60.
54. See Giuseppe Di Cesare, Marzio Gerbella, and Giacomo Rizzolatti, "The Neural Bases of Vitality Forms," *National Science Review* 7 (2020): 202–13; see also Giacomo Rizzolatti et al., "The Neural Bases of Tactile Vitality Forms and Their Modulation by Social Context," *Scientific Reports* (2021).
55. Di Cesare, Gerbella, and Rizzolatti, "The Neural Bases of Vitality Forms," 202, cf. with reference to tactile vitality forms Giacomo Rizzolatti et al., "The Neural Bases of Tactile Vitality Forms and Their Modulation by Social Context."
56. Cf. Giada Lombardi and Giuseppe Di Cesare, "From Neuroscience to Art: The Role of 'Vitality Forms' in the Investigation of Multimodality," *Multimodality: On the Sensually Organized and at the Same Time Meaningful and Socio-culturally Relevant Potential of Artistic Works*, ed. Martina Sauer and Christiane Wagner: *Art Style*, special issue 10, no. 1 (September 2022). Giada Lombardi, Martina Sauer, and Giuseppe Di Cesare, "An Affective Perception: How 'Vitality Forms' Influence Our Mood," *Atmosphere and Mood. Two Sides of The Same Phenomenon*, ed. Martina Sauer and Zhuofez Wang: *Art Style*, special issue 10, no. 2 (March 2023).
57. Stern, *The Interpersonal World of the Infant*, 157–61, 159. For a detailed discussion of performing arts, see Stern, *Forms of Vitality*, 75–101.

58. On the methodological and interpretive consequences of this consideration, cf. Martina Sauer, "Affordance as a Method in Visual Cultural Studies Based on Theory and Tools of Vitality Semiotics: A Historiographic and Comparative Study of Formal Aesthetics, Iconology, and Affordance Using the Example of Albrecht Dürer's Christ Among the Doctors from 1506," *Material Image: Affordances as a New Approach to Visual Culture Studies*, ed. Elisabeth Günther and Martina Sauer: *Art Style*, special issue 7 (2021): 11–37.
59. Langer, *Mind*, vol. 1, 127, and Langer, *Feeling and Form*, 306–25, 319.
60. Langer, *Mind*, vol. 1, 225.
61. Langer, *Feeling and Form*, 392–415, 399.
62. Langer, *Mind*, vol. 1, 153–97, 164.
63. Ibid.
64. Langer states, "The secret of the 'fusion' is the fact that the artist's eye sees in the nature, and even in human nature betraying itself in action, an inexhaustible wealth of tensions, rhythms, continuities and contrasts which can be rendered in line and color; and those are the 'internal forms' which the 'external forms'—paintings, musical or poetic compositions or any other works of art—express for us" (Langer, *Mind*, vol. 1, 73–106, 87).
65. Langer, *Mind*, vol. 1, 73–106, 87.
66. Hanna Arendt, ed., *Walter Benjamin, The Work of Art in the Age of Mechanical Reproduction*, 1936 (New York: Schocken/Random House, 1969), II, line 26.
67. Cf. Martina Sauer, "Visualität und Geschichte," 39–60.
68. Cf. Langer, *Mind*, vol. 1, 230.

BIBLIOGRAPHY

Arendt, Hanna, ed. *Walter Benjamin: The Work of Art in the Age of Mechanical Reproduction*. 1935/6. Translated by Harry Zohn. New York: Schocken/Random House, 1969. Available online: https://www.marxists.org/reference/subject/philosophy/works/ge/benjamin.htm (accessed September 14, 2021).

Bredekamp, Horst and Marion Lauschke, eds. *John M. Krois: Bildkörper und Körperbilder. Schriften zur Verkörperungstheorie ikonischer Formen*. Berlin: Akademie Verlag, 2011.

Cassirer, Ernst. *An Essay on Man: An Introduction to a Philosophy of Human Culture*. New York: Doubleday & Company, [1944] 1953.

Cassirer, Ernst. *The Philosophy of Symbolic Forms*, vol. 3: *The Phenomenology of Knowledge*. Translated by Ralph Mannheim. 10th edn. New Haven, CT: Yale University Press, [1929] 1979.

Di Cesare, Giuseppe, Marzio Gerbella, and Giacomo Rizzolatti. "The Neural Bases of Vitality Forms." *National Science Review* 7 (2020): 202–13.

Freedberg, David, and Vittorio Gallese. "Motion, Emotion and Empathy in Esthetic Experience." *Trends in Cognitive Sciences* 11, no. 5 (2007): 197–203.

Gallese, Vittorio. "Embodied Simulation: Its Bearing on Aesthetic Experience and the Dialogue between Neuroscience and the Humanities." *Gestalt Theory* 41, no. 2 (2019): 113–28. Available online: https://www.researchgate.net/publication/334722167_Embodied_Simulation_lts_Bearing_on_Aesthetic_Experience_and_the_Dialogue_Between_Neuroscience_and_the_Humanities (accessed September 14, 2021).

Gallese, Vittorio, and Cinzia Di Dio. "Neuroesthetics: The Body in Esthetic Experience." In *Encyclopedia of Human Behavior*, edited by V. S. Ramachandran, 1–7. Amsterdam: Elsevier, 2012. Available online: https://www.researchgate.net/publication/279718809_Neuroesthetics_The_Body_in_Esthetic_Experience (accessed September 14, 2021).

Langer, Susanne, K. *Feeling and Form: A Theory of Art*. 4th edition. London: Routledge & Kegan, [1953] 1967.

Langer, Susanne, K. *Mind: An Essay on Human Feeling*, vol. 2. Baltimore and London: Johns Hopkins University Press, 1972.

Langer, Susanne, K. *Mind: An Essay on Human Feeling*, vol. 1. Baltimore and London: Johns Hopkins University Press, [1967] 1985.

Langer, Susanne K. *Philosophy in a New Key: A Study in Symbolism of Reason, Rite, and Art*. 6th edn. New York: American Library, [1942] 1954.

Lombardi, Giada, and Giuseppe Di Cesare. "From Neuroscience to Art: The Role of 'Vitality Forms' in the Investigation of Multimodality." Special issue on "Multimodality," edited by Martina Sauer and Christiane Wagner: *Art Style*, 10, no. 1 (September 2022): 11–23. Available online: https://zenodo.org/record/7020465 (accessed July 16, 2023).

Lombardi, Giada, Martina Sauer, and Giuseppe Di Cesare. "An Affective Perception: How 'Vitality Forms' Influence Our Mood." Special issue on *Atmosphere and Mood. Two Sides of The Same Phenomenon*, edited by Martina Sauer and Zhuofez Wang: *Art Style*, 10, no. 2 (March 2023): 127–39.

Müller, Ulrich. "The Context of the Formation of Heinz Werner's Ideas." In *Heinz Werner and Developmental Science: Path in Psychology*, edited by Jaan Valsiner, 25–54. New York: Kluwer Academic, 2005.

Rizzolatti, Giacomo, et al. "Action Recognition in the Premotor Cortex." *Brain* 119 (1996): 593–609.

Rizzolatti, Giacomo, et al. "The Neural Bases of Tactile Vitality Forms and Their Modulation by Social Context." *Scientific Reports* 11, no. 1 (2021): art. 9095. Available online: https://doi.org/10.1038/s41598-021-87919-z.

Rizzolatti, Giacomo, et al. "The Neural Correlates of 'Vitality form' Recognition: An fMRI Study." *Social Cognitive and Affective Neuroscience* 9 (2013): 951–60.

Sauer, Martina. "Affordance as a Method in Visual Cultural Studies Based on Theory and Tools of Vitality Semiotics. A Historiographic and Comparative Study of Formal Aesthetics, Iconology, and Affordance Using the Example of Albrecht Dürer's Christ among the Doctors from 1506." Special Issue on *Material Image: Affordances as a New Approach to Visual Culture Studies*, edited by Elisabeth Günther and Martina Sauer: *Art Style* 7 (2021): 11–37. Available online: https://doi.org/10.5281/zenodo.6371528.

Sauer, Martina. "Ästhetik und Pragmatismus. Zur funktionalen Relevanz einer nicht-diskursiven Formauffassung bei Cassirer, Langer und Krois." *Image* 20 (2014): 49–69. Available online: https://doi.org/10.11588/artdok.00003196.

Sauer, Martina. "Bildkraft und Tatkraft': Zum Verhältnis von ästhetischer Erfahrung und Technik im Anschluss an Cassirer, Langer und Krois." *Techne – poiesis – aisthesis*, edited by Birgit Recki: *Kongress-Akten, Deutsche Gesellschaft für Ästhetik* 3 (2018): 1–25. Available online: https://doi.org/10.11588/artdok.00003688.

Sauer, Martina. *Faszination—Schrecken. Zur Handlungsrelevanz ästhetischer Erfahrung anhand Anselm Kiefers Deutschlandbilder*. 2nd edn. 2012. Reprint with introduction. Heidelberg: arthistoricum.net, 2018. Available online: https://doi.org/10.11588/arthistoricum.344.471.

Sauer, Martina. "From Aesthetics to Vitality Semiotics—From l'art pour l'art to Responsibility. Historical Change of Perspective Exemplified on Josef Albers." In *BildGestalten. Topographien medialer Visualität*, edited by Lars C. Grabbe, Patrick Rupert-Kruse, and Norbert M. Schmitz, 185–205. Marburg: Büchner, 2020. Available online: https://doi.org/10.11588/artdok.00007276.

Sauer, Martina. "Visualität und Geschichte. Bilder als historische Akteure im Anschluss an Verkörperungstheorien." In *Jenseits des Illustrativen. Visuelle Medien und Strategien politischer Kommunikation*, edited by Niels Grüne and Christian Oberhauser, 39–60. Göttingen: V&R unipress, 2015. Available online: https://doi.org/10.11588/artdok.00004293.

Stern, Daniel, N. *Forms of Vitality: Exploring Dynamic Experience in Psychology, the Arts, Psychotherapy, and Development*. Oxford: University Press, 2010.

Stern, Daniel N. *The Interpersonal World of the Infant: A View from Psychoanalysis and Developmental Psychology*. New York: Basic Books, 1985.

Uexküll, Jacob v. *Umwelt und Innenwelt der Tiere*. Berlin: J. Springer, 1909. Available online: https://www.biodiversitylibrary.org/item/15902#page/5/mode/1up (accessed August 27, 2021).

Werner, Heinz. *Comparative Psychology of Mental Development*. Adapted 1940. Translated by Edward Ballar Garside. 2nd edn. New York: International University Press, 1957. Available online: https://archive.org/details/in.ernet.dli.2015.138929/page/n11/mode/2up (accessed November 4, 2021).

Werner, Heinz. *Einführung in die Entwicklungspsychologie*. 4th edn. Munich: Barth, [1926] 1959.

Whitehead, Alfred N. *Symbolism: Its Meaning and Effect*. New York: Little and Ives Company, 1927.

CHAPTER SIXTEEN

Virtual Powers in Susanne K. Langer's Theory of Dance and its Application in Postcolonial Hong Kong

EVA KIT WAH MAN

SUSANNE K. LANGER'S ILLUSORY WORLD OF DANCE

In her book *Feeling and Form*, Susanne Langer writes, "[dance] can harbor no raw material, no things or facts, in its illusory world. The virtual form must be organic and autonomous and divorced from actuality. Whatever enters into it does so in radical artistic transformation: its space is plastic, its time is musical, its themes are fantasy, its actions symbolic."[1] How can this virtual space of dance be read except through contemplation? Do we apprehend only form? If so, how can a dance critic respond to questions of social reality or culture? What does Langer really mean in her discourse on dance?

To expand on Langer's understanding of dance, I use case studies from my recent work on contemporary dance choreographers in postcolonial Hong Kong.[2] These dance "authors" and performers build and develop their choreographies across time and within their social reality, and in the form of these choreographies, audiences perceive a powerful emotional import and interactivity. Here I focus on the case study of Helen Lai in illustrating Langer's theory of "layers of illusion." She said that the second of these layers contains devices that support the overall creation or enhance its expressiveness.[3] I argue that the artistic effects present in contemporary dance performances are rich in meaning when viewed through the lens of Langer's concept of "primary illusions." Langer's theory of dance suggests that a simple illusion, e.g., pure space or pure time, is distinctly present within the more complex illusion of dance. The case study of Lai's work considered here exemplifies this theory, as it involves a sudden revelation of emotive import (by stressing a formal aspect and abstracting it) that makes the *feeling*-content of the work cited apparent. Here, I offer a further reflection on the issue of contextualizing Langer's notion of the "virtual form" of dance.

DANCE AS A FIELD OF VIRTUAL POWERS

Langer describes the world of dance as a field of "virtual powers," noting that "no art suffers more misunderstanding, sentimental judgment, and mystical interpretation than the art of dancing."[4] This confusion relates to the very nature of dancing—what it expresses, what it creates, and how it is related to other art forms, to the artist, and to the material world. In response, Langer calls for a return to an intuitive appreciation of dance, which society has largely rejected. The appreciation of dance, she argues, should be as direct and natural as the enjoyment of any other art form.

Langer suggests that to intuitively appreciate dance, one must acknowledge that the art form has its own primary illusion. Dance is a form of movement involving actual life gestures, which signals its creators' desires, intentions, expectations, and demands. These gestures can be consciously controlled, so they may also be elaborated into a system of assigned and combinable symbols, effectively becoming a genuine discursive language.[5] In modern dance in particular, the dancer seems to directly present his/her emotions, and this self-portraiture, which reflects the dancer's experiences of and feelings about the real world, becomes a motif. Langer claims that "the appearance of movement as gesture requires only its emanation from a center of living force; strangely enough, a mechanism 'come to life' [*sic*] intensifies this impression, perhaps by the internal contrast it presents."[6] In this way, the mystic force of the art form becomes even more apparent.

The aesthetic and cultural assumptions underlying Langer's idea of virtual powers draw attention to dancers' leaping, whirling, and kicking, their seemingly struggling and stamping piston-like steps, and other of their carried away movements. Langer further suggests that dance's magic is projected onto spectators to heal, purify, or inspire.[7] She emphasizes the dancer's physical body, specifically the musculature. Langer suggests that "the dancer knows well that the lines composed by his body form the illusionary forces, even for a solo dancer, the rhythmic play of the muscle, the freedom in which impulses go along in complete and intended movements."[8] Accordingly, we are able to differentiate the world of dance from reality and understand the transformation of artistry and symbolism of movement in dance. In the words of Mary Wigman, quoted by Langer,

> all dance construction arises from the dance experience which the performer is destined to incarnate and which gives his creation its true stamp. The experience shapes the kernel, the basic accord of his dance existence around which all else crystallizes. Each creative person carries one's own characteristic theme. It is waiting to be aroused through experience and completes itself during one whole creative cycle in manifold radiations, variations and transformations.[9]

Langer closely follows her teacher Ernst Cassirer's theory of culture and art as systems of symbols when she presents dance forces as virtual powers. Langer echoes Cassirer's notion of "animal symbolicum," positing that human beings are essentially living creatures using and living with symbols and that dance expresses vital forces and emotions in symbolic ways. Langer emphasizes a holistic approach to dance, which also covers the issue of identity. Dance expresses forms of feeling in a stream of tensions and resolutions, and it conveys emotions, mood, and personal existence in an indefinite interplay of tensions.[10] I echo this conceptualization of the art form in my study of six contemporary dance choreographers in Hong Kong.

For Langer, dance choreographers and dancers transform their art into the articulation of symbols in organic diversity. They do not have to experience the emotions embedded in their works in real life; rather, they skillfully utilize dance to objectify their works. Langer elaborates on this point by referring to the theory of intuition, in particular art-making and artistic appreciation as subjective activities in the mastery and understanding of the art form. Cognition of form is intuitive and the intuition relates parts in an organic wholeness. The qualities of distinctness, congruence, correspondence of forms, contrast, and synthesis are in a total gestalt, which can be grasped only through direct insight or intuition. Relevant examples, for me, are Mondrian's formalistic composition in his abstract paintings and dance. Langer points out that meaning in dance always emerges through logical intuition or insight.[11] To understand dance is to fully apprehend the denotations of the emotions it conveys, its expressiveness, and its form through intuition. Crucially, Langer concludes that there is no failure of expression in dance.[12]

Cassirer suggests that the significance of symbols lies in their inclusion of thoughts and ideas that lie beyond the artist's personal experience and history. Criticism of this argument is also applicable to the proposition that dance is symbolic. In work published in the 1920s, Heidegger challenges Cassirer's philosophy of symbolic forms by invoking the limits of the human body, arguing that such physicality, not just thoughts, concepts, and ideas, contributes to the make-up of our world. In response, the posthumous work *The Metaphysics of Symbolic Forms* of Cassirer acknowledges the phenomenon of vitality, including the human body, and modifies his proposition that the foundation of cultural philosophy is the study of the phenomenology of perception.[13] The Cassirer–Heidegger debate had an important influence on Cassirer's student, Langer, and contributed to her theory of symbolism. The consideration of vitality, emotion, and expressiveness became effective theoretical tools in Langer's understanding of dance as expression of vital forces.

PHILOSOPHY OF DANCE AND ITS LAYERS

Langer notes that the philosophical significance of dance stems from two fundamental sources: the primary illusion and the basic abstraction in which the illusion is created and shaped.[14] Langer believes that the primary illusion is created with the first touch—or, in the case of dance, with the first motion, whether performed or merely implied. This motion, as a physical reality and therefore "material" component of the art, must be transformed, as "everything becomes expression, gesture."[15] For Langer, inspired by Wigman, gesture is the basic abstraction whereby the dance illusion is made and organized as vital movement, at once subjective and objective, personal and public, willed (or evoked) and perceived.

Langer suggests that in "real" life (outside the world of dance) gestures function as signals or symptoms of our desires, intentions, expectations, demands, and feelings. At the same time, like any language, they can be consciously controlled and elaborated into a system of assigned and combinable symbols, rendering gesture a genuine discursive language. Gesture in dance is always spontaneously expressive, by virtue of its form.[16] When gesture is imagined, it becomes an artistic element, a possible dance-gesture; eventually, it becomes a free symbolic form of will, which may be used to convey ideas of emotion and combined with or incorporated into other virtual gestures, to express physical and mental tensions.[17]

Langer further argues that the spontaneously gestic character of dance motions is illusory. In her words, "the primary illusion of dance is a virtual realm of Power—not actual, physically exerted power, but appearances of influence and agency created by virtual gesture ... The prototype of these purely apparent energies is not the 'field of forces' known to physics, but the subjective experience of volition and free agency."[18] She notes that in dance, the actual and virtual aspects of gesture are mingled in complex ways; it is the dancer's emotion that turns movement into dance-gesture.[19] Langer further suggests that it is the conceptualization of a feeling that allows the dancer's body to symbolize it.[20] In other words, through dance-gesture, something is revealed, articulated, and made manifest by the symbol. In her words, "everything illusory, and every imagined factor (such as a feeling we imagine ourselves to have) which supports the illusion, belongs to the symbolic form; the feeling of the whole work is the 'meaning' of the symbol, the reality which the artist has found in the world and of which he wants to give his fellow men a clear conception."[21]

Langer delves deeper into the nature of dance by referring to the virtuality that is typically employed in the making of symbols, alongside Cassirer's concept of "mythical consciousness," which is structurally the same as artistic consciousness.[22] In *Language and Myth*, Cassirer states that "mythology is inevitable, it is natural, it is an inherent necessity of language … it is in fact the dark shadow which language throws upon thought … Mythology, in the highest sense, is the power exercised by language on thought in every possible sphere of mental activity."[23]

Although Langer wrote an in-depth preface to Cassirer's theory of language and myth in 1945, the book containing this preface, *Language and Myth*, was published only later, in 1946. In her preface, Langer emphasizes Cassirer's account of the constitutive character of symbolic renderings in the making of "experience." Cassirer uses the Kantian doctrine, which stipulates that the mind is constitutive of the external world, to explain how the world is experienced, in addition to the mere fact that humans experience the external world at all.[24] The prevalence of myth in early cultures and its persistence in religious thought are given renewed significance through Cassirer's constructive philosophy, according to which sense data come together to form our experience of objects. Cassirer's influence is evident in Langer's statement that

> the mythic mind never perceives passively, never merely contemplates things; all its observations spring from some acts of participation, emotion and will … Only where this vital feeling is stirred from within, where it expresses itself as love or hatred, fear or hope, joy or sorrow, mythic imagination is roused to the pitch of excitement at which it begets a definite world of representations.[25]

Langer touches on the meaning and position of objectivity and describes dance as expressing moments in vivid phases. Cassirer said,

> there is the utmost tension between the subject and its object, the outer world; when external reality is not merely viewed and contemplated, but overcomes a man in sheer immediacy, with emotions of fear or hope, terror or wish fulfillment: then the spark jumps somehow across, the tension finds release, as the subjective excitement becomes objectified and confronts the mind as a god or a daemon.[26]

Cassirer further elaborates, "[the dancing moment] is merged with its object in an indissoluble unity … The potential between 'symbol' and 'meaning' is resolved; in place of a more or less adequate 'expression,' we find a relation of identity, of complete congruence between 'image' and 'object,' between the name and the thing."[27] Similar to the world presented in mythology, dance in every form is capable of changing: it reconfigures itself at every instant. "There is no clear division between mere 'imagining' and 'real' perception, between wish and fulfillment, between image and object," Langer concludes.[28]

Cassirer's *An Essay on Man* echoes Langer's philosophy of dance:

> The world of myth is a world of actions, of forces, of conflicting powers. In every phenomenon of nature, its [mythic consciousness] sees the collision of these powers. Mythical perception is always impregnated with these emotional qualities. Whatever

is seen or felt is surrounded by a special atmosphere—an atmosphere of joy or grief, of anguish, of excitement, of exultation or depression.[29]

Langer calls dance a field of virtual powers, saying that "there are no actualities left in it, no untransformed materials, but only elements, living beings, centers of force, and their interplay."[30] Again, there is no differentiation between the virtual and real worlds, but total absorption, assimilation, and appropriation. If there are primary and secondary illusions in dance, then its secondary illusion assimilates the whole phenomenon of dance into the realm wherein the given illusion is primary. The secondary illusion is "an art of time, poetry and drama," in Langer's words.[31] Like other art forms, dance composition or choreography is constructive and imaginative. Indeed, as Langer notes, "it springs from an idea of feeling, a matrix of symbolic form and grows organically."[32] What is distinctive about its secondary layer is that "each creative person carries his own characteristic theme, waiting to be aroused through experience and completes itself during one whole creative cycle in manifold radiations, variations and transformations."[33]

Discussing the relationship between art and audiences, Langer suggests that art, including dance, possesses a social element, whereby the meaning of an artwork is established through its audience's intuitive interpretation. She states that "only in so far as the work is objective, the feeling it exhibits becomes public; it is always bound to its symbol. The effect of this symbolization is to offer the beholder a way of conceiving emotion; and that is something more elementary than making judgments about it."[34]

According to Langer, the viewer experiences excitement when encountering an artwork. This excitement is not personal but shared by the various audience members, indicating the collective nature of the human experience. A work of dance inspires us to formulate both perceptions of feeling and conceptions of visual and audible reality; i.e., "it gives us forms of imagination and forms of feeling, inseparably."[35] Langer's idea of "intuitive anticipation" suggests that audiences have certain expectations of dance before the curtain goes up, including expectations of style, which refers to the choreographer's choice of form and structural expression.

Langer notes that the "energy of art imposes itself on men, and becomes for them the plastic standard of the period ... All the art works of an epoch end by resembling the most energetic, the most expressive, and the most typical works of the period."[36] Covering three decades of contemporary dance (1980–2010) in postcolonial Hong Kong, my case studies illustrate the expressiveness symbolized through dance and show how well-known choreographies mirror changes in Hong Kong society over several decades.

I use Langer's notion of "expressive force" to understand choreographic works. There is no general principle or theory of expressiveness; choreographers must find their own ways to express and symbolize their ideas and emotions as a whole. As a critic, I see and comprehend the overall form of the work, which is closely linked with feelings, vital forces, sensitivity, muscle regulation, mentality, and external and internal interaction, as Langer suggests.[37] However, I also consider the social environment, and I find that dance is equally linked with hope and despair, indictment, and compromise, as expressed in the works themselves. I echo Langer's assertion that art is a public possession, because the formulation of "felt life" is at the core of a culture. Once life is symbolized by its setting, the world seems important and beautiful and is intuitively "grasped."[38] I am inspired by Langer's claim that great artists have often struggled for expression, only that the urgency of their ideas caused them to develop every vestige of talent until it rose to their demands.[39]

HONG KONG DANCE CHOREOGRAPHY (1980–2010): THE CASE OF HELEN LAI

The Taiwan dance scholar Ya-ping Chen reviews three of Hong Kong choreographer Helen Lai's representative works: *Revolutionary Pekinese Opera* (1997); *Tales of Two Cities—Hong Kong. Shanghai. Eileen Chang* (2010); and *Her Story* (2007). Linking the dance stage with the city of Hong Kong, Chen draws attention to the multi-dimensional narratives of Lai's productions and concludes that her method is that of "volumetric choreography," which has strong parallels with the three-dimensionality and vitality of the city of Hong Kong.[40]

Chen concludes her comments on Lai's volumetric choreography with reference to the hybrid culture of Hong Kong, particularly the fluctuation and intertwining of multiple ideologies and body languages in the media. These body languages are related to the marginal positions of colonial citizens, women, and dancers, and also to the tension created by the patriarchy culture of China in Hong Kong. Lai's work echoes Langer's conceptualization of dance, specifically the idea that body movements are filled with

FIGURE 16.1 *Soledad* by Helen Lai (2015). Photograph by Ringo Chan. Courtesy of CCDC City Contemporary Dance Company, Hong Kong.

symbols representing the frustration of some of the Hong Kong citizens living in the ex-colony, facing the pressures of living and social constraints. Furthermore, beyond movement, we need to consider the words, signs, sounds, rhythms and original music clips in Lai's productions, which connect and interact to form an organic whole, as Chen suggests.[41]

The narrative position and perspective of the repeating multi-tones in Lai's dance, which she deliberately uses to interrupt the harmonious flow of the performance, makes symbolization a multifaceted and "living" organism. The stage performance reflects Langer's ideas of forms and feelings in movements, which at times serve to communicate interpersonal entanglement, including the sense of fracture and personal loss one may experience in a relationship. For example, the dance *Tales of Two Cities—Hong Kong. Shanghai. Eileen Chang* features several main characters and their relationships. The imagery provides insights into their complicated emotional relationships through a montage of dance movements, in which the love and hatred within each relationship are extended to encounters in Chang's other stories. As Chen suggests, Lai integrates literary symbols in Chang's well-known novels, in addition to historical events and Chang's own words, into her evocation of congested symbols and meanings in the postcolonial city and its gender politics.[42]

HELEN LAI'S CHOREOGRAPHY IN *SOLEDAD* (2015)

In revisiting Lai's choreographic work *Soledad* (2015), I find that the close relationship between the primary illusion of the gestures with the dance composition, which assembles people's actual and volumetric actions, corresponds to Hong Kong's "Umbrella Movement" of 2014. During this social movement, many Hong Kong citizens who asked for broader representations in the Chief Executive election process occupied the city's Central District and clashed with the Hong Kong police. The event became known as the Umbrella Movement because the protesters used yellow umbrellas to protect themselves from the police force. The protest scenes were captured by the media and broadcast throughout the seventy-nine days of the movement, until the protesters' camps were dismantled.

I draw parallels between this protest movement and *Soledad*, the performance of which featured patterns of physical movement and props that were restaged to project feelings of outrage, with Langer's notion of "expressive force" created by the dancers which stirred up her notion of "intuitive anticipation" in the audience. This was a fully intuitive process, not requiring further analysis or conceptual cognition.

Soledad was produced in 2015, with Helen Lai as the choreographer and Peter Stuart responsible for the music, art, and narration. This particular production was performed by the City Contemporary Dance Company of Hong Kong. The story of *Soledad* is based on Gabriel García Márquez's celebrated novel *One Hundred Years of Solitude*. When viewing the performance, it is clear that Lai deliberately chose not to recount the complexities of Márquez's novel in a literal way. Instead, Lai focused on solitude as enacted through the individual roles in the dance and the social context. The dance reflected human relations and the magical, illusory scenes abstracted from Márquez's novel, and individual readings and interpretations of the work. As such, the performance was divided into separate episodes. Similar to the novel, *Soledad* explores the characters' solitude, with Lai basing the characters on those she encountered in real life, including workers on strike, insistent protesters, people standing in the pouring rain, and those involved in lustful and incestuous relationships. During the performance, it became clear that the audience could easily relate the stories to real occurrences, especially audience members who had experienced similar demonstrations and other social situations.

Between reality and illusion, Peter Stuart provided narration for the performance, which was distantly inspired by the source novel. Stuart, a British and a Hong Kong citizen, a performer and a novelist himself, acted primarily as a narrator reminding people of García's awarded novel. He sometimes acts also as an observer and a commentator in *Soledad*, appearing in the dance delivering readings of the novel when at times. His entries blended well with the dancers' performance: their movements evoked the characters in the novel. It is noted that Lai's choreography and movement arrangement are deviated from the original order of the narration of the novel, aiming for illusory perception. In her commentary, Lai focuses on the body contact and human relations in the performance, as the choreography was tailor made for the individual dancers, reflecting their personalities and temperaments. The integration of reality and illusion in *Soledad* was realized through experiments and improvisation in all aspects, including choreography, relational dynamics, original music, set design, and costumes.

A description of individual scenes from the performance showed three cartoon-like characters dancing solo on the stage at the beginning. They did not interact with each other, but the rhythms of their movements were all in harmony with the music. They then disappeared into the darkness. A male dancer appeared, waiting for a stack of chairs to be installed. He came forward to touch the furniture, which reminded him of a place he had inhabited in the past.

Pairs of dancers ran in from the darkness to dance intimate duets. The pairs were each composed of one male and one female dancer. They were struggling to engage with each other and became entangled in love and in conflict, unable to separate but knowing that they would end up in solitude. Similar scenes appeared a number of times, performed by different pairs of dancers, showing the struggles of heterosexual relations. The background changed accordingly: sometimes it was a falling building and sometimes it was Stuart burying items in sand and clay.

Stuart recited his own poems, inspired by *A Hundred Years of Solitude*. These pieces were filled with metaphors and allegories, describing wars and crises, although no one in the audience seemed to notice them.

Groups of dancers then took over the stage and each of them expressed their fears and frustrations. Their heads came together and they curved their bodies and fought off invisible attacks before they fell, one at a time. Patches of light were moving across the floor and the dancers began to awaken and stand up to find their own paths in solitude, with their bodies casting long shadows on the stage floor.

The tone of the dance changed when sand began to fall from the ceiling onto the stage. The sand represented rainfall. The dancers, each holding a yellow umbrella, came together then quickly dispersed, moving their umbrellas individually. The dancers then fell to the ground, leaving only one umbrella in motion until it was moved offstage. Some dancers grouped together to build a "yellow sea." After Stuart had recited one of his poems, the umbrella-holding people began to stand up and dance in a group with the same rhythm, speaking in chorus lines and finally turning the umbrellas upside down. They retreated and then moved forward as if resisting an invisible big rainstorm, murmuring the number of days it had taken to reach this point. Weak bodies were dragged across the stage; they were suppressed until they were motionless. Stuart then appeared on stage to read his last piece. The dance ended when he dropped his paper to the ground and removed his hat.

Nearly all of the scenes in *Soledad* were dimly lit, evoking an emotional response in the audience. Indeed, the scenes focusing on emotional topics and challenging relationships were extremely touching. These scenes attempted to communicate common experiences of individual solitude in various life situations and incidents, in particular heartbreak and despair. The scenes near the end of the performance undoubtedly reminded the audience, particularly Hong Kong citizens, of their social reality, reported daily in the news. The umbrella holders evoked the protesters in the city's Central and Wan Chai districts in 2014 and their desperate but strong resistance against the police.

In "A Critical Bicultural Pedagogy of Dance: Embodying Cultural Literacy," Darde and Cronini posit that dance in socially oppressed societies reflects citizens' struggle for emotional, psychological, physical, and spiritual survival, as well as creating a space for individual and communal empowerment, even when dance practices are commodified by the dominant social force.[43] Although their study focuses on cultural suppression and literacy, the development of voice and expression in creating the conditions for emancipatory social agency is evident. I echo the saying that dance can be a means through which one builds one's own identity, shaped by personal and social histories and the distinctive engagement with one's surroundings.[44] People in Hong Kong who had experienced the occupation of the city's Central District in 2014 would probably have recognized every step and movement that the dancers performed with yellow umbrellas in *Soledad*.

Soledad as a dance echoes the four key features that Darde and Cronini propose in their pedagogy of dance for cultural awareness. In the authors' words, a dance work can be anchored in the histories, cultural knowledge, lived experiences, and understandings of everyday life in a particular social context. Dance can link the sociopolitical development of the critical consciousness of social agency. It can nurture the emancipatory expression of identity, community, and self-determination that reinforces a deep sensitivity to the issues, concerns, and practices that link the subjects of a dance and its audiences. Dance can lead viewers to adopt a humanizing ethos of life that requires consciousness of political commitment and cultural integrity.[45]

We see evidence of this dance's communal and incarnated knowledge communicated through the dancers' collective interactions, communicating empowerment, in the yellow umbrella scenes, in contrast to the solos and duets. In the group scenes we see acts of resistance, which are meaningful symbols of resistance and liberation. Darde and Cronini eloquently state that dance allows the body to release trauma through the communal process of movement, fostering new ways to experience (and build) the world.[46] Dance is where "power, authority, [and the] community relationship are affected, rearranged and affirmed."[47] A "necessary synergy and solidarity"[48] was built among most citizens in Hong Kong during the social protest in the mid-2010s.

LANGER'S NOTION OF VIRTUAL SPACE AND POWERS IN CONTEXT

The questions raised at the beginning of this chapter are as follows. How can Langer's virtual space of dance be read? Is it only through contemplation? Do we apprehend only form? If so, how can a dance critic respond to questions related to the work in its social reality or culture? Langer's notion of virtual power provides inspiration to address these questions.

The response should begin by referring to Langer's suggestion of layers of illusion in contemplating dance. The illusions begin with gesture, which, to Langer, is more than vital movement. Langer regards gesture as at once subjective and objective, personal and public, willed (or evoked) and perceived. She points out that, in real life, gestures function as signals or symptoms of our desires, intentions, expectations, demands, and feelings, but they can also be consciously controlled and elaborated into a system of assigned and combinable symbols, by virtue of their form.[49]

The primary illusion of dance is a virtual realm of powers, created by virtual gestures, that involves the subjective experience of volition.[50] Langer further posits that in dance, the actual and virtual aspects of gestures are mingled in complex ways. *Soledad*, in which one can see numerous solos and duets containing gestures, is no exception to Langer's theory. It is the dancer's emotions expressed through gestures that turn movement into dance-gesture.[51] Langer further posits that it is the conception of a feeling that disposes the dancer's body to symbolize it.[52] In other words, through the dancer's movements, something actual is revealed, articulated, and made manifest by the symbol. In the performance of *Soledad* considered here, the conception, the symbols and the virtual powers then at the higher level, were rendered possible through the events and images that happened in social reality and in the news, and could immediately be captured and assimilated by the audience. Helen Lai's choreography produced possible dance-gestures in a free symbolic form of will to convey the individual feelings of solitude and social emotions of certain groups. These groups simultaneously used colors and symbolic costumes, and designed unique movements and gestures to articulate their political identities. In Langer's terms, these layers or levels are all illusory: a gesture as physical reality is transformed into art with imagination. The second level relates to the devices that support the total creation and enhance the expressiveness of the gestures.

Soledad, a dance composed by Helen Lai, exemplifies Langer's notion of the "virtual form of dance." It supports Langer's argument that the art form expresses feelings in a stream of tensions and resolutions, placing emotions, mood, and personal existence in an indefinite interplay.[53] In addressing whether we only apprehend the form of dance, Langer suggests that the cognition of form is intuitive in the organic wholeness of all relatedness; distinctness, congruence, correspondence of forms, contrast, and synthesis are in a total gestalt, which can be grasped only by direct insight.[54] Indeed, the above-mentioned performance of *Soledad* succeeded in conveying these effects.

Langer delves deeper into the nature of dance when she refers to the virtuality in the making of symbol and in Cassirer's term, "the mythical consciousness." According to Langer, dance springs from an idea of feeling, a matrix of symbolic form and grows organically.[55] Langer further posits that audiences experience "intuitive anticipation" of dance, in that they anticipate elements of the performance before the curtain goes up, including the style, which reflects the formal and structural choices of the choreographer.[56] "Intuitive anticipation" was a strong feature of the focal performance of Lai's *Soledad*, particularly when we saw the dancing bodies stretching and restraining under the yellow umbrellas, which had already become a political symbol representing the social movement in Hong Kong in the 2010s. This echoes Langer's belief that art is a public possession, the formulation of "felt life" as the core of a culture.[57]

The intertwining of repeating multi-tones in Lai's dance, including in the solos, duets, and group scenes, makes symbolization in art a multifaceted and living organism. These multi-tones echo Langer's ideas relating to forms and feelings, illusory layers, virtual powers, and intuitive anticipation. Lai's work extends Langer's idea of illusory layers in dance to social reality, in that the feelings expressed in dance are deliberately created symbolic forms. They are multi-dimensional adoptions of semiotics in multiple propositions, like all dance works in other particular social, political, and cultural contexts.

NOTES

1. Susanne K. Langer, *Feeling and Form* (New York: Charles Scribner's Sons, 1953), 204.
2. See Eva K. W. Man, ed., *Six Contemporary Dance Choreographers in Hong Kong 1980s–2010s: History, Aesthetics and Cultural Identities* (Hong Kong: The International Association of Theatre Critics, 2019).
3. Langer, *Feeling and Form*, 205.
4. Ibid., 169.
5. Ibid., 174–5.
6. Ibid., 181.
7. Cf., ibid., 199.
8. Ibid., 197.
9. Ibid., 206.
10. Ibid., 372.
11. Ibid., 379.
12. Ibid., 381.
13. Ernst Cassirer, *The Philosophy of Symbolic Forms*, vol. 4: *The Metaphysics of Symbolic Forms* (New Haven, CT: Yale University Press, 1998).
14. Langer, *Feeling and Form*, 169.
15. Ibid., 174.
16. Cf., ibid., 175.
17. Cf., ibid., 174.
18. Ibid., 175.
19. Cf., ibid., 180.
20. Cf., ibid., 181.
21. Ibid., 182.
22. Ibid., 184.
23. Ernst Cassirer, *Language and Myth*, trans. Susanne K. Langer in 1945 (New York: Dover, 1946), 5.
24. Susanne K. Langer, "Preface," in Cassirer, *Language and Myth*, vii–x.
25. Susanne K. Langer, "On Cassirer's Theory of Language and Myth," in *The Philosophy of Ernst Cassirer*, ed. Paul Schilpp (Evanston, IL: The Library of Living Philosophers, 1949), 396. Langer refers here to Cassirer's then untranslated book *Philosophie der symbolischen Formen, Teil 2: Das mythische Denken* (Berlin: Bruno Cassirer, 1925), 90.
26. Cassirer, *Language and Myth*, 33.
27. Ibid., 58.
28. Langer, "On Cassirer's Theory of Language and Myth," 397. Here, Langer again refers to Cassirer's *Philosophie der symbolischen Formen* 2, 48–9.
29. Ernst Cassirer, *An Essay on Man* (New Haven, CT: Yale University Press, 1962), 76–7.
30. Langer, *Feeling and Form*, 187.
31. Ibid., 205.
32. Ibid.
33. Ibid., 206.
34. Ibid., 394.

35. Ibid., 397.
36. Ibid., 400.
37. Ibid., 407.
38. Ibid., 409.
39. Ibid. 408.
40. Ya-ping Chen, "Hong Kong as Method: Narratives, Body and Search of Identities in Helen Lai's Work," in *Six Contemporary Dance Choreographers in Hong Kong 1980s–2010s: History, Aesthetics and Cultural Identities*, ed. Eva K. W. Man (Hong Kong: The International Association of Theatre Critics, 2019), 90.
41. Ibid., 82–3.
42. Ibid., 88–9.
43. Antonia Darde and Sharon Cronini, "A Critical Bicultural Pedagogy of Dance: Embodying Cultural Literacy." *Portuguese Journal of Education* 31 (2018): 27.
44. Darde and Cronini, "A Critical Bicultural Pedagogy of Dance," 29.
45. Ibid., 30.
46. Ibid., 33.
47. Yvonne Daniels, *Dancing Wisdom: Embodied Knowledge in Haitian Vodou, Cuban Yoruba, and Bahian Candomble* (Urbana and Chicago, IL: University of Illinois Press, 2005), 55.
48. Darde and Cronini, "A Critical Bicultural Pedagogy of Dance," 36.
49. Cf., Langer, *Feeling and Form*, 175.
50. Cf., ibid.
51. Cf., ibid., 180.
52. Cf., ibid., 181.
53. Cf., ibid., 372.
54. Cf., ibid., 379.
55. Cf., ibid., 205.
56. Cf., ibid., 400.
57. Cf., ibid., 409.

BIBLIOGRAPHY

Cassirer, Ernst. *An Essay on Man*. New Haven, CT: Yale University Press, 1962.

Cassirer, Ernst. *Language and Myth*. Translated by Susanne Langer. New York: Dover, 1946.

Cassirer, Ernst. *Philosophie der symbolischen Formen*. Teil 2. Das mythische Denken. 1st edn. Berlin: Bruno Cassirer, 1925.

Cassirer, Ernst. *The Philosophy of Symbolic Forms*, vol. 4: *The Metaphysics of Symbolic Forms*. Edited by John Michael Krois and Donald Phillip Verene. New Haven, CT: Yale University Press, 1998.

Chen, Ya-ping. "Hong Kong as Method—The Body Narrative in Helen Lai's Dance Productions." In *Six Contemporary Dance Choreographers in Hong Kong 1980s–2010s: History, Aesthetics and Cultural Identities*, edited by Eva K. W. Man, ch. 2. Hong Kong: The International Association of Theatre Critics, 2019.

Daniels, Yvonne. *Dancing Wisdom: Embodied Knowledge in Haitian Vodou, Cuban Yoruba, and Bahian Candomble*. Urbana and Chicago, IL: University of Illinois Press, 2005.

Darde, Antonia, and Sharon Cronini. "A Critical Bicultural Pedagogy of Dance: Embodying Cultural Literacy." *Portuguese Journal of Education* 31 (2018): 26–41.

Man, Eva K. W. *Cross-Cultural Reflections on Chinese Aesthetics, Gender, Embodiment and Learning*. Singapore: Springer Nature PTE Co, 2020.
Man, Eva K. W. ed. *Six Contemporary Dance Choreographers in Hong Kong 1980s–2010s: History, Aesthetics and Cultural Identities*. Hong Kong: The International Association of Theatre Critics, 2019.
Langer, Susanne K. *Feeling and Form: A Theory of Art*. New York: Charles Scribner's Sons, 1953.
Langer, Susanne K. "On Cassirer's Theory of Language and Myth." In *The Philosophy of Ernst Cassirer*, edited by Paul Schilpp, vol. 6. Evanston, IL: The Library of Living Philosophers, 1949.
Schilpp, Paul, ed. *The Philosophy of Ernst Cassirer*. Evanston, IL: The Library of Living Philosophers, 1949.

CHAPTER SEVENTEEN

"Virtual Acts" as a Langerian Approach to Performance Art

CHRISTOPHE VAN EECKE

VIRTUAL ACTS: A PROPOSAL

What exactly is it that artists "do" when they "do" performance art? This question has long frustrated critics. RoseLee Goldberg observes that "by its very nature, performance art defies precise or easy definition beyond the simple declaration that it is live art by artists."[1] Consequently, critics have tended to group performance art together with other body-related art practices under such container terms as Body Art,[2] Live Art,[3] or the more poetical "fleshworks."[4] But while it may be true that "the distinctions between performance art and other new media are now quite blurred, calling for new terminologies, for fresh ways to describe 'performance,'"[5] we need a robust concept of performance art if we are to move beyond generalities. It is at this juncture that Langer proves helpful. In *Feeling and Form*, Langer argues that every major art form creates what she calls its own primary illusion, which is that art form's specific presentational form. The primary illusion identifies how, for example, what is presented in a painting or a sculpture is significantly different in its formal structure from what is offered in a work of literature, a dance, or a film. The notion of "illusion" is key because it indicates that, for Langer, all artistic forms are virtual. This means that every work of art is first and foremost "given to our perception."[6] In the visual arts, for example, the primary illusion creates what Langer calls "virtual space" because in painting and drawing spatial relations are rendered in a two-dimensional plane: they create "a picture space that exists for vision alone. Being only visual, this space has no continuity with the space in which we live; it is limited by the frame ... The created virtual space is entirely self-contained and independent."[7] Along similar lines, Langer argues that "virtual time" is the primary illusion of music because "musical duration is an image of ... the passage of life"[8] while literary fiction establishes the primary illusion of "virtual life"[9] and architecture creates "an ethnic domain."[10]

Following this model, I propose "virtual action" as the primary illusion of performance art. This concept is made up of two terms: performances are claimed to be *acts*, and these acts are said to be *virtual*. Both terms require some clarification.

Langer introduces a concept of act in the first volume of *Mind* (1967), where she uses the term in a biological context to refer to "any unit of activity"[11] in an organism and to the "elements in the continuum of a life."[12] This means that her notion of act is not immediately helpful for a discussion of performance art because it is not directly concerned with deliberate human activity. I will therefore borrow Hannah Arendt's concept of action from her book *The Human Condition* (1958). Action, says Arendt, is "the only activity that goes on directly between [people] without the intermediary of things or matter, correspond[ing] to the human condition of plurality."[13] To act is to do something: it is to take an initiative in the world and in the process "beginning something new," which is what Arendt calls "the human condition of natality."[14] The newness set in motion through action is both unexpected, because any person can take an initiative that "cannot be expected from whatever may have happened before,"[15] and unpredictable, because all actions have a "boundlessness"[16] in their possible consequences or outcomes. This means that action is the most political human activity for Arendt, namely the one most closely connected to the condition of living together. Its very boundlessness makes it an open-ended process: every act may engender a new (unpredictable) act from someone else. Because it is the business of all of humankind, so many different parties and interests are involved that the outcomes of the process can never be predicted or predetermined. Furthermore, nothing that is achieved in politics (or in action) is ever final, if only because subsequent actors can always undo or change what was previously established. To set something in motion in the public realm therefore always involves a leap into the unknown, for as soon as the initial push is given the process is out of one's hands: other hands join in, and other interests become involved. As Arendt says, action has the structure of narrative, and "its full meaning can reveal itself only when it has ended."[17]

It is the precarious open-endedness of action that provides a conceptual bridge to performance art, which typically involves acts performed in public and where the doing of the act is the work of art. As Kristine Stiles points out, "in performance, artists present and represent themselves in the process of *being and doing*, and these acts take place in a cultural context for a public to witness."[18] Furthermore, performance art is often also engaged art, which provides a second link to political action. Few performance artists ever perform merely to entertain, and they typically expect their work to grate or provoke and make people think, and perhaps even change something in the world. But if performance art is very much like action, it is not the thing itself: it is still a work of art. This means that performances are first and foremost given to our perception, to be seen and to be thought about. They are not the real world of pragmatic day-to-day existence: like all art, they signify *something about* the world through the presentation of actions. It is in this sense that performance art consists of *virtual* acts, and it is the virtuality of the event that now needs further unpacking.

MAGIC CIRCLES AND WILLING PARTICIPANTS

It is counterintuitive to suggest that the acts performed in performance art are virtual rather than real, especially because so much performance art has a physical immediacy that makes it intensely and even uncomfortably real for those present. We therefore need to get a firm grasp of what is meant by its virtuality. Although Langer never wrote about performance art, her discussion of dance provides interesting elements that help to make the virtuality of performance art more concrete. Langer argues that "all dance motion is gesture ... *Gesture* is the basic abstraction whereby the dance illusion is made and organized."[19] Langer distinguishes between gesture and gesticulation. "Gesticulation, as part of our actual behavior, is not art. It is simply vital movement."[20] In dance, such vital movement is choreographed into a deliberate pattern. This turns gesticulation into gestures, and these are virtual: rather than involuntary or pragmatic expressions of the hustle and bustle of everyday life or the vital needs of our organism they are deliberate enactments of these movements. In this sense, gesture is an abstraction of gesticulation: while gesticulation is simply part of the patterns of life, gesture makes artistic use of gesticulation for the purpose of expressing the significance of these patterns. This is especially clear in types of modern dance that take everyday movements as the basis for their choreography.

For our present purpose, which is to illuminate how performance art is virtual, it is interesting that Langer further writes that

> the primary illusion of dance is a virtual realm of Power—not actual, physically exerted power, but appearances of influence and agency created by virtual gesture. In watching a collective dance—say, an artistically successful ballet—one does not see people running around; one sees the dance driving this way, drawn that way, gathering here, spreading there ... In a *pas de deux* the two dancers appear to magnetize each other; the relation between them is more than a spatial one, it is a relation of forces; but the forces ... really do not exist physically at all. They are dance forces, virtual powers.[21]

This description of the virtual powers of dance illuminates how the virtual realm of the dance work is created. As Langer explains, "the dance creates an image of nameless and even bodiless Powers filling a complete, autonomous realm, a 'world'."[22] This means that, in their performance of the dance, the dancers "transform the stage for the audience as well as for themselves into an autonomous, complete, virtual realm."[23] The establishment of such a self-contained virtual realm is a principle that can be observed throughout the history of the performing arts, and it is especially in evidence in the creation of a so-called magic circle in street theater such as the medieval mummers' play. To create an open space for the performance, several actors would walk around in an expanding circle, pushing the audience back. Creating this circular area ensures that "the action is set apart from the spectators yet made accessible to all of them."[24] Furthermore, "clearing the magic circle for the players ... is also a physical necessity, as anyone who has performed with an unstructured audience will tell you."[25]

The audience for performance art is often similarly unstructured, especially when an artist performs in a public space or in an open walk-in area where the audience will have considerable freedom to move around (including the freedom to leave or re-enter the venue). But if the magic circle is used in theatrical performance to create a virtual realm that is separate from the real world of the audience, then performance art often relies on an inverse dynamic: to the extent that audience response and even audience involvement are often integral to a performance its magic circle will have to *include* rather than exclude the audience. If, in a traditional theatrical performance, the magic circle could be said to coincide with the edge of the proscenium, where the footlights demarcate the border between fiction and reality (with the magic circle of the stage illuminated and the real world of the audience cloaked in darkness), then in many performances the magic circle includes both the entire space in which the performance takes place and everyone inside that space. This space need not have clear physical borders: it is the dynamic of audience involvement that constitutes the magic circle. Just as dancers can seem to magnetize each other, so the audience of a performance is included in and magnetized by the virtual force field of the work. Many performances are in fact not complete *as a work* without an audience and its response. This is quite a unique feature of performance art. After all, most works of art are completed *before the audience comes in*. This obviously holds for paintings and novels, which are completed before we get to see them, but also for many of the performing arts. This is the function of rehearsals: performers must typically master or internalize the complete work (the text of a play, the score of a symphony, the choreography of a ballet, etc.) *before* the audience enters so that they may give a satisfying performance of the work once the audience is present. The audience itself does not figure into the creation of the work: the play is performed *for* the audience, not *with* it. What is presented to the audience is the achieved work. Before the curtain goes up, the complete work is already finished in the minds and memories of the performers, who are *ready to perform* it again, and again, and again (and it is this readiness that also allows performers to improvise in a non-trivial way).

In much performance art, however, *the work cannot be complete before the audience comes in*, and many major works of performance art have been performed only once—and they certainly do not become part of the repertoire as classic plays or symphonies do, to be performed anywhere at any time by any group of artists with sufficient dedication to train themselves to readiness for the performance. A remarkable consequence of this dynamic is that the audience of a performance is truly a virtual audience. By this I mean that the people attending a performance consent to becoming participants in the work and not just a conventional audience that keeps its observational distance: they become part of what is created. Consider Paul McCarthy's piece *Hot Dog* (1974), a masochistic work of self-degradation in which the artist stuffed his mouth with hot dogs to the point of suffocation. Barbara Smith, an artist who attended the performance, reported its effect on members of the audience as they watched McCarthy "struggling with himself, trying to prevent his own retching. It is apparent that he is about to vomit … Should he vomit he might choke to death, since the vomit would have no place to go. And should any one of us vomit, we might trigger him to do likewise."[26] This is performance art as a hostage situation: like the dancers in the magnetizing *pas de deux*, McCarthy and the audience are locked together in a perverse choreography of assaulted bodies. The audience's acute and deeply uncomfortable awareness of its own stomach, raging to vomit yet knowing it should not, is at the heart of what the performance signifies. It would simply not make sense to perform this piece without (the response of) an audience because *without the audience there is no piece*. Throughout this process, however, the audience is a willing participant: everyone is free to walk out. This is not incidental: it is key to the artistic success of performance art.

HARM'S WAY: BLURRED BOUNDARIES AND SPURIOUS ACTS

Having suggested "virtual action" as the primary illusion of performance art, we can now explore the question of the concept's usefulness for practical criticism. Let us begin by addressing the question of what happens if performance art oversteps the boundaries of the magic circle of virtuality and invades the real world beyond that magic line. Interestingly, this is exactly what happens in several classic pieces of performance art. Consider Chris Burden's notorious piece *Deadman* (1972), of which the artist himself explained that

> at 8 pm I lay down La Cienega Boulevard and was covered completely with a canvas tarpaulin. Two 15-minute flares were placed near me to alert cars. Just before the

flares extinguished, a police car arrived. I was arrested and booked for causing a false emergency to be reported. A trial took place in Beverly Hills. After three days of deliberations, the jury failed to reach a decision and the judge dismissed the case.[27]

This performance raises several ethical questions, for, in performing it, Burden implicated and even endangered the lives of many people who never consented to being part of his event. What if a passing car had run over Burden's body? What if a driver coming upon the obstacle had swerved and caused a collision? If we use virtuality as a formal test to determine whether *Deadman* is a successful work of performance art, we might argue that the artist took a reckless and cynical risk with the lives of non-consenting others, thereby breaking the bounds of art. *Deadman* then emerges as a spurious work of art for which it is difficult to offer any kind of ethical or artistic justification. An ethical person might be forgiven for not considering it art at all.

A similar problem pertains, but in a much more subtle way, to one of the most canonical pieces of performance art, namely Marina Abramović's *Rhythm 0* (1974), for which the artist allowed the audience to do with her body as they wished using a range of objects including a gun, a saw, needles, and scissors. "By the end of the performance all her clothes had been sliced off her body with razor blades, she had been cut, painted, cleaned, decorated, crowned with thorns and had had the loaded gun pressed against her head. After six hours the performance was halted by concerned spectators."[28] In a very thoughtful discussion of the piece, Peggy Phelan correctly observes that *Rhythm 0* "placed performance art squarely in the ongoing post-war conversation about the ethics of the act: what does it mean to act when full knowledge of the consequence of your act cannot be known in advance?"[29] In this respect, the piece certainly created a situation in which the open-endedness of action, in Arendt's sense, was key. But although the audience for *Rhythm 0* was doubtless a willing participant, and although Abramović had stated that she would accept full responsibility for its outcome, I suggest that the performance was not sufficiently bounded. If the performance had been taken to its uttermost conclusion by one of the participants, it would have had tremendous real-life consequences that extended far beyond the context of the performance. The hypothetical audience member who shot the artist would likely have found that the artist's presumed assumption of full responsibility did not protect them from criminal prosecution. This means that the magic circle was broken, and it was broken on principle: the performance cheated on the terms on which it negotiated the audience's participation.

Rhythm 0 involved a reckless endangerment of both the artist and her participatory audience. There is no doubt that *Rhythm 0* was a momentous piece and possibly a life-changing experience for those present. It has certainly had a revolutionary impact on performance art practices. But something's being momentous, and life-changing, and revolutionary does not necessarily make it good or interesting art. Impact and (artistic) significance are different qualities. Like *Deadman*, *Rhythm 0* is flawed and even objectionable as a work of art because it spills over from self-regarding performance piece to other-regarding real-life event.

THEATER OF LIFE

As a counterexample of a significant work of performance art that does not cheat on its contract with its audience, but which also illuminates the key features of performance art as virtual action, I propose Christoph Schlingensief's notorious performance *Bitte liebt*

Österreich (2000).[30] For this piece, Schlingensief installed several containers in front of the Vienna Opera, which is the cultural heart of the city and of Austria. For the duration of a week, Schlingensief housed twelve purported asylum seekers inside these containers. Their lives inside the containers were broadcast live online and, following the model of the television program *Big Brother*, which was hugely popular at the time, the audience was allowed to vote out two asylum seekers every day. These were then allegedly deported. On top of the containers, Schlingensief had a large sign installed which read "*Ausländer raus*" or "foreigners out," which was a direct reference to the extreme-right party FPÖ (Austrian Freedom Party), which was part of the Austrian government coalition and had been elected on a radical anti-immigration agenda. With *Bitte liebt Österreich*, Schlingensief decided to take the party and the people who had voted for it at their word by putting the party's program into action. Unsurprisingly, *Bitte liebt Österreich* proved incredibly controversial. The press, including the right-wing press that had supported the FPÖ, widely condemned it, exposing its own hypocrisy in the process. Some people demonstrated against the performance, while others showed up to express their support, not just for the performance but also for the presumed deportation of the asylum seekers. At one point leftist activists stormed the containers to destroy the "*Ausländer raus*" sign (it was subsequently replaced) and "liberate" the asylum seekers.

How did Schlingensief's container project operate as a virtual as opposed to a real act? Georg Seesslen has pointedly remarked that "the essence of this art installation [was] its radical proximity to reality."[31] This seems to me to be correct: Schlingensief achieved extreme proximity to the real, but not reality, and the performance's play with the real and the fictional was key to its success. The widely differing responses to the performance suggest that people were unclear about the ontological status of the event: was it real or not? As Jens Roselt has pointed out, the template of *Big Brother*, on which Schlingensief modeled his performance, borrowed its aesthetic from naturalism in late nineteenth-century theater, where the audience was allowed to peep into seemingly real situations through the fourth wall of the theater. Naturalist drama tried to present drama on stage "as if" it was real, playing with the line between the realistic (that which looks real) and the authentic (that which is real).[32] *Bitte liebt Österreich* cleverly traded on the same distinction between realism and authenticity. For while many people thought the event was real, the mediated nature of the event, staged in a public space and with the continuous presence of cameras and the press (not to mention the online livestream), clearly signaled to all passers-by that they were witnessing a staged event. Schlingensief clearly stated (and continually shouted through his megaphone) that the whole event was merely a theatrical performance.

Yet it did not really matter whether the asylum seekers were real or actors (or real asylum seekers being paid to perform), and the containers might as well have been empty for Schlingensief's purposes. The key element was the "as-if"-factor of the event, whose instability functioned as a screen onto which all societal and political stakeholders could and did project their own aims, fears, and ideologies. Schlingensief himself called it "a mirroring technique"[33] while Peter Sloterdijk dubbed it "a kind of event-management, which is to say an attempt to build the explanation directly into the event."[34] The ingenuity of Schlingensief's intervention lay in the fact that it seemed impossible to respond to the performance without somehow becoming implicated in or tainted by it. Nobody who did respond to it, however, could claim not to be aware of the conditions of publicity under which their response took place. Those who chose to make public statements, shout at cameras, demonstrate, or charge the containers were doing so in full knowledge of the fact that they were participating in a public event, that it was part of a theatrical festival, and that their acts were being recorded. They may not have fully realized the precise nature of the event, but that was both Schlingensief's point *and not his fault*.

In this respect, the liminality of the performance, balancing on the edge between reality and illusion, can be short-circuited with Arendt's notion of action in an interesting way. As Arendt argued regarding political action, it is impossible for anyone to have full knowledge of or control over the consequences of their acts because their full significance typically only becomes clear in retrospect. To the extent that the public space is constantly full of events whose full scope we cannot understand and to which we may or may not respond, Schlingensief did not put his audience in a situation that was in any way structurally different from our everyday presence in the public realm. In this sense, encountering and deciding to publicly respond to (and become implicated in) Schlingensief's container performance is not dissimilar to agreeing to a "vox pop" interview by a television journalist who hails one in the street. The news is now daily flooded with random citizens giving their impromptu opinions on just about any topic. Like the people hysterically screaming at Schlingensief's containers, these people may regret their fifteen seconds of television exposure when they see their own imprudent words or angrily distorted faces played back on the nation's screens. In that sense, the terms of the contract in Schlingensief's action were clear: this is the public realm, this is an event, and you can respond if you wish, but there is no guarantee about the outcome or consequences of your intervention. In allowing his audience the freedom to respond he never promised them they could control what their responses would mean and how they would be perceived.

In that sense, Schlingensief's action was a completely virtual act because it duplicated the structure of public acts yet had no real-life effects beyond those that we typically expect either from real public acts or from art, which are changes in our relations with ourselves and with others. Those actions that did have more far-reaching real-life consequences were neither orchestrated nor invited by Schlingensief, although perhaps silently hoped for. Here we might think of the storming of the containers, during which people might really have gotten hurt, and for which the police were immediately called. But Schlingensief neither invited nor encouraged it, and therefore cannot be held accountable for it. No artist is accountable for the irresponsible, violent, or imprudent responses of members of the public to their work. With his container project, Schlingensief cleverly staged a perfect double of the public realm, a *theatrum mundi* of the Austrian political mindset of the time. On this stage, all the men and women were merely players, but in playing their parts they revealed themselves. This is the ultimate paradox of Schlingensief's container action and the ultimate measure of its artistic success: while nothing about it was *real*, everything about it was *authentic*. Its authenticity lay in our own response to it. Its mirror revealed us to ourselves. It was the ultimate coup in audience participation, achieved without the artist's (false) promise to relieve us of responsibility for our acts.

ENVOI: ART AND LIFE

I have proposed "virtual acts" as a concept for thinking about performance art, about what it does, and how, and what this means. In this sense, my discussion has been exploratory: the range of examples has necessarily been limited and the discussion has doubtless raised more questions than it has been able to address. But the aim was not to resolve all issues and questions: my intention was merely to propose a conceptual shift in the way we think about performance art by offering a Langerian definition of what performance art might be, and how it might best be understood. The most important conclusion to draw from my analysis would seem to be the significance of virtuality in performance art: the need for the work to have a presentational form given primarily to our perception. As the uniquely liminal example of Schlingensief's container project shows, there are many ways in which art can meaningfully break through the fourth wall, engage the audience, and blur the lines between the real and the fictional without fully rupturing them and falling back into the realm of everyday life, as did happen in *Deadman* and *Rhythm 0*. For Langer, this virtuality was a necessary condition for something to succeed as art. In this respect, it may be helpful to recall that, in theater as much as in all the other arts, the magic circle functions as a safe space (to use a fashionable term) for both presentation and reflection. Because there is a frame (real or metaphorical) around a work of art it can present the world to our perception and invite reflection without swallowing us up. There is room for contemplation. When art spills over into life, or when real life intrudes upon art (because a performance is not sufficiently bounded by virtuality), the distinction between the two ultimately becomes meaningless. In holding the mirror up to life, performance art can make us think about who we are, and how we act, without peril.

NOTES

1. RoseLee Goldberg, *Performance Art: From Futurism to the Present* (New York: Thames & Hudson, 2001), 9.
2. See Lea Vergine, *Body Art and Performance: The Body as Language*, 1974 (Milan: Skira, 2000) and *The Artist's Body*, ed. Tracey Warr (London: Phaidon, 2000).
3. See *Live: Art and Performance*, ed. Adrian Heathfield (London: Tate Publishing, 2004).
4. Amelia Jones, "Working the Flesh: A Meditation in Nine Movements," in *Live: Art and Performance*, ed. Adrian Heathfield (London: Tate Publishing, 2004), 134.
5. RoseLee Goldberg, "One Hundred Years," *Live: Art and Performance*, ed. Adrian Heathfield (London: Tate Publishing, 2004), 180.
6. Susanne K. Langer, *Feeling and Form: A Theory of Art Developed from "Philosophy in a New Key"* (London: Routledge and Kegan Paul, 1953), 86. For more on Langer's notion of virtuality, see Thomas Leddy in Chapter 18 in the present volume.
7. Langer, *Feeling and Form*, 72.
8. Ibid., 109.
9. Ibid., 262.
10. Ibid., 95. See Christophe Van Eecke, *Only Connect: Five Exercises in Aesthetics* (Breda: Lokaal 01, 2011), 56–98 for a general presentation of Langer's philosophy of art.
11. Susanne K. Langer, *Mind: An Essay on Human Feeling*, vol. 1 (Baltimore: Johns Hopkins University Press, 1967), 202.
12. Langer, *Mind*, vol. 1, 261. See Martina Sauer's contribution in Chapter 15 in the present volume for more on Langer's notion of act.
13. Hannah Arendt, *The Human Condition* (Chicago: University of Chicago Press, 1998), 7.
14. Ibid., 9.
15. Ibid., 178.
16. Ibid., 191.
17. Ibid., 192.
18. Kristine Stiles, "Performance," in *Critical Terms for Art History* (Chicago: University of Chicago Press, 2003), 75.

19. Langer, *Feeling and Form*, 174.
20. Ibid., 175.
21. Ibid., 175–6.
22. Ibid., 190.
23. Ibid., 204. Eva Kit Wah Man also discusses Langer's ideas on dance in Chapter 16 in this volume.
24. Alan Brody, *The English Mummers and Their Plays: Traces of Ancient Mystery* (Philadelphia: University of Pennsylvania Press, 1970), 17.
25. Meg Twycross, "The Theatricality of Medieval English Plays," in *The Cambridge Companion to Medieval English Theatre* (Cambridge: Cambridge University Press, 2008), 61.
26. Quoted in Tracey Warr, ed., *The Artist's Body* (London: Phaidon, 2000), 104.
27. In Fred Hoffman, ed., *Chris Burden* (London: Thames & Hudson, 2007), 158.
28. Tracey Warr, ed., *The Artist's Body* (London: Phaidon, 2000), 125.
29. Peggy Phelan, "On Seeing the Invisible," in *Live: Art and Performance*, ed. Adrian Heathfield (London: Tate Publishing, 2004), 19.
30. The performance was documented in Paul Poet's excellent documentary *Ausländer Raus! Schlingensiefs Container* (2001) and in the book *Schlingensiefs Ausländer raus*, ed. Matthias Lilienthal and Claus Philip (Frankfurt a.M: Suhrkamp, 2000).
31. Georg Seesslen, "Der Populist, der Provokateur, der fremde Blick und das eigene Bild," in *Schlingensiefs Ausländer raus: Bitte liebt Österreich*, ed. Matthias Lilienthal and Claus Philipp (Frankfurt a.M: Suhrkamp, 2000), 254.
32. See Jens Roselt, "Big Brother: Zur Theatralität eines Fernsehereignisses," in *Schlingensiefs Ausländer raus: Bitte liebt Österreich*, ed. Matthias Lilienthal and Claus Philipp (Frankfurt a.M: Suhrkamp, 2000), 74.
33. In Lilienthal and Philipp, eds., *Schlingensiefs Ausländer raus*, 224.
34. Ibid., 225.

BIBLIOGRAPHY

Arendt, Hannah. *The Human Condition*. 2nd edn with an introduction by Margaret Canovan. Chicago: University of Chicago Press, [1958] 1998.

Ausländer Raus! Schlingensiefs Container (2000), [Film] Dir. Paul Poet, Austria.

Beadle, Richard and Alan J. Fletcher, eds. *The Cambridge Companion to Medieval English Theatre*. 2nd edn. Cambridge: Cambridge University Press, 2008.

Brody, Alan. *The English Mummers and Their Plays: Traces of Ancient Mystery*. Philadelphia: University of Pennsylvania Press, 1970.

Goldberg, RoseLee. "One Hundred Years." In *Live: Art and Performance*, edited by Adrian Heathfield, 176–80. London: Tate Publishing, 2004.

Goldberg, RoseLee. *Performance Art: From Futurism to the Present*. Revised and expanded edn. New York: Thames & Hudson, 2001.

Heathfield, Adrian, ed. *Live: Art and Performance*. London: Tate Publishing, 2004.

Hoffman, Fred, ed. *Chris Burden*. London: Thames & Hudson, 2007.

Jones, Amelia. "Working the Flesh: A Meditation in Nine Movements." In *Live: Art and Performance*, edited by Adrian Heathfield, 132–43. London: Tate Publishing, 2004.

Langer, Susanne K. *Feeling and Form: A Theory of Art Developed from "Philosophy in a New Key."* London: Routledge and Kegan Paul, 1953.

Langer, Susanne K. *Mind: An Essay on Human Feeling*, vol. 1. Baltimore: Johns Hopkins University Press, 1967.
Lilienthal, Matthias and Claus Philipp, eds. *Schlingensiefs Ausländer raus: Bitte liebt Österreich*. Frankfurt a.M.: Suhrkamp, 2001.
Nelson, Robert S. and Richard Shiff, eds. *Critical Terms for Art History*. 2nd edn. Chicago: University of Chicago Press, 2003.
Phelan, Peggy. "On Seeing the Invisible." In *Live: Art and Performance*, edited by Adrian Heathfield, 16–27. London: Tate Publishing, 2004.
Roselt, Jens. "Big Brother: Zur Theatralität eines Fernsehereignisses." In *Schlingensiefs Ausländer raus: Bitte liebt Österreich*, edited by Matthias Lilienthal and Claus Philipp, 70–8. Frankfurt a.M.: Suhrkamp, 2000.
Seesslen, Georg. "Der Populist, der Provokateur, der fremde Blick und das eigene Bild." In *Schlingensiefs Ausländer raus: Bitte liebt Österreich*, edited by Matthias Lilienthal and Claus Philipp, 241–66. Frankfurt a.M.: Suhrkamp, 2000.
Stiles, Kristine. "Performance." In *Critical Terms for Art History*. 2nd edn, edited by Robert S. Nelson and Richard Shiff, 75–97. Chicago: University of Chicago Press, 2003.
Twycross, Meg. "The Theatricality of Medieval English Plays." In *The Cambridge Companion to Medieval English Theatre*, 2nd edn, edited by Richard Beadle and Alan J. Fletcher, 26–74. Cambridge: Cambridge University Press, 2008.
Van Eecke, Christophe. *Only Connect: Five Exercises in Aesthetics*. Breda: Lokaal 01, 2011.
Vergine, Lea. *Body Art and Performance: The Body as Language*. 1974. Milan: Skira, 2000.
Warr, Tracey, ed. *The Artist's Body*. London: Phaidon, 2000.

CHAPTER EIGHTEEN

Susanne K. Langer, Everyday Aesthetics, and Virtual Worlds

THOMAS LEDDY

Susanne Langer did not invent the idea of virtual reality, yet she inspired it by way of her development of the idea of "virtual world" in *Feeling and Form* (1953). David Chalmers, in a recent book, recognizes this, but defines a "virtual world" more narrowly than Langer as "an interactive and computer-generated space."[1] It was actually Jaron Lanier who coined the term "virtual reality" in the late 1980s. In the 1960s Ivan Sutherland of MIT had discussed a virtual world seen through a headset but, as Lanier puts it, the term "virtual world" came from Langer "who was using it as a way to think about modernist painting."[2] Lanier thought of "virtual reality" as meaning "moving beyond the headset experience to include some other elements" including your body as an avatar that could pick up things or be social with other avatars.[3] This, of course, is very different from what Langer refers to as a "virtual world." And yet, we might visualize her virtual worlds as somewhat like virtual realities.

The first virtual reality headset was created in 1968 by Sutherland and his student, Bob Sproull. However, the term "virtual reality" first appeared in Stanley Weinbaum's 1935 science-fiction story "Pygmalion's Spectacles."[4] Likewise, it is noteworthy that Antonin Artaud described the illusory nature of characters and objects in the theater as "la réalité virtuelle," and the 1958 English translation of his work marked the second published use of the term "virtual reality."[5]

Starting with *Feeling and Form*, Langer used "virtual" and "virtual world," but not "virtual reality." She used these terms in relation to several arts, including painting, sculpture, architecture, and dance. She spoke not only of virtual worlds but of virtual objects, space, kinetic volume, time, gesture, powers, and memory. Chalmers credits her with inventing the term "virtual object,"[6] although he sees virtual objects as real, whereas Langer sees them as illusion: for Chalmers, gamers who have a sense of being in a virtual space really are in that place.[7] However, in what follows I will support Langer's idea of illusion.

Unlike "virtual reality," no technology is required for actualization of a Langerian virtual world. She introduces the idea of "virtual world" after developing the more primary idea of "virtual." Her first use of it is with regard to architecture. For Langer, a virtual world in architecture is a matter of virtual space, what she calls an "ethnic domain," which she sees as being at the center of a virtual world. This is an illusory world, an illusion of "self-contained, self-sufficient, perceptual space."[8]

This illusion, I will maintain, can also animate everyday life phenomena. But this is only accomplished by way of seeing as an artist does. Within the current volume, Chapter 10 by Robert Innis is particularly helpful in understanding how this might work.[9] He emphasizes imaginative seeing as well as feeling, quoting from Langer, "It is perception molded by imagination that gives us the outward world we know … [and] by virtue of our thought and imagination we have not only feelings but a *life of feeling* [which is] a stream of tensions and resolutions,"[10] and it is within this stream that we live. Langer speaks of this imaginative seeing in terms of having the eye of the artist: "The artist's eye sees in nature, and even in human nature betraying itself in action, an inexhaustible wealth of tensions, rhythms, continuities and contrasts."[11] Further, "all sensitivity bears the stamp of mentality."[12] Innis concludes that "Langer considered even the primary world of senses to be primarily a domain of symbols, that is, experiential configurations with 'symbolic pregnancy,' a form of significance exhibited in the luring or repelling affective tones or felt qualities of objects and experiential situations."[13]

Moreover, unlike many philosophers of art, Langer stresses the creative process of the artist. It is in the studio, and in the artist's daily movements through the world, that we can see the transition from art to everyday aesthetics, and the possibility of applying the notion of "virtual" from one to the other. I will return to this later.

Here I need to explain my larger project and the direction from which I am coming to Langer's aesthetics. Since the 1990s, I have been working on providing theoretical underpinnings for a new subdiscipline within philosophical aesthetics on the par with the aesthetics of nature and the aesthetics of art. This new subdiscipline, which came in part out of environmental aesthetics, is called "everyday aesthetics." Everyday aesthetics deals with aesthetic properties and experiences in everyday life, for example in the home, the workplace, the walk in the neighborhood, and increasingly in virtual online environments, for example in social media contexts. The best and most current explanation of this new field is Yuriko Saito's *Stanford Encyclopedia of Philosophy* article.[14]

Pragmatist philosopher John Dewey has widely been considered the grandfather of everyday aesthetics,[15] and yet, my argument here is that Langer, who often saw herself as an opponent of Dewey, also provides important, untapped resources for developing this new field, particularly in her notion of "virtual worlds" and other related concepts. The thrust of my own approach to everyday aesthetics has been to stress the extraordinary in the ordinary, i.e., the way in which seeing the everyday aesthetically is modeled on seeing in the way that an artist sees their subject matter and their materials in the creative process. This has been in opposition to the dominant approach to the subdiscipline in which the ordinariness of the ordinary is stressed as also a deep disconnect between the aesthetics of art and aesthetics of everyday life.[16] I see Langer as a significant ally in this debate.

Langer may seem at first a poor fit for everyday aesthetics since she deliberately dissociates virtuality, so central to her aesthetic theory, from the everyday. Unlike Dewey, for whom there is continuity between art and everyday life,[17] Langer sees the virtual as discontinuous with everyday life. In this respect, she might be more like Clive Bell. Langer actually adopts Bell's idea of "significant form"[18] as her own when she writes, "'Significant Form' (which really has significance) is the essence of every art; it is what we mean by calling anything 'artistic.'"[19] Although Bell is traditionally thought to have denied any expressive function to art, Langer notes that Bell allows the possibility "that created form moves us so profoundly because it expresses the emotion of its creator."[20]

Langer believes both that art is autonomous and that the aesthetic is closely associated with art. It is thus not surprising that she is dismissive of everyday aesthetic phenomena. She considers music, for example, radically distinct from pleasurable non-musical sound. She thinks "somatic influences," e.g., the pleasures of sound, taste, and smell, are not part of the aesthetic and that exploiting these influences is more "self-indulgence" than "spiritual triumph."[21] She strictly separates "mere epicures" (of whom connoisseurs of fine food would be an example) from "artists" who, she believes, are "torchbearers of culture" and "inspired creators."[22] She holds that if the only function of music was to stimulate and soothe us, pleasing our ears as gourmet foods please our palates, it might be popular but "not culturally important."[23]

Yet although her dissociation of art from "mundane environment"[24] seems to disallow everyday aesthetics, she *expands* the domain of art to include such inhabitants of our everyday worlds as textiles and pots. Illusion (which, for her, is the goal of art) does not require representation: in a textile, a pot, or a sonata the "air of illusion ... exists as forcibly as in the most deceptive picture."[25] This expansion of the concept of illusion allows that *other* objects, not normally called "art," could have such an air. Other things can give the impression that an illusion enfolds them so as to give a feel of "detachment from actuality" and of "otherness."[26] She limits this feel to products of human arts. However, as I shall argue, one could have a sense of detachment and otherness in perception itself.

As mentioned above, Langer places a lot of emphasis on the creative process of the artist. But the art studio as place of creative work poses a problem for a strict separation between the world of art and the world of everyday things. Langer holds that the object of creation is an "image." And yet, an image is created out of ordinary things. Langer herself realizes a painting "is an image, created for the first time out of things that are not imaginal, but quite realistic—canvas or paper, and paints."[27] Something has to happen to canvas, paint, etc., so that they can be used to create an image. Perhaps it happens by the way the artist looks at them as he or she uses them in the creative process.

For Langer, the difference between image and reality is functional. She observes that "real objects, functioning in a way that is normal for images, may assume a purely imaginal status."[28] This is why works of art that are non-representational can have an illusory quality. But the idea can also be applied to *any* real object, for example a urinal in the eyes of Duchamp. Of course, Langer also says that "the true power of the image lies in the fact that it is an abstraction, a symbol, the bearer of an idea."[29] However, an object or scene not intentionally made to be the bearer of an idea can be an "image" in her sense if it functions as such.

It makes sense, then, to see the canvas and paper *seen by the artist* as having potential for creation and hence as part of a virtual space. An "image" can be created even out of things that have no meaning prior to use: *real objects (such as paint and canvas) may assume an imaginal status in the creative process*, a process that allows for the creation of a virtual reality.

So, what then is an image? Langer says that a building, for example, "becomes an image when it presents itself purely to our vision, i.e., as a sheer visual form instead of a locally and practically related object."[30] She asserts that "[w]e abstract its appearance from its material existence."[31] It then becomes "simply a thing of vision," is detached from its actual setting, and "acquires a different context."[32] Further, an image as "something that exists only for perception, abstracted from the physical and causal order" is the creation of an artist.[33] This all seems to take it very much away from the everyday. And yet, a building, for example a church or a post office, is something people experience every day.

For Langer, things of everyday life can be perceived either practically or as illusion. When the second occurs, it is *as though* they are art. One could even say (although Langer would not) that in seeing the building this way its practical relations are themselves transformed and *carried up into* the aesthetic realm. Material existence is never completely distinct from imaginative experience since material existence can be imaginatively perceived and worked. For Langer, the "image" "arises from the process of arranging colors on a surface." It is "created, not just gathered and set in new order."[34] When it emerges, canvas and paint seem to cease to exist, or at least become difficult to perceive.[35]

Langer's idea of "semblance" (related to Schiller's "*Schein*") also poses a problem for a strict separation of art from everyday life. She speaks of liberating perception and conception from practical purposes so that the mind can "dwell on the sheer appearances of things."[36] One can focus on the mere appearance of the chair or telephone as long as the practical is bracketed. Hence, the aesthetic is not limited to the realm of art. A vase can, like fine art, seem to be given to one sense alone.[37] A concentration on appearances can also give a sense of illusion. Thus, the "unreality" of art "tinges even perfectly real objects like pots, textiles, and temples."[38] Further, being surrounded by good visual forms in the everyday is good training for perception of great pictures.[39] Good taste in everyday life is continuous with (and dynamically interactive with) good taste in art. Bringing in the emotions of life, and not just a special aesthetic emotion, which is the only emotion Bell allows for art, also softens the distinction between art and everyday life.

Langer informs us that a good work of art clarifies and presents feelings proper to forms and colors the painter has seen.[40] A Langerian everyday aesthetics is possible, then, because one can do to everyday things what one can do in making art, i.e., one can "estrange it from actuality." And one can do this by creating "a realm of illusion, in which it functions as *Schein*."[41]

Langer wants an uncoupling of art from practical life [42] whereas Dewey wants art and life brought together again.[43] Yet, when talking about decoration, Langer sounds very much like Dewey. She believes that craftsmanship can be art and, when it is, the elements of abstraction, plastic freedom and expressiveness are present. Thus, for her, as for Dewey, there is no legitimate distinction between higher and lower arts. All involve "creation of forms expressive of human feeling."[44] She observes that pure design, which is found throughout the world, is not to be taken lightly, since it is associated with fitness and formalization. Good decoration makes the surface more visible and unifies it.[45] Several of the black and white plates in *Feeling and Form* are used to illustrate this and related points, for instance an image of the face of a Maori man with tattoos.[46]

In support of the idea that elementary forms of decoration are based instinctively on principles of perception, and that organization of the visual field comes directly to expression, Langer quotes, surprisingly (given her earlier attacks on Dewey), Albert Barnes, a close friend and ally of Dewey. Barnes sees the appeal of decorative beauty as satisfying a need to perceive freely and agreeably. However, unlike Barnes, she insists that decoration has "emotional import" and that, like all created form, its function is to "impregnate" and transform perception.[47] In support of this she refers to such diverse things as Chinese embroideries and Mexican pots.[48]

It may seem, contra Dewey, that Langer denies continuity between art and life since, as she says, virtual space is "a self-contained, total system."[49] Indeed, she directly attacks his position not by name, but through a footnote that refers back to her first criticism of him. Langer rejects the idea of a similarity between perception of things as a painter sees

FIGURE 18.1 Wiremu Pātara Te Tuhi. Ref. 1/2-051849-F. Alexander Turnbull Library, Wellington, New Zealand. /records/22361561. Dated around 1880. Original photograph of plate III in Susanne K. Langer's *Feeling and Form* (1953) with the caption, "The immediate effect of good decoration is to make the surface, somehow, more *visible*."

them and perception in practical life. She holds that, for the pragmatist (such as Dewey) "'Creation' becomes a somewhat pretentious word to apply to the modifications an artist may make in the appearance of things by selection and emphasis."[50] And yet, she believes certain artists can see the actual world in an art-like way, which does seem to be a matter of selection and emphasis. She says that, for Cézanne, "the transformation of natural objects into pictorial elements took place *in his seeing*," and this is why he thought he painted exactly what was there. He attributed to the object seen the properties he found in virtual space.[51] In a sense, he lived in virtual space.

This is an important concession to everyday aesthetics, since, if Cézanne could do this on a grand scale, it could happen in a much more limited way for *everyone*. Just as artists and art-writers can speak of the art object as living,[52] things in the real world can be animated in perception. Painting and its effect on us must be based on some more fundamental pre-perceptual activity in which we all participate. Cézanne would not have his effect on us if we did not also have this capacity in some way.

Similarly, one could argue that aesthetic perception of nature and of the phenomena of everyday life is also a form of animation; i.e., seeing a non-living thing *as if* alive, or seeing some living thing *as* more alive. The piece of trash, seen aesthetically on the side of the road, takes on life—becomes *as if* a living thing. The ability to see the world in this way makes art possible.

Langer does not fail to see this when it happens in the studio. She speaks of it as "the process of transforming the actual datum, the canvas, or paper surface, into a virtual space, creating the primary illusion of artistic vision."[53] But this can also be done when the actual datum is a scene taken in during a walk. Moreover, Van Gogh's painted chair will seem alive, as Langer observes,[54] *because* Van Gogh can see the actual chair as animated: transforming that experience into the medium of paint intensifies the animation. Pottery, too, can be "living"[55] in the sense of expressing "living form."[56]

Admittedly, there are discontinuities as well as continuities between art and life. The virtual objects and spaces created in perception in everyday aesthetic experience are not communicable directly. They lack both the intensity and the semantic and syntactic density—to use Nelson Goodman's language in *Languages of Art* (1976)—of the virtual worlds of art. Whereas Langer tells us that "[t]he primary illusion of virtual space comes at the first stroke of brush or pencil that concentrates the mind entirely on the picture plane and neutralizes the actual limits of vision,"[57] nothing like that happens in the aesthetics of everyday life.

But Langer is wrong when she says that artists and art lovers have no need to cultivate the "aesthetic attitude." She sees this attitude as merely a selecting of sense data from the actual world and contemplating them as qualitative. However, the aesthetic attitude is better seen as the attitude taken when looking at the world in the way an artist does, for example, the attitude taken by Cézanne as he sees ordinary things animated, becoming "virtual" before his eyes.[58]

A similar issue is raised with regard to materials observed in the studio during the creative process. Langer distinguishes elements, as factors in the illusion, and therefore "virtual" themselves, from materials, which are merely actual. As she puts it, "Paints are materials, and so are the colors they have in the tube or on the palette; but the colors in a picture are elements determined by their environment." As such, they can create tensions in the picture. Colors in the paintbox are materials that, as she avers, "lie side by side in their actual, undialectical materialism."[59] Yet paints are not just handled by the artist: they are also seen. And they can be seen with or without the artist's vision. If they were just dead material things in the artist's eyes, they would never be able to take on the role of elements in the painting. The process is not one of moving directly from dead matter to living form but of taking something already animated and concentrating and intensifying it in the creative process. Contra Langer, this gradual transformation of paints into painterly elements probably is, actually, dialectical.

Langer is right that the painter is not just taking and arranging elements, that she is creating a virtual space, and even that she brackets from the realm of the practical. But this does not mean that art is, in the manner of Bell, completely isolated from life. Bell argues that art has nothing to do with our everyday emotions: it only has to do with that special aesthetic experience that we get from appropriately apprehending something with "significant form." For Langer, art expresses the emotions of life, not just the aesthetic emotion. To be sure, it does not express the particular emotion of the particular artist. Rather, it expresses the forms of sentience as the composer or artist understands them. [60] So the composer is not interested in expressing his/her own feelings but rather what he/she "knows about" the inner life of people in a world of things. The idea is to make a statement about human sensibility through art as a symbol. But this means that there is a relation between art and life, which Bell would deny. This is why Langer can quote from Matisse, with approval, something that Bell could never have sanctioned: "When I see the Giotto frescoes at Padua I do not trouble to recognize which scene of the life of Christ I have before me, but I perceive instantly the sentiment which radiates from it and which is instinct in the composition in every line and color. The title will only serve to confirm my impression."[61] For a formalist like Bell the title can only be a distraction.

Langer, in discussing individual art forms, continues to undercut the radical discontinuity between art and life that she elsewhere maintains. As I have observed, she already begins to move away from the formalists simply by insisting that art has to do with the emotions of life, and not just with a rarefied "aesthetic emotion." Borrowing from Bell, she holds that music, and art generally, is "significant form." However, for her, the significance of music (and art) "is that of a symbol, a highly articulated sensuous object, which by its dynamic structure can express the forms of vital experience which language is peculiarly unfit to convey."[62] So "significant form" is not here tied to the special "ecstatic" aesthetic experience in the way it is for Bell. Bell says that this experience is like that of a mathematical discovery, or like religious experience.[63] It is transcendent in a Platonic-like way. But, for Langer, significant form is tied to the actual structures of our emotions. And, as I have argued, these may also be manifested in everyday perception.

We have seen how Langer's discussions of painting and decoration have sometimes opened up paths for everyday aesthetics. But, so far, our emphasis has been on image and form. It is, however, through the concept of a virtual world that a true reconciliation between art and everyday life can be found. We find this first in her discussion of painting and sculpture and then, even more, when she discusses architecture, where she re-establishes continuity between art and life at least on one level.

The illusion created in visual art is a "virtual scene" a "space opposite the eye and related directly and essentially to the eye."[64] This is contrasted to what is before us in the actual world. And yet the actual world can be *taken as* a virtual scene. When reality is taken as virtual it glows with what Langer, at one point, calls "aura." This happens when she speaks of a vase or a building as having an "aura of illusion" based on its detachment from actuality.[65] To paint a scene one has to see it as a virtual scene, as symbolic. The person who sees the world with the artist's eyes, I am arguing, similarly, sees it, or aspects of it—or just things in it—as virtual.

Sculpture, as with architecture and painting, creates what Langer calls virtual space. However, this space does not stop at the physical boundaries of the art object. The volume created "is more than the bulk of the figure." First, it is a space made visible. Second, its area is greater than what the figure takes up. In emphasizing the negative spaces and the spaces around the sculpture, Langer expands the virtual space that she otherwise seeks to keep separate from everyday life. For instance, she says that "[t]he figure itself seems to have a sort of continuity with the emptiness around it, however much its solid masses may assert themselves as such." The space itself has a "vital form" that continues that of the figure.[66]

The space around the sculpture is brought into the world of the work. It can be constituted as physical space and as virtual space. Langer sees physical space as discursive, logical, scientific, and *not* the space in which we live. By contrast, virtual space is phenomenological. And what is of importance here is not the relation between physical and virtual space but the dynamic relation between *two* phenomenological spaces: the space we actually inhabit and the virtual space of the work.

Langer could argue that the sculpture does not really move into the realm of the everyday even though it goes beyond the physical object. The empty space it "commands" is part of the sculptural volume.[67] Yet her analysis of art places it closer to life as we live it than Bell's in that she is committed to "significant form" as having to do with the actual structure of our emotions, which are tied to our lives.

Further, although Langer tries to keep life and art radically separate, the value of art, in her view, is that it reflects or expresses something about us as live organisms. She observes, like Dewey, that living organisms maintain themselves, resist change and seek to preserve their structure.[68] They must have certain forms or cease to exist. Following Aristotle, she stresses that life has "telos": an acorn strives to become an oak. Although a sculpture is not actually organic it gives us "semblance of living form."[69] Here is another connection between art and life that would allow for an aesthetics of everyday life: they both exhibit telos.

Sculpture, for Langer, is "virtual kinetic volume, created by—and with—the semblance of living form." When sculpture makes "tactual space visible"[70] it enters even more intimately into contact with life. The "semblance of kinetic volume" which is sculpture overlays actual volumes in lived experience, dominating surrounding space. Langer quotes the German-American patron of the arts Bruno Adriani sympathetically on the similarities between sculptural space and the construction of the world by the self. And yet the sculptor creates sculptural space to symbolize the universe. Thus, unlike Adriani, she holds that we do not identify the space which centers in the statue with our own environment. Sculptural space is autonomous, and this autonomy allows the created world both to be objective and to be an image of our surrounding space. Thus, on her view, the sculpture is at the center of its own space, and we of ours. It remains possible, however, that our own "restricted space" can be virtual as well.[71]

Of architecture, Langer says it "creates the semblance of that World which is the counterpart of Self," and that "[i]t is the total environment made visible."[72] However, in this case, as opposed to sculpture, it is not clear how the "environment made visible" is to be distinguished from the actual environment. The introduction of "Self" may indicate a phenomenological point of view. We are talking about the world as perceived by a self. And, as she observes, in the case of architecture, the Self is collective, which means that we perceive the world according to certain shared worldviews: the World is communal.[73] But if this is so then the World really is the world of everyday experience. Thus, continuity is re-established. The only world that is cut off is the world understood not phenomenologically but discursively, i.e., the world as understood by science.

Langer also insists that "the actual environment of a being is a system of functional relations," and therefore a virtual environment "is a symbol of functional existence."[74] Yet is the phenomenological world in which we live and act itself a virtual world? And is not the world surrounding each work of architecture, basically most of the world in which city and town-dwellers live, one that is infused with the spirit of architecture? But if that is true then "the everyday," if only in cities and other architectural places, is a collection of virtual spaces constituted by various architectural entities.

Is Langer right? Answering this question is difficult for an avowed Deweyan like myself. Langer is clearly saying some of this in opposition to Dewey. She plainly wants to overcome the idea of continuity between art and everyday life. I think, however, that Deweyans could learn from Langerians and vice versa. I agree that art creates a world of illusion or, rather, each artwork creates its own illusion. Speaking of architecture as creating virtual space seems to help. So too with sculpture. In architecture, you are both walking in a real space and also transported into another space, for example when walking through Frank Lloyd Wright's *Fallingwater* (1935–9). Dewey speaks of refinement and intensification of experience, and he, like Langer, is very aware of how visual art both excludes all of the other senses, and also incorporates them indirectly, so that in seeing a painting of oranges you can sometimes almost smell the oranges.[75] One refines and intensifies ordinary experiences by way of creating a virtual reality.

CONCLUSION

In this chapter I have argued that reading Langer provides resources towards the construction of an aesthetics of everyday life. My own approach to aesthetics is essentially Deweyan, and thus I stress continuities over discontinuities between art and life. Langer's opposition to Dewey might predict failure in this. However, there are several points in her analysis that soften the separation she seems to set up between art and life. These include (1) her expansion of the concept of art to include items of everyday life as virtual images; (2) her understanding of art in terms of the creative process, in which the artist must take certain objects in the actual world as virtual; (3) her recognition that the artist needs to find symbolism (and emotional meaning) in the world he or she seeks to portray; (4) her understanding of sculpture as expanding virtual reality into the surrounding space; and (5) her understanding of architecture as setting up an overlay between the actual and the virtual worlds. With the rise of digital "virtual reality," inspired in part by Langer's notion of "virtual world," we live increasingly in surroundings that offer their own bifurcation, i.e., between screen world and actual world, the world in which we do all the things we did before the rise of screens and of that related space seen through technological goggles as "virtual reality." From a Deweyan standpoint, this bifurcation too needs to be overcome as that between art and life, so that the analogue virtual worlds so important to us are not drained of their significance.

NOTES

1. David J. Chalmers, *Reality+: Virtual Worlds and the Problems of Philosophy* (New York: W. W. Norton, 2022), 191.
2. Jaron Lanier, *Dawn of the New Everything: Encounters with Reality and Virtual Reality* (New York: Henry Holt, 2017), 42.
3. Ian Evenden, "The History of Virtual Reality," *Science Focus*, March 17, 2016.
4. Daniel Cooke, "When Was Virtual Reality Invented?" *Pebble Studios: Blog*, August 17, 2017.
5. Antonin Artaud, *The Theater and its Double* (New York: Grove Press, 1958), 49.
6. Chalmers, *Reality+*, 192.
7. Ibid., 205.
8. Susanne K. Langer, *Feeling and Form: A Theory of Art Developed from Philosophy in a New Key* (New York: Charles Scribner's Sons, 1953), 95.
9. See Robert Innis's Chapter 10 in this volume, "Psychological Dimensions, Cultural Consequences, and their Breakings in Susanne K. Langer's Symbolic Mind."
10. Innis, this volume Chapter 10, 150; Langer, *Feeling and Form*, 372.
11. Langer, *Feeling and Form*, 372.
12. Langer, *Philosophy in a New Key: A Study in the Symbolism of Reason, Rite, and Art* (Cambridge, MA: Harvard University Press, 1942), 90.
13. Innis, this volume, Chapter 10, 152.
14. Yuriko Saito, "Aesthetics of the Everyday," *The Stanford Encyclopedia of Philosophy*, ed. Edward N. Zalta (2021). See also Saito, *Everyday Aesthetics* (Oxford: Oxford University

Press, 2007); Thomas Leddy, *The Extraordinary in the Ordinary: The Aesthetics of Everyday Life* (Peterborough, ON: Broadview, 2012); Thomas Leddy, "Resolving the Tension of Everyday Aesthetics in a Deweyan Way," in *American Aesthetics Today: Theory and Practice*, ed. Walter Gulick and Gary Slater (Albany, NY: SUNY Press, 2020).

15. See John Dewey, *Art as Experience* (New York: Minton, Balch and Company, 1934).
16. Cf. Saito, *Everyday Aesthetics.*
17. Dewey, *Art as Experience*, 3.
18. See Clive Bell, *Art* (New York: Capricorn Books, 1958).
19. Langer, *Feeling and Form*, 24.
20. Ibid.
21. Ibid., 28.
22. Ibid.
23. Ibid.
24. Ibid., 45.
25. Ibid.
26. Ibid., 46.
27. Ibid.
28. Ibid., 47.
29. Ibid., 27.
30. Ibid., 47.
31. Ibid.
32. Ibid.
33. Cf. ibid.
34. Ibid.
35. Cf. ibid., 48.
36. Ibid., 49.
37. Cf. ibid.
38. Ibid., 50.
39. Cf. ibid., 53.
40. Cf. ibid., 58.
41. Ibid., 59.
42. Ibid., 60.
43. Cf. John Dewey, *Art as Experience* (New York: Minton, Balch and Company, 1934), 81.
44. Langer, *Feeling and Form*, 60.
45. Ibid., 61.
46. Langer, *Feeling and Form*, plate III.
47. Ibid., 62.
48. Cf. ibid., 63.
49. Ibid., 75.
50. Ibid., 77.
51. Cf. ibid., 78.
52. Ibid., 79.
53. Ibid., 80.
54. Cf. ibid., 81.
55. Ibid., 82.
56. Ibid., 66.
57. Ibid., 84.
58. Cf. ibid.

59. Ibid., 84–5.
60. Ibid., 28.
61. Here, Langer quotes Matisse in *Feeling and Form*, 83. Strangely, Langer does not give a citation, but the quote is found in Henri Matisse, "Notes of a Painter," in *The Problems of Aesthetics: A Book of Readings*, ed. Eliseo Vivas and Murray Krieger (New York: Rinehart, 1953), 260.
62. Langer, *Feeling and Form*, 32.
63. Cf. Bell, *Art*, 27, 54.
64. Langer, *Feeling and Form*, 86.
65. Ibid., 46. Langer only uses the term "aura" once in *Feeling and Form*. I did not know this when I featured the same term in virtually the same sense in my book *The Extraordinary in the Ordinary*, ch. 4.
66. Langer, *Feeling and Form*, 88.
67. Ibid.
68. Cf. Ibid.
69. Ibid., 89.
70. Ibid., 89–90.
71. Cf. ibid., 91–2.
72. Ibid., 98.
73. Ibid.
74. Ibid., 91.
75. Cf. Dewey, *Art as Experience*, 125.

BIBLIOGRAPHY

Artaud, Antonin. *The Theater and its Double*. Translated by Mary C. Richard. New York: Grove Press, 1958.

Bell, Clive. *Art*. 1914. New York: Capricorn Books, 1958.

Chalmers, David J. *Reality+: Virtual Worlds and the Problems of Philosophy*. New York: W. W. Norton, 2022.

Cooke, Daniel. "When Was Virtual Reality Invented?" *Pebble Studios: Blog*, August 17, 2017. Available online: https://www.pebblestudios.co.uk/2017/08/17/when-was-virtual-reality-invented/ (accessed August 15, 2022).

Dewey, John. *Art as Experience*. New York: Minton, Balch and Company, 1934.

Evenden, Ian. "The History of Virtual Reality." *Science Focus*, March 17, 2016. Available online: https://www.sciencefocus.com/future-technology/the-history-of-virtual-reality/ (accessed December 28, 2020).

Goodman, Nelson. *Languages of Art: An Approach to a Theory of Symbols*. Indianapolis: Hackett Publishing Co., 1976.

Lanier, Jaron. *Dawn of the New Everything: Encounters with Reality and Virtual Reality*. New York: Henry Holt, 2017.

Langer, Susanne K. *Feeling and Form: Feeling and Form: A Theory of Art Developed from Philosophy in a New Key*. New York: Charles Scribner's Sons, 1953.

Langer, Susanne K. *Philosophy in a New Key: A Study in the Symbolism of Reason, Rite, and Art*. Cambridge, MA: Harvard University Press, 1942.

Leddy, Thomas. "Dewey's Aesthetics." *The Stanford Encyclopedia of Philosophy*, Summer 2020 edn. Edited by Edward N. Zalta. Stanford, CA: University of Stanford.

Available online: https://plato.stanford.edu/archives/sum2020/entries/dewey-aesthetics/ (accessed October 30, 2021).

Leddy, Thomas. *The Extraordinary in the Ordinary: The Aesthetics of Everyday Life*. Peterborough, ON: Broadview, 2012.

Leddy, Thomas. "Resolving the Tension of Everyday Aesthetics in a Deweyan Way." In *American Aesthetics Today: Theory and Practice*, edited by Walter Gulick and Gary Slater, 111–34. Albany, NY: SUNY Press, 2020.

Matisse, Henri. "Notes of a Painter." In *The Problems of Aesthetics: A Book of Readings*, edited by Eliseo Vivas and Murray Krieger, translated by Alfred H. Barr, 255–61. New York: Rinehart, 1953. Available online: https://archive.org/details/problemofaesthet0000unse/ (accessed October 28, 2021).

Saito, Yuriko. "Aesthetics of the Everyday." In *The Stanford Encyclopedia of Philosophy*, edited by Edward N. Zalta, 2021. Available online: https://plato.stanford.edu/archives/spr2021/entries/aesthetics-of-everyday/ (accessed October 30, 2021).

Saito, Yuriko. *Everyday Aesthetics*. Oxford: Oxford University Press, 2007.

EPILOGUE

"That she cannot be catalogued": Tracing Susanne K. Langer's *Fortuna*

CAROLYN BERGONZO

In addition to a half dozen digital humanities projects and editorial initiatives intent on rectifying the lack of diversity in the Western philosophical canon, two group biographies—within a year—on Elizabeth Anscombe, Philippa Foot, Mary Midgley, and Iris Murdoch, new books on Hannah Arendt and Simone De Beauvoir, and reparative compendia like *The Philosopher Queens: The Lives and Legacies of Philosophy's Unsung Women* (London: Unbound, 2020) signal a return to the pioneering feminist scholarship of the 1980s and 1990s that challenged women's exclusion from standard histories of philosophy and wrote more inclusive histories into the discipline.[1] Among these recent efforts, projects like this handbook make the case for a thinker's capaciousness and complexity by employing her ideas in the present towards creative, occasionally unexpected ends. Not only is Langer, as a prolific American philosopher who cut a bold path through the twentieth century, an historically significant figure; her philosophy, for a variety of thinkers, artists, and writers, remains patently alive.

Set against what can seem like overly simplified narratives of exclusion, feminist histories of philosophy that chronicle the reception and transformation of a philosopher's ideas and standing over time can offer a robust accounting of a life in thought and its multiple afterlives. Sarah Hutton gives us a useful term for this approach, that of tracing a philosopher's *fortuna*—her fortune and misfortune, rediscoveries and reappraisals of her work, and the myriad private or public provocations it elicits decades after its creation.[2] By documenting these shifting contexts of reception, historians of the discipline can avoid tendencies to pull forward only those threads that align with current perspectives or to project present values into the past. Such an approach also embraces the admixture of institutional impasse and access, prejudice, and personality that converge with the ever-shifting ground of philosophical context to determine a figure's reception and legacy. If the publication of this handbook, following Adrienne Dengerink Chaplin's *The Philosophy of Susanne Langer: Embodied Meaning in Logic, Art and Feeling* (London: Bloomsbury, 2019) and the founding of the Susanne K. Langer Circle, mark a resurgence in Langer scholarship, one wonders: what has brought us here? Why now? What is next?

In other words, what of Susanne K. Langer's *fortuna*? Langer does not fit into an easily plotted narrative in the history of philosophy; her *not fitting* is precisely what makes her a restlessly provocative figure in the present. To recruit Langer to a single school of philosophy risks narrowing her to the point of unrecognizability. Perhaps she can be said to belong only to a methodological type: a systematic philosopher who expanded the viewing frame of the discipline during a century of narrowing adherence to logical analysis in order to illuminate those murkier, purportedly irrational, realms of human activity: art, ritual, myth, dreams. Through rather conventional means of philosophical inquiry, Langer pursued a radical project. Working and re-working generative concepts from late nineteenth- and early twentieth-century developments in a range of disciplines, Langer assembled a theory that intertwined more aspects of human experience than discursive thought alone. In this pursuit, she toppled the perceived hierarchy of rationality into a plural horizon of semantics under the new key of human cognition: symbolization. Philosophers and historians alike have documented how Langer's theory faced critiques of haziness by fellow philosophers who saw her as taking logic too far afield from its mathematical roots. I am not alone in suggesting that this very wandering, by which Langer turned toward "the whole creative process of ideation, metaphor, and abstraction that makes human life an adventure in understanding," is precisely what gives her work such resonance in the present.

Langer's expansive theory, originally presented in *Philosophy in a New Key* in 1942, met with remarkable success among a general readership, especially after its release as a 35 cent Mentor paperback in 1948. The developmental psychologist Howard Gardner has detailed the formative impact the book had on him and the generation of students who encountered Langer's slim, curious paperback in the 1960s.[3] Presenting a theory of symbolic transformation in which language was only one vehicle for the construction and conveyance of meaning, Langer admitted more dimensions of human experience into the sanctum of philosophy than the dominant logic would allow. This opening-out struck a chord with readers. The text was translated into many languages and cropped up in course readings and anthologies throughout the 1970s. In a 2020 dossier on "Undead Texts" in *Public Culture*, historian Joel Isaac traces *Philosophy in a New Key*'s "unusual trajectory through postwar intellectual culture" and offers a persuasive account of Langer's underacknowledged influence on the work of ethnologist Clifford Geertz.[4] Taken together, Gardner and Isaac's histories of the book's reception give us a fuller understanding of Langer's influence. Both Gardner and Isaac wonder at Langer's exclusion from the philosophical canon given the popularity of *Philosophy in a New Key* and the pervasive influence of the new key she synthesized for the lay-reader and philosopher alike. To echo Isaac's rallying cry regarding Langer's under-credited influence on Geertz: few outside of a devoted circle of Langer enthusiasts have made much of this exclusion; I think more of us should.

"That she cannot be catalogued," Gardner suggests, "may explain why she has escaped certain honors." Another possible explanation for her exclusion is, of course, her gender, which—both Gardner and Isaac address. For her popularity and for her unwavering self-conception as a philosopher, one might expect Langer to loom large in feminist histories of twentieth-century American philosophy. However, a large part of feminist scholarship in the history of philosophy has focused, importantly, on surfacing the contributions of early-modern women, whose theories—previously ignored or embedded in works of fiction or correspondence—are not as ready-to-hand as Langer's mass-produced paperback. Even as scholars look closer afield to figures from the twentieth century, Langer remains largely absent. Mary Ellen Waithe excluded Langer from the fourth volume of her *History of Women Philosophers* (1900–present) on the grounds that Langer's work, ten years after her death, was still readily accessible.[5] It is notable, by this reasoning, that prominent twentieth-century figures like Hannah Arendt, Simone de Beauvoir, and Simone Weil are profiled in the volume. Additionally, Langer is not mentioned in Sophia Connell's recent article, "The Lost Women of Early Analytic Philosophy," likely because Langer is recognized, if at all, for her philosophy of art rather than for her early contribution to logic. Fortunately, scholars like Sander Verhaegh and Jeanne Peijnenburg are resurfacing Langer's engagement with the logical empiricists and her contribution to the development of symbolic logic in the States in the 1920s and 1930s.[6]

Langer's own views on gender complicate efforts to claim her as a feminist trailblazer in a male-dominated discipline. Often referred to as the first "academic woman philosopher" in the US, Langer had a contested relationship both with the academy and with the "woman" qualifier. She held a series of temporary teaching appointments at institutions across the country, only landing a permanent position at Connecticut College, then a women's college, in 1954, at the age of fifty-eight. On its face, this might seem clear evidence of institutional exclusion. However, Dengerink Chaplin, quoting Donald Dryden, suggests that Langer might have deliberately sought temporary or part-time teaching appointments so that she could prioritize her own research and writing, which were supported by grants from the Rockefeller Foundation and the Edgar J. Kaufmann Charitable Trust in the decades after *Philosophy in a New Key*'s publication. In order to better understand Langer's relationship with the academy, we need a fuller accounting both of her experiences at, and between, institutions, and of her designs on how to build a fulfilling life in thought.

What comes across clearly to Langer scholars is that she disliked being described as a "woman philosopher." She turned down invitations to speak about the status of women or to be included in anthologies of texts written by women. As she wrote to *The World Who's Who of Women* in 1976, rejecting the offer to be included in what she called their "feminist inventory": "I am a scholar, not a 'woman scholar.'"[7] In a letter at around the same time to historian Bruce Kuklick, she conveyed the same disinterest in being included in any list "cataloguing the achievements of women as women" (emphasis hers).[8] Kuklick went on to include Langer in the final pages of the final appendix ("Women Philosophers at Harvard") of his book *The Rise of American Philosophy: Cambridge, Massachusetts, 1860–1930*. At the same time as Langer resisted inclusion in feminist compendia, her archives reveal their fair share of sexism, with well-positioned Harvard philosophers—Langer's own teachers—begrudgingly acknowledging her brilliance. Most familiar among those who study Langer is William Ernest Hocking's backhanded endorsement of the *Philosophy in a New Key* manuscript: "I am prejudiced against books on philosophy by women; according to this prejudice no woman could write a book as good as she has written."[9] Less well known is a letter by Ralph Barton Perry to the President of Radcliffe College in which he praises Langer as among the greatest philosophers of the time, then goes on to critique her personality. According to Perry, Langer is insufficiently warm and too self-asserting. He ascribes the latter to her precarious employment. (Fortunately, her divorce from Harvard historian William L. Langer in 1942, according to Perry, left her mentally unscathed.)[10] The letter is emblematic of gender stereotypes in the academy, where a woman's intellectual tenacity, prioritization of her own work, and refusal to be first and foremost a pleasant social agent are marks against her, while the same characteristics are endured, if not championed, in men. Perry himself mounts ample counterevidence of Langer's skills as an instructor—well liked by her tutees, a brilliant but occasionally inscrutable lecturer—which show his critique of her personality to be all the more irrelevant. Although it is unclear whether Langer was aware of these specific sentiments, the prejudices they convey certainly impacted Langer's experience of academia have narrowed the professional opportunities available to her.

Langer's desire to be recognized first and foremost as a philosopher, and not as a woman, is a familiar refrain—if seemingly remarkable for its appearance amidst the robust feminist activism of the 1970s—among women writers and thinkers who did not want to be reduced to their demographic. In a 1975 letter to then-president of Radcliffe College Matina Horner, Langer expresses her hope that Radcliffe would continue to support women's scholarship across disciplines, rather than primarily support feminist-oriented research. She reflects on her life as an early mother finishing her doctoral work at Radcliffe and her commitment, during that time, to pursuing her own intellectual project. She then arrives at what strikes me as a perplexing calculus—if she had spent her time fighting for her right to pursue philosophy, she explains, then she would not have had the time or resources to do the work of philosophy. This is perplexing to me because if there had not been activists a generation earlier fighting for her right to pursue an education, then her formal pursuit of philosophy from within the College would not have been viable. In her resistance to the idea that the merit of one's work could be overshadowed by one's identity, Langer projects a misplaced optimism.

Yet with her unwavering self-conception as a philosopher, her belief in the significance of her own work, and her independent spirit, Langer can certainly be described as a feminist force. In the last decades of her life, as her health deteriorated and her efforts to finish her three-volume *Mind: An Essay on Human Feeling* intensified, she was a paragon of refusal, turning down invitations to speak or review others' work in order to prioritize her own writing. What is more, as Iris van der Tuin notes in her chapter, Langer's work surfaces a "universe of women." Whereas Langer's significant male influences have been studied at length—most comprehensively in Dengerink Chaplin's book—not much has been made of the matrix of women cited in Langer's index cards and published volumes, especially *Mind*. Among these are philosopher of aesthetics Ivy G. Campbell-Fisher, zoologist Libbie Hyman, founder of paleoneurology Tilly Edinger, spiritualist and scholar of Tibet Alexandra David-Néel, and philosopher Pepita Haezrahi. At a key moment in volume 1 of *Mind*, Langer makes a special recognition of Ivy Campbell-Fisher's papers on the complex logic of art, as "the only work that I know which centers on this topic, and to which I can subscribe almost without reservation."[11] If one examines Langer's life, one learns that her closest friends and thought-partners were predominantly women, among them: her older sister Ilse Knauth Dunbar; her childhood friend, the artist and illustrator, Helen Sewell; Anna ("Nancy") Ward Perkins, a medical doctor in rural New York to whose house Langer would cart her index cards; and colleagues at Connecticut College—Rosemary Parks, Berenice Wheeler, Dorothy Richardson. Our record of these women and of the exchanges they shared with Langer is scant, yet worth our attention.[12]

To trace Langer's *fortuna*, we need to follow the threads of her philosophy as they intertwine beyond disciplinary frames, even now. We need to be perplexed by the story of her career and her story of feminism to create more intricate narratives of women's lives in thought. We deserve our histories to be as complex as the figures they seek to recover. If Langer resists the narrative that I would have for her, if she complicates my affinity with or even affection for her, then she can exist as something more—as an expansive and unfixable intellectual figure; and as a person, ineluctably real.

NOTES

1. Digital humanities projects and editorial initiatives include Project Vox, Extending New Narratives, In Parenthesis, History of Women Philosophers and Scientists, and Oxford New Histories of Philosophy. The two Somerville quartet biographies are *Metaphysical Animals: How Four Women Brought Philosophy Back to Life* (New York: Knopf Doubleday, 2022), and *The Women Are Up to Something* (Oxford: Oxford University Press, 2021).
2. Sarah Hutton, "'Context' and 'Fortuna' in the History of Women Philosophers: A Diachronic Perspective," *Methodological Reflections on Women's Contribution and Influence in the History of Philosophy* (Cham: Springer, 2020).
3. Howard Gardner, "*Philosophy in a New Key* Revisited: An Appreciation of Susanne Langer," *Art, Mind, and Brain: A Cognitive Approach to Creativity* (New York: Basic Books, 1982).
4. Joel Isaac, "Susanne K. Langer, *Philosophy in a New Key* (1942)," *Public Culture* 32, no. 2 (91) (May 1 2020): 355–61.
5. Mary Ellen Waithe, *A History of Women Philosophers*, vol. 4: *Contemporary Women Philosophers, 1900–Today* (Dordrecht/Boston: Kluwer Academic Publishers, 1995).
6. See Verhaegh's article in this handbook as well as "Susanne Langer and the American Development of Analytic Philosophy," in *Women in the History of Analytic Philosophy* (Cham: Springer, 2022) and Jeanne Peijnenburg and Sander Verhaegh's "Analytic Women | Aeon Essays," Aeon. August 1, 2023. Available online: https://aeon.co/essays/the-lost-women-of-early-analytic-philosophy (accessed 19 August 2023).
7. Susanne Langer to *The World Who's Who of Women*, November 1, 1976, Susanne Langer Papers, 1895–1985 (MS Am 3110), Correspondence, Box 19, Houghton Library, Harvard University, Cambridge, MA.
8. Susanne Langer to Bruce Kuklick, March 16 1973, Susanne Langer Papers, 1895–1985 (MS Am 3110), Correspondence, Box 2, Houghton Library, Harvard University, Cambridge, MA.
9. Max Hall, *Harvard University Press: A History* (Cambridge, MA: Harvard University Press, 1986), 80. Cited in Dengerink Chaplin's *The Philosophy of Susanne K. Langer* (London: Bloomsbury, 2019), 47.
10. Ralph Barton Perry to Bernice Cronkhite, December 22, 1953, Radcliffe College Student Files, RG XXI, Series I, Susanne Knauth Langer, Identifier: Box 70, Schlesinger Library, Radcliffe Institute, Harvard University, Cambridge, MA.
11. Susanne K. Langer, *Mind: An Essay on Human Feeling*, vol. 1 (Baltimore: Johns Hopkins University Press, 1967), 85.
12. Gerard A. Finin's biography, *A Good and Noble Thing: The Pioneering Life and Service of Anna Ward Perkins, M.D.* (Rensselaerville, NY: The Rensselaerville Historical Society, 2018), provides invaluable insight into Langer's friendship with Perkins.

BIBLIOGRAPHY

Connell, Sophia. "The Lost Women of Early Analytic Philosophy." Blog of the American Philosophy Association (September 7, 2022). Available online: https://blog.apaonline.org/2022/09/07/the-lost-women-of-early-analytic-philosophy/ (accessed July 16, 2023).

Dengerink Chaplin, Adrienne. *The Philosophy of Susanne Langer: Embodied Meaning in Logic, Art and Feeling.* London: Bloomsbury, 2019.

Finin, Gerard A. *A Good and Noble Thing: The Pioneering Life and Service of Anna Ward Perkins, M.D.* Rensselaerville, NY: The Rensselaerville Historical Society, 2018.

Gardner, Howard. "*Philosophy in a New Key* Revisited: An Appreciation of Susanne Langer." In *Art, Mind, and Brain: A Cognitive Approach to Creativity*, 45–54. New York: Basic Books, 1982.

Hall, Max. *Harvard University Press: A History*. Cambridge, MA: Harvard University Press, 1986.

Hutton, Sarah. "'Context' and 'Fortuna' in the History of Women Philosophers: A Diachronic Perspective." In *Methodological Reflections on Women's Contribution and Influence in the History of Philosophy*. Cham: Springer, 2020.

Isaac, Joel. "Susanne K. Langer, *Philosophy in a New Key* (1942)." *Public Culture* 32, no. 2 (91) (May 1, 2020): 355–61.

Kuklick, Bruce. *The Rise of American Philosophy: Cambridge, Massachusetts, 1860–1930*. New Haven, CT: Yale University Press, 1979.

Langer, Susanne K. *Philosophy In A New Key. A Study In The Symbolism Of Reason, Rite And Art*. New York: New American Library, 1954.

Langer, Susanne K. *Mind: An Essay on Human Feeling*, vol. 1. Baltimore: Johns Hopkins University Press, 1967.

Langer, Susanne K. to Bruce Kuklick, March 16, 1973, Susanne Langer Papers, 1895–1985 (MS Am 3110), Correspondence, Box 2, Houghton Library, Harvard University, Cambridge, MA.

Langer, Susanne K. to *The World Who's Who of Women*, November 1, 1976, Susanne Langer Papers, 1895–1985 (MS Am 3110), Correspondence, Box 19, Houghton Library, Harvard University, Cambridge, MA.

Lipscomb, Benjamin J. Bruxvoort. *The Women Are Up to Something: How Elizabeth Anscombe, Philippa Foot, Mary Midgley, and Iris Murdoch Revolutionized Ethics*. Oxford: Oxford University Press, 2021.

Mac Cumhaill, Clare, and Rachael Wiseman. *Metaphysical Animals: How Four Women Brought Philosophy Back to Life*. New York: Knopf Doubleday, 2022.

Peijnenburg, Jeanne, and Sander Verhaegh. 2023. "Analytic Women | Aeon Essays." Aeon. August 1, 2023. Available online: https://aeon.co/essays/the-lost-women-of-early-analytic-philosophy (accessed 19 August 2023)

Perry, Ralph Barton to Bernice Cronkhite, December 22, 1953, Radcliffe College Student Files, RG XXI, Series I, Susanne Knauth Langer, Identifier: Box 70, Schlesinger Library, Radcliffe Institute, Harvard University, Cambridge, MA.

Verhaegh, Sander. "Susanne Langer and the American Development of Analytic Philosophy." In *Women in the History of Analytic Philosophy*. Cham: Springer, 2022.

Waithe, Mary Ellen. *A History of Women Philosophers*, vol. 4: *Contemporary Women Philosophers, 1900–Today*. Dordrecht/Boston: Kluwer Academic Publishers, 1995.

INDEX